# Smart and Power Grid Systems – Design Challenges and Paradigms

## RIVER PUBLISHERS SERIES IN POWER

*Series Editors:*



The "River Publishers Series in Power" is a series of comprehensive academic and professional books focussing on the theory and applications behind power generation and distribution. The series features content on energy engineering, systems and development of electrical power, looking specifically at current technology and applications.

The series serves to be a reference for academics, researchers, managers, engineers, and other professionals in related matters with power generation and distribution.

Topics covered in the series include, but are not limited to:

- Power generation
- Energy services
- Electrical power systems
- Photovoltaics
- Power distribution systems
- Energy distribution engineering
- Smart grid
- Transmission line development

For a list of other books in this series, visit www.riverpublishers.com

# Smart and Power Grid Systems – Design Challenges and Paradigms

**Editors**

**Kolla Bhanu Prakash**

K.L. Deemed to be University, India

**Sanjeevikumar Padmanaban**

Aarhus University, Denmark

**Massimo Mitolo**

Irvine Valley College, USA

**Published 2023 by River Publishers**
River Publishers
Alsbjergvej 10, 9260 Gistrup, Denmark
www.riverpublishers.com

**Distributed exclusively by Routledge**
605 Third Avenue, New York, NY 10017, USA
4 Park Square, Milton Park, Abingdon, Oxon OX14 4RN

*Smart and Power Grid Systems – Design Challenges and Paradigms* / by Kolla Bhanu Prakash, Sanjeevikumar Padmanaban, Massimo Mitolo.



Routledge is an imprint of the Taylor & Francis Group, an informa business

ISBN 978-87-7022-672-1 (print)
ISBN 978-10-0079-514-1 (online)
ISBN 978-1-003-33955-7 (ebook master)

# Contents

**Preface** xv

**List of Figures** xxiii

**List of Tables** xxxi

**List of Contributors** xxxiii

**List of Abbreviations** xxxvii

**1 Power Electronics in Smart Grid** 1
*Padmanaban Sanjeevikumar, Mohammad Zand, Morteza Azimi Nasab, Mohsen Eskandari, Tina Samavat, Alireza Jahangiri, and Mohammad H. Moradi*
1.1 Introduction . . . 2
1.2 Smart Grid, Components and Advantages . . . 2
1.2.1 Structure of Photovoltaic Intelligent Charging Station (PVCS) Based on SST Solid State Transformer . . . 3
1.3 Power Electronic Converters in Smart Grids . . . 5
1.4 Application of Power Electronic Technology in Smart Grid . 6
1.4.1 The Application of DC-AC and DC-AC Converters in SG . . . 7
1.4.2 The Application of HVDC Technology in Smart Grid 8
1.4.3 The Application of FACTS Technology in Smart Grid 8
1.5 CPS Attacks Mitigation Approaches on Power Electronic . . 9
1.5.1 Architecture of Digitally-Controlled Power Electronic CPSs . . . 10
1.5.2 CPS Protection Vulnerabilities and the Applications of Power Electronics . . . 14
1.6 Conclusion . . . 14
1.6.1 Topology Detection and Cyber Attack . . . 15

1.6.2 Vulnerability Analysis of Cyber-attacks in the Control of Voltage Source Converters . . . . . . . . 18
References . . . . . . . . . . . . . . . . . . . . . . . . 18

**2 Power Electronics in HVDC Transmission Systems 21**
*Mehdi Abbasipour, Xiaodong Liang, and Massimo Mitolo*
2.1 Introduction . . . . . . . . . . . . . . . . . . . . . . . 22
2.2 HVDC Transmission Systems . . . . . . . . . . . . . . . . 22
2.2.1 Brief Overview on HVDC Transmission Technologies 22
2.2.1.1 Back-to-Back HVDC transmission . . . . 23
2.2.1.2 Point-to-Point HVDC transmission . . . . 23
2.2.1.3 Multi-Terminal HVDC grids . . . . . . . 24
2.2.2 HVDC Configurations . . . . . . . . . . . . . . . . 27
2.2.2.1 Monopolar . . . . . . . . . . . . . . . . 27
2.2.2.2 Bipolar . . . . . . . . . . . . . . . . . . 28
2.2.2.3 Homopolar . . . . . . . . . . . . . . . . 30
2.2.2.4 Hybrid . . . . . . . . . . . . . . . . . . 31
2.2.3 Power Electronics Converters in HVDC Transmission Systems . . . . . . . . . . . . . . . . . . . . 31
2.3 Power Converters . . . . . . . . . . . . . . . . . . . . . 33
2.3.1 Voltage Source Converters . . . . . . . . . . . . . 33
2.3.1.1 Two-level VSCs . . . . . . . . . . . . . . 34
2.3.1.2 Multilevel converters . . . . . . . . . . . 35
2.3.1.2.1 Monolithic multilevel converters 36
2.3.2 Current Source Converters . . . . . . . . . . . . . 48
2.3.2.1 Multipulse CSCs . . . . . . . . . . . . . 49
2.3.2.2 Modular current source converters . . . . 51
2.3.2.2.1 Power electronics current source SMs . . . . . . . . . . . . . . 51
2.3.2.2.2 Conventional MCSCs . . . . . . 53
2.3.2.2.3 Modular multilevel current source converters . . . . . . . . . . . . 54
2.3.2.2.4 Hybrid MCSCs . . . . . . . . . 55
2.3.3 Hybrid Current and Voltage Source Converters . . . 56
2.4 DC/DC Converters . . . . . . . . . . . . . . . . . . . . . 58
2.4.1 Isolated DC/DC Converters . . . . . . . . . . . . . 59
2.4.1.1 Flyback/Forward-based . . . . . . . . . . 59
2.4.1.2 DAB . . . . . . . . . . . . . . . . . . . 60
2.4.1.2.1 Two-level DAB . . . . . . . . . 60

2.4.1.2.2 Cascaded DAB multilevel converter . . . 61
2.4.1.2.3 DAB-MMC . . . 62
2.4.2 Non Isolated DC/DC Converters . . . 67
2.4.2.1 DC autotransformer . . . 68
2.4.2.2 Transformerless . . . 68
2.4.2.2.1 Resonant DC/DC converters . . 68
2.5 DC Power Flow Controllers . . . 72
2.5.1 SDC-PFC . . . 73
2.5.2 CDC-PFCs . . . 74
2.5.3 IDC-PFCs . . . 75
2.6 Conclusion . . . 76
References . . . 77

**3 Optimal Multi-Objective Energy Management of Renewable Distributed Integration in Smart Distribution Grids Considering Uncertainties** **83**
*M. Zellagui, N. Belbachir, S. Settoul, and C. Z. El-Bayeh*
3.1 Introduction . . . 85
3.2 Uncertainty Modeling of RDG Source . . . 87
3.2.1 Modeling of Load Demand Uncertainty . . . 87
3.2.2 Modeling of Solar DG Uncertainty . . . 87
3.2.3 Wind Turbine DG Uncertainty Modeling . . . 90
3.3 Multi Objective Indices Evaluation . . . 91
3.3.1 Multi Objective Indices . . . 91
3.3.2 Equality Constraints . . . 92
3.3.3 Distribution Line Constraints . . . 92
3.3.4 RDG Constraints . . . 93
3.4 Distribution Test System . . . 93
3.5 Analysis Results and Comparison . . . 93
3.6 Conclusion . . . 110
References . . . 111

**4 Security Challenges in Smart Grid Management** **117**
*S. Nithya, K. Vijayalakshmi, and M. Parimala Devi*
4.1 Introduction . . . 118
4.2 The Demand for a Smart Grid . . . 118
4.3 Benefits of Smart Grid . . . 119
4.4 Smart Grid Operation . . . 120
4.5 Smart Grid Security Challenges . . . 121

4.6 Literature Review . . . . . . . . . . . . . . . . . . . . . . . 121
4.7 Key Points that Require Special Attention . . . . . . . . . . . 122
4.7.1 Requirements for Data and Information Security . . 122
4.7.2 Extensive use of "Smart" Devices . . . . . . . . . . . 122
4.7.3 Grid Perimeter and Physical Security . . . . . . . . . 123
4.7.4 Protocols of a Legacy and (in) Secure Communication . . . . . . . . . . . . . . . . . . . . 123
4.7.5 Many Stakeholders and Synergies with other Services . . . . . . . . . . . . . . . . . . . . . . . 124
4.7.6 A Lack of Clarity about the Smart Grid Concept and its Security Requirements . . . . . . . . . . . . . . 125
4.7.7 Lack of Awareness among Smart Grid Stakeholders: 125
4.7.8 Supply Chain Security . . . . . . . . . . . . . . . . . 125
4.7.9 Encourage the Interchange of Risk, Vulnerability, and Threat Information . . . . . . . . . . . . . . . . 126
4.7.10 International Cooperation . . . . . . . . . . . . . . 126
4.7.11 Utility Security Management . . . . . . . . . . . . . 127
4.8 Smart Grid Security Policies . . . . . . . . . . . . . . . . . 127
4.8.1 Confidentiality . . . . . . . . . . . . . . . . . . . . . 127
4.8.2 Integrity . . . . . . . . . . . . . . . . . . . . . . . . 127
4.8.3 Availability . . . . . . . . . . . . . . . . . . . . . . 127
4.8.4 Accountability . . . . . . . . . . . . . . . . . . . . 128
4.9 Corrective Strategies to Improve Smart Grid Protection . . . 128
4.10 Important Areas to Safeguard the Grid . . . . . . . . . . . . 128
4.10.1 Powerful Digital Identities . . . . . . . . . . . . . . 128
4.10.2 Mutual Verification . . . . . . . . . . . . . . . . . . 129
4.10.3 Encryption . . . . . . . . . . . . . . . . . . . . . . 129
4.10.4 Continuous Security Updates . . . . . . . . . . . . . 130
4.11 Conclusion . . . . . . . . . . . . . . . . . . . . . . . . . . 131
References . . . . . . . . . . . . . . . . . . . . . . . . . . 132

**5 Differential Protection Scheme along with Backup Blockchain System for DC Microgrid 135**
*E. Fantin Irudaya Raj, K. Manimala, and M. Appadurai*
5.1 Introduction . . . . . . . . . . . . . . . . . . . . . . . . . 136
5.2 DC Microgrids . . . . . . . . . . . . . . . . . . . . . . . . 139
5.2.1 Various Power Sources in DC Microgrids . . . . . . 140
5.2.2 Energy Storage Systems in DC Microgrids . . . . . 141
5.2.3 Power Converters used in DC Microgrids . . . . . . 141

5.3 Challenges in Protection of Smart Grid . . . . . . . . . . . . . 141
5.3.1 Protection from Cyber-Attacks . . . . . . . . . . . . . 142
5.3.2 Converters with a Low Tolerance . . . . . . . . . . . 142
5.3.3 Inefficacy of AC Circuit Breakers . . . . . . . . . . . 142
5.3.4 Fault Current in both Directions . . . . . . . . . . . 142
5.4 Cyber Attacks . . . . . . . . . . . . . . . . . . . . . . . . . 142
5.4.1 Network security cyber attacks . . . . . . . . . . . . 143
5.4.2 GOOSE and SV Messages . . . . . . . . . . . . . . . 144
5.5 Blockchain . . . . . . . . . . . . . . . . . . . . . . . . . . . 145
5.6 Blockchain-Based DC Microgrid Protection Approach . . . 146
5.6.1 Differential Fault Identification . . . . . . . . . . . 146
5.6.2 Blockchain-based Backup and Protection System . . 149
5.7 Results and Discussion . . . . . . . . . . . . . . . . . . . . 149
5.7.1 Detecting and Isolating the Faults without Considering Cyber-Attack . . . . . . . . . . . . . . . 150
5.7.2 Detecting and Isolating Faults when considering a Cyber-Attack . . . . . . . . . . . . . . . . . . . . . 151
5.8 Conclusion . . . . . . . . . . . . . . . . . . . . . . . . . . . 153
References . . . . . . . . . . . . . . . . . . . . . . . . . . . 153

**6 Planning Active Distribution Systems Using Microgrid Formation 157**
*Shah Mohammad Rezwanul Haque Shawon, Xiaodong Liang, and Massimo Mitolo*
6.1 Introduction . . . . . . . . . . . . . . . . . . . . . . . . . . 157
6.2 Step 1: Defining the Objectives . . . . . . . . . . . . . . . . 159
6.3 Step 2: Defining the Microgrid Topology . . . . . . . . . . . 162
6.4 Step 3: System Modeling . . . . . . . . . . . . . . . . . . . 163
6.4.1 Modeling of DGs . . . . . . . . . . . . . . . . . . . . 163
6.4.1.1 Dispatchable DG model . . . . . . . . . . 163
6.4.1.1.1 Diesel engine model . . . . . . . 163
6.4.1.1.2 Micro turbine model . . . . . . 164
6.4.1.2 Nondispatchable DG model . . . . . . . . 164
6.4.1.2.1 PV modeling . . . . . . . . . . . 165
6.4.1.2.2 Wind turbine modeling . . . . . 166
6.4.2 Load Modeling . . . . . . . . . . . . . . . . . . . . . 167
6.4.3 Energy Storage System Modeling . . . . . . . . . . . 168
6.5 Step 4: Network Optimization . . . . . . . . . . . . . . . . 168
6.5.1 Power Flow . . . . . . . . . . . . . . . . . . . . . . . 171

6.5.2 Demand Response . . . . . . . . . . . . . . . . . . 172
6.6 Switch Placement . . . . . . . . . . . . . . . . . . . . . . . 172
6.6.1 Objectives of Switch Placement . . . . . . . . . . . 174
6.6.2 Methodology of Switch Placement . . . . . . . . . . 176
6.6.3 Discussion . . . . . . . . . . . . . . . . . . . . . . . 177
6.7 Conclusion . . . . . . . . . . . . . . . . . . . . . . . . . . 177
References . . . . . . . . . . . . . . . . . . . . . . . . . . 178

**7 Overview on Reliability of PV Inverters in Grid-connected Applications 185**
*Ranjith kumar Gatla, N.V Prasad K, P. Sridhar, Jianghua Lu, and Devineni Gireesh kumar*
7.1 Introduction . . . . . . . . . . . . . . . . . . . . . . . . . 186
7.2 Power Converters for PV Systems . . . . . . . . . . . . . . 188
7.3 Basic Principles of Reliability . . . . . . . . . . . . . . . . 191
7.3.1 Failure Rate . . . . . . . . . . . . . . . . . . . . . . 191
7.3.2 Mean Time to Failure (MTTF) . . . . . . . . . . . . 192
7.3.3 Mean Time to Repair (MTTR) . . . . . . . . . . . . 192
7.3.4 Mean Time Between Failure (MTBF) . . . . . . . . 193
7.4 Power Module Reliability . . . . . . . . . . . . . . . . . . . 193
7.4.1 Reliability Analysis of IGBT Module . . . . . . . . 194
7.5 Capacitor Reliability . . . . . . . . . . . . . . . . . . . . . 197
7.6 Lifetime Estimation Methods . . . . . . . . . . . . . . . . . 199
7.6.1 Parts Stress Method . . . . . . . . . . . . . . . . . . 199
7.6.2 Lifetime Prediction Methods of Power Devices . . . 200
7.6.2.1 Coffin-Manson Lifetime Model . . . . . . 200
7.6.2.2 Coffin-Manson-Arrhenius Lifetime Model . . . . . . . . . . . . . . . . . 200
7.6.2.3 Norris-Landzberg Lifetime Model . . . . 200
7.6.2.4 Semikorn Lifetime Model . . . . . . . . . 201
7.6.2.5 Bayerer Lifetime Model . . . . . . . . . . 201
7.6.3 Lifetime Prediction Methods of DC-Link Capacitors . . . . . . . . . . . . . . . . . . . . . 202
7.7 Conclusion . . . . . . . . . . . . . . . . . . . . . . . . . . 203
References . . . . . . . . . . . . . . . . . . . . . . . . . . 204

**8 Energy Storage 211**
*Sanjeevikumar Padmanaban, Mohammad Zand, Morteza Azimi Nasab, Mohamadmahdi Shahbazi, and Heshmatallah Nourizadeh*

8.1 Introduction . . . . . . . . . . . . . . . . . . . . . . . . . 211
8.2 Installed Capacity in the World . . . . . . . . . . . . . . . . 213
8.3 Application of Energy Storage Devices . . . . . . . . . . . . 214
8.4 Classification of Energy Storage Devices . . . . . . . . . . . 215
8.4.1 A Variety of Storage Technologies in the Supply Chain to Consume Electricity . . . . . . . . . . . . 216
8.4.2 Electrical storage technologies . . . . . . . . . . . . 216
8.5 Superconducting Magnetic Storage (SMES) . . . . . . . . . . 219
8.6 Mechanical Storage Method . . . . . . . . . . . . . . . . . . 220
8.6.1 Storage Pump . . . . . . . . . . . . . . . . . . . . . 220
8.6.2 Compressed Air Storage . . . . . . . . . . . . . . . . 221
8.6.3 Flight Wheel Storage . . . . . . . . . . . . . . . . . 223
8.7 Thermal Storage Method . . . . . . . . . . . . . . . . . . . 224
8.7.1 Reasonable Thermal Energy Storage Systems . . . . 225
8.7.2 Sensible thermal energy storage systems . . . . . . . 226
8.8 Chemical Storage Method . . . . . . . . . . . . . . . . . . . 226
8.8.1 Chemical Storage Systems with Internal Storage . . 226
8.8.1.1 Hydrogen storage system (HES) . . . . . 226
8.8.1.2 GAT power system: Artificial natural gas methanation . . . . . . . . . . . . . . . . . 228
8.8.1.3 Current batteries . . . . . . . . . . . . . . 228
8.8.2 Chemical Storage Systems with External Storage . . 229
8.8.2.1 Lithium-ion battery . . . . . . . . . . . . 229
8.8.2.2 Lead-acid battery . . . . . . . . . . . . . 230
8.8.2.3 High-temperature batteries (sulfur - sodium) . . . . . . . . . . . . . . 231
8.8.2.4 Nickel (nickel-cadmium) battery . . . . . 232
8.8.2.5 Status of energy storage technologies . . . 232
8.9 Storage Cost . . . . . . . . . . . . . . . . . . . . . . . . . . 232
8.10 Criteria for Determining Appropriate Energy Storage Technologies . . . . . . . . . . . . . . . . . . . . . . . . . . 233
References . . . . . . . . . . . . . . . . . . . . . . . . . . 234

**9 A Comprehensive Review of Techniques for Enhancing Lifetime of Wireless Sensor Network** **237**
*Raj Gaurang Tiwari, Alok Misra, Ambuj Kumar Agarwal, and Vikas Khullar*
9.1 Introduction . . . . . . . . . . . . . . . . . . . . . . . . . 238
9.1.1 Scalability . . . . . . . . . . . . . . . . . . . . . . 238

9.1.2 Routing . . . . . . . . . . . . . . . . . . . . . . 238
9.1.3 Quality of Service . . . . . . . . . . . . . . . . . . 239
9.1.4 Safety Measures . . . . . . . . . . . . . . . . . . . 239
9.1.5 Energy/Power Preservation . . . . . . . . . . . . . . 239
9.1.6 Node Collaboration . . . . . . . . . . . . . . . . . . 239
9.1.7 Interoperation . . . . . . . . . . . . . . . . . . . . 239
9.2 Intricacy while Deployment of Manet . . . . . . . . . . . 240
9.3 Wireless Sensor Networks . . . . . . . . . . . . . . . . . 240
9.3.1 Sensor Network Communication Architecture . . . . 241
9.4 Coverage Problem in Sensor Network . . . . . . . . . . . 242
9.5 Lifetime Maximization of Wireless Sensor Networks . . . . 243
9.6 Overview of Optimization Techniques Employed to Maximize . . . . . . . . . . . . . . . . . . . . . . . 246
9.7 Conclusion . . . . . . . . . . . . . . . . . . . . . . . . 248
References . . . . . . . . . . . . . . . . . . . . . . . . 248

**10 Soft Open Points in Active Distribution Systems** **253**
*Md Abu Saaklayen, Xiaodong Liang, Sherif O. Faried, and Massimo Mitolo*
10.1 Introduction . . . . . . . . . . . . . . . . . . . . . . . 253
10.2 Basic Concept of SOP . . . . . . . . . . . . . . . . . . . 254
10.2.1 Benefits of SOPs . . . . . . . . . . . . . . . . . . . 256
10.3 Comparison of SOPs with other Power Electronic Devices . 258
10.4 Principle and Modeling of SOPs in Active Distribution Networks . . . . . . . . . . . . . . . . . . . . . . . . 260
10.4.1 Mathematical Modeling of SOPs in Active Distribution Networks . . . . . . . . . . . . . . . . . 263
10.5 Classification of Existing SOP Configurations . . . . . . . . 264
10.5.1 Two-Terminal Soft Open Points . . . . . . . . . . . 264
10.5.2 Multi-Terminal Soft Open Points . . . . . . . . . . . 265
10.5.3 Soft Open Points with Energy Storage . . . . . . . . 266
10.5.4 DC Soft Open Points . . . . . . . . . . . . . . . . . 267
10.6 Planning for Sizing and Placement of SOPs in Distribution Networks . . . . . . . . . . . . . . . . . . . . . . . . 267
10.6.1 SOP Coordinated Optimization . . . . . . . . . . . 270
10.6.1.1 SOP Coordinated Optimization in Balanced Distribution Networks . . . . 270
10.6.1.2 SOP Coordinated Optimization in Unbalanced Distribution Networks . . . 271

10.7 Operation of SOPs in Distribution Networks . . . . . . . . . . 272
10.7.1 Operation of SOPs under Normal Conditions . . . . 272
10.7.1.1 Control Block Diagram for Power Flow Control Mode Operation of SOPs . . . . . 273
10.7.2 Operation of SOP during Abnormal Condition (Supply Restoration) . . . . . . . . . . . . . . . . . . . . 275
10.7.2.1 Control Block Diagram for Supply Restoration Mode of SOPs . . . . . . . . . . . . 276
10.7.2.2 Supply Restoration Approaches Using SOPs . . . . . . . . . . . . . . . . . . . . 276
10.8 Conclusion and Future Research Direction . . . . . . . . . . 279
References . . . . . . . . . . . . . . . . . . . . . . . . . . . 279

**11 Future Advances in Wind Energy Engineering 287**
*Biswajit Mohapatra*
11.1 Introduction . . . . . . . . . . . . . . . . . . . . . . . . . . 288
11.1.1 Airborne Wind Energy . . . . . . . . . . . . . . . . . 288
11.1.1.1 Ground-Gen airborne wind energy systems . . . . . . . . . . . . . . . . . . . . 288
11.1.1.2 Fly-Gen airborne wind energy systems . . 290
11.2 Offshore Floating Wind Concepts . . . . . . . . . . . . . . . 291
11.2.1 Floating hybrid energy platforms . . . . . . . . . . . 292
11.3 Smart Rotors Technology . . . . . . . . . . . . . . . . . . . 293
11.3.1 Passive and active control systems . . . . . . . . . . 294
11.3.2 Degree of development, challenges and potential of smart rotors . . . . . . . . . . . . . . . . . . . . . . 294
11.4 Wind Turbine with TIP Rotors . . . . . . . . . . . . . . . . 296
11.5 Multi Rotor Wind Turbine . . . . . . . . . . . . . . . . . . 298
11.6 Diffuser Augmented Wind Turbines . . . . . . . . . . . . . 299
11.7 Other Small Wind Turbine Technologies . . . . . . . . . . . 301
11.8 Wind Induced Energy Harvesting from Aeroelastic Phenomena . . . . . . . . . . . . . . . . . . . . . . . . . . 302
References . . . . . . . . . . . . . . . . . . . . . . . . . . . 305

**Index 309**

**About the Editors 311**

# Preface

The Smart Grid represents an unprecedented opportunity to move the energy industry into a new era of reliability, availability, and efficiency that will contribute to our economic and environmental health. During the transition period, it will be critical to carry out testing, technology improvements, consumer education, development of standards and regulations, and information sharing between projects to ensure that the benefits we envision from the Smart Grid become a reality. Today, an electricity disruption, such as a blackout, can have a domino effect, that is, cause a series of failures that can affect banking, communications, traffic, and security. This is a serious threat in winter, when users can be left without heat. A smarter grid will add resiliency to our electric power system and make it better prepared to address emergencies such as severe storms, earthquakes, large solar flares, and terrorist attacks. Because of its two-way interactive capacity, the Smart Grid will allow for automatic rerouting when equipment fails or outages occur. This will minimize outages and their effects when they do happen. Smart Grid technologies can detect and isolate the faulty portion of the system, containing the power outage before it becomes large-scale blackouts. The new technologies will also help ensure that electricity resumes quickly and strategically, after, for example, an emergency-routing of electricity to emergency services. In addition, the Smart Grid will take greater advantage of customer-owned power generators to produce power when it is not available from utilities. By combining these "distributed generation" resources, a community could keep its health center, police department, traffic lights, phone system, and grocery stores operating during emergencies. The Smart Grid is an effective way to address an aging energy infrastructure that needs to be upgraded or replaced. The book shows that Smart Grids can address energy efficiency issues, bring increased awareness of the connection between electricity use and the environment, improve national security of energy systems, enhancing the resilience of the power grid and make it more resistant to natural disasters and attack.

Detailed abstracts of each chapter given below:-

**Power electronics in smart grid**

The high penetration of emerging power electronics-based distributed energy resources into electric grids has raised control, management, and stability issues. Besides, electric grids have been converted to multi-input multi-output cyber-physical systems (CPSs), which need advanced communication infrastructure to be monitored and optimally controlled. The smart grid (SGs) concept has been introduced as the platform that facilitates the solution to the challenges of modern electric grids. Monitoring and control capacities, automatic grid control, and active consumer participation in the energy industry further expand the SGs influence. The important features of SGs have been achieved by combining power system technologies with emerging power electronics and digital telecommunications technologies. Since almost all the distributed energy resources (DERs) technologies, such as renewable energy resources, micro sources, and energy storage systems, use power electronics interfaces, it is vital to match the characteristics of the DERs with the grid requirements. On the other hand, power electronics provide more flexibility, controllability, and better power quality, which are essential factors in SGs. However, with the rapid progression of physical systems in power electronics applications to interface DERs combined with cyber frameworks, the threat of cyber-attacks has a significant impact on SG security and performance. A comprehensive study of the vital role of power electronics in SGs, such as CPSs to provide sustainable/reliable energy resources, is provided in this chapter. Cyber-attacks in power electronics systems are briefly discussed.

**Power Electronics in HVDC Transmission Systems**

High Voltage Direct Current (HVDC) transmission are widely used for bulk power transmission, asynchronous connection, and marine power transmission. The principle of HVDC transmission relies on AC to DC power conversion and vice versa. Generally, power electronics converters in HVDC transmission systems can be classified into three major groups: a) power converters, b) DC/DC converters, and c) DC power flow controllers. Power converters convert AC power to DC power and vice versa. DC/DC converters (or named DC transformers) provide DC voltage matching; they can also be used to divide large Multi-Terminal HVDC (MT-HVDC) grids into several smaller protection zones, regulate DC voltage, isolate faults, and connect Bipolar/Monopolar configurations. DC power flow controllers (DC-PFCs) or DC current flow controllers (DC-CFCs) control power flow in HVDC transmission systems, especially in MT-HVDC grids. In this chapter, power electronic converters and inverters, commonly used in HVDC transmission systems, are summarized through extensive literature review, and their various topologies are introduced.

**Optimal Multi-Objective Energy Management of Renewable Distributed Generator Integration in Smart Distribution Grids Considering Uncertainties**

The energy supplies problem is crucial. Daily improvements are implemented to obtain the optimization of the power system configuration and the power generation. Renewable Distributed Generators (RDGs) are one of the best solutions to realize those improvements. The optimal placement and sizing of RDG sources in the Smart Distribution Grid (SDG) are considered a trendy problem, which, due to the high complexity, can usually be solved based on various approaches and algorithms. The presence of RDGs in the SDG can provide many benefits and advantages. These benefits may be generally summarized in power losses minimization, voltage profiles improvement, system load-ability and reliability growth, system security, and protection enhancement. To achieve these benefits, RDGs should be optimally located and sized based on various objective functions.

A recent nature-inspired metaheuristic approach, defined as the Marine Predators Algorithm (MPA), is employed, which is based on various foraging strategies among optimal encounter rates policy in biological interaction and ocean predators. This algorithm was utilized to optimize many types of RDG units to obtain the optimal location and size of Photovoltaic Distributed Generator (PVDG) and Wind Turbine Distributed Generator (WTDG) units into the SDG. This has been performed by taking into consideration the uncertainties of the power generated from the RDG, as well as the load demand variation during each of the day's hours.

This chapter proposes new Multi-Objective Indices (MOI), which simultaneously minimize five technical indices based on the power losses, the voltage deviation, and the over current protection system of the SDG. The chosen algorithm is validated on different standards IEEE 33-bus, and 69-bus distribution grids for the purpose of testing its efficiency, where also three cases of RDGs' allocation were studied. The convergence characteristics reveal that the MPA is effectively a quick technique that may provide the best solutions in a small number of iterations, compared to other algorithms, such as Particle Swarm Optimization (PSO), Ant Lion Optimization (ALO), Grey Wolf Optimizer (GWO), Grasshopper Optimization Algorithm (GOA), and

Moth Flame Optimizer (MFO) algorithms. The optimal allocation of RDG identifies the suitable results satisfying the permissible voltage limits and power loss minimization. After the installation of both RDGs, power losses are minimized, the voltage profile improves, and so does the over current protection system. The simulation results confirm the feasibility of optimal power planning. The results reveal that the optimal integration of the WTDG units based on the chosen algorithm was the best choice over the PVDG units, which led to the minimization of the expected APL, RPL, VD, OT, and CTI of different test systems.

**Security Challenges in Smart Grid Management**

The transformation from the power grid to the smart grid has become one of the greatest technological evolutions of the last few years. The reliability of electric supply is the most crucial feature for both developing and developed countries. The smart grid enables the reduction of the emission of carbon, thanks to the integration of renewable energy resources. The smart grid is a system that monitors and manages energy use through a network of computers and power infrastructures. The smart grid combines different domains and must withstand natural disasters, intentional attacks, hackers, transportation, and storage. It also includes Smart meter problems such as privacy invasion, reliability, overcharging, hacking and other health issues. To overcome the above issues traditional security analysis approaches such as certification and internal quality assurance are essential, but they fall short when it comes to critical systems. To evaluate smart grid systems, industry and government must be innovative. Even though services such as Google Power Meter are opt-in, customers have little control over how power data is used by utility providers. To address these issues with smart grid applications, various algorithms can be proposed.

**Differential Protection Scheme along with Backup Blockchain System for DC Microgrid**

The advances in Intelligent Smart Meters, the Internet of Things (IoT), and communication technologies converge into Smart Electric Grids. In the conventional grid, the communication is unilateral. In a Smart Electric Grid platform, all system members, from different generating units and different consumers, can bilaterally communicate in real time via modern technologies. At the same time, the new Smart Electric Grid has some challenges. There might be a chance for personal information leakage, cyber-attacks; managing small customers or prosumers is another crucial challenge in Smart

Grids. Introducing Blockchain technology to the Smart Electric Grid is an attractive solution to these problems. The Blockchain platform comprises cryptographic security measures, a decentralized consensus mechanism, and a distributed ledger. In this chapter, protection schemes of the DC microgrids based upon Blockchain technology are discussed in detail. Compared with AC microgrids, DC microgrids are more efficient and are more suited for renewable energy sources, thanks to the reduction in conversion stages and the absence of skin effect. The differential protection scheme along with a backup blockchain system for DC microgrid is discussed. Detection and isolation of fault with and without cyber-attack are explained with simulation results. Differential protection schemes detect and isolate the high impedance faults, and the blockchain backup system enhances the trustworthiness of the proposed scheme by mitigating the communication failure impact.

**Planning Active Distribution Systems Using Microgrid Formation**

In this chapter, the planning of active distribution systems through microgrid formation is discussed. There are four critical steps involved: 1) defining the objectives, 2) defining the microgrid topology, 3) system modelling, and 4) network optimization. The objectives of such planning problems in the literature are extensively reviewed, dominant and popular objectives used are demonstrated. The modelling of microgrid components, such as dispatchable and non-dispatchable DGs, load, and energy storage, is reviewed and summarized in the chapter. As critical tools, optimization techniques in planning are summarized. The switch placement in distribution networks is also discussed in this chapter.

**Overview on Reliability of PV Inverters in Grid-connected Applications**

In terms of reliability, the PV inverter is the weakest part of grid-connected Photovoltaic (PV) systems. This chapter presents an overview of the reliability of PV inverters in grid-connected applications. The discussion includes different PV inverter configurations for grid-connected systems, basic principles of reliability, and the importance of reliability evaluation in PV inverters. Finally, the chapter presents different lifetime estimation methods available to evaluate the reliability of the most frequent failure components in the PV inverter, which are power devices and dc-link capacitors. The main objective of this review is to provide the basics of reliability research in PV inverters.

**Energy Storage**

Energy continues to be a pivotal element of worldwide development. Due to the oil price volatility, depletion of fossil fuel resources, global warming, local pollution, geopolitical tensions, and growth in energy demand, alternative energies, renewable energies, and effective use of fossil fuels have become much more critical than at any time in history. Current and future markets in fossil fuels are subject to volatile price changes in oil and natural gas. National and international energy/ environmental crises and conflicts are combining to motivate a dramatic shift from fossil fuels to reliable, clean, and efficient fuels. Renewable energy resources such as wind and solar energies cannot produce power steadily since their power production rates change with parameters such as seasons, months, days, hours.

All of these challenges require using some storage device to develop viable power system operation solutions. There are different types of storage systems with different costs, operation characteristics, and potential applications. Understanding these is vital for the future design of power systems, whether for short-term transient operation or long-term generation planning.

Due to inefficiency and increasing reduction of fossil fuel resources, optimal use of generation capacity and excess power storage can improve power networks' performance and reduce generation capacity. For this purpose, in recent decades, energy storage systems with different motives to improve the performance of the power system have been considered. In this chapter, the upgraded equipment of several technologies such as (electrical storage, mechanical storage, thermal storage, and chemical storage) is reviewed, and their various characteristics are analyzed. This review includes their application, classification, ability and storage properties, the current state of the industry, and future installation possibilities. This chapter also examines the main features of energy-saving technologies suitable for renewable energy systems.

## A Comprehensive Review of Techniques For Enhancing Lifetime Of Wireless Sensor Network

A sensor is a device that detects an event and turns it into an electrical, mechanical, or another form, signal. In diverse application areas like phones, electronics and electronic devices, mechanical devices, industries, etc., sensors can be utilized. There is a large variety of sensors, such as seismic, optical, magnetic, infrared, thermal, acoustic, and radar, which can detect a variety of ambient parameters. There are several resource constraints on a sensor, like storage, energy, communication, and computation capabilities. A sensor network can be constructed by combining identical or diverse sensors. In the wireless sensor network, physical information is sensed by a set of independent sensors and also transmitted to main locations through the network. Armed forces applications, such as the supervision of the battlefield, are motivated by the development of sensor networks. As of today, we can find their applications in process control, device status monitoring, health monitoring through body area networks, civil and disaster management, environmental, and commercial applications, etc. These networks can also be utilized to supervise and analyze tornado/storm movements, to monitor the temperature of the area surrounding a volcano, and wild animal behavior. In this chapter, we present an inclusive analysis of the recent literature on prolonging the lifetime and coverage area of sensor networks.

**Soft Open Points in Active Distribution Systems**

In this chapter, the Soft Open Point (SOP), an emerging power electronics device in distribution networks, is introduced. SOPs can be installed to replace the normally open points (NOPs) in distribution networks to achieve the flexible connection between feeders and the upgrade from radial to closed-loop configuration. Under normal operation, SOPs can provide active power flow control, reactive power compensation and voltage regulation; under fault conditions, SOPs offer post-fault supply restoration. In this chapter, the basic concept of the SOP and its benefits are explained; a comparison between SOPs and other power electronics devices used in distribution networks is provided; the modeling method of SOPs, along with their placement and sizing in active distribution networks are reviewed.

**Future Advances in Wind Energy Engineering**

Wind energy is a renewable and non-conventional energy source, which is the result of uneven heating up of earth surface by the sun. It is clean, inexhaustible, indigenous energy resources. In recent years, wind energy has become one of the most important and promising sources of renewable energy, which demands additional transmission capacity and better means of maintaining system reliability. Advances in wind energy engineering have been resulted in better reliability, controllability, low cost and less maintenance. Some of the advances are variable speed wind turbines, airborne wind energy, offshore floating concepts, smart rotors, wind-induced energy harvesting devices, blade tip-mounted rotors, unconventional power transmission systems, multi-rotor turbines, alternative support structures, and modular high voltage direct current generators, innovative blade manufacturing techniques, diffuser-augmented turbines and small turbine technologies.

**Advances in energy supervisory and management of distributed energy resources**

The roadmap for the transition to a decentralised, democratised, and highly data driven power systems must pass through the tree dimensions of the Energy Trilemma prism, i.e., energy efficiency – energy security – environmental sustainability and must achieve a balance among its three competing aims, i.e., economics, politics and the environment. Distributed energy resources (DERs) will support the transformation toward smart grids which makes use of computational intelligence in an integrated fashion.

In this chapter, we will explain the state-of-the-art and illustrate the future trends of the pro-active control of the DERs by the use of artificial intelligence tools that are able to implement predictive modeling of the DERs behavior in order to implement distributed management and prognostic approaches.

We will consider full deep learning approaches based on recurrent models for data aggregation and multivariate prediction. A reduction of such models will be illustrated in terms of randomized layers and/or shallow recurrent architectures, in order to consider a possible implementation in the DER's embedded hardware for real-time control and forecasting. We will illustrate centralized approaches as well as innovative distributed learning approaches. Several examples and results will be proposed through the whole chapter.

**Dr. Kolla Bhanu Prakash**

**Dr. Sanjeevikumar Padmanaban**

**Dr. Massimo Mitolo**

# List of Figures


| | | |
|---|---|---|
| **Figure 1.1** | SST-based scheduling for an example of energy-intensive smart grid with integrated parking charge and photovoltaic f . . . . . . . . . . . . . . . . . . | 3 |
| **Figure 1.2** | Micro grid architecture design . . . . . . . . . . . | 6 |
| **Figure 1.3** | Grid-connected power electronic converters in smart grids . . . . . . . . . . . . . . . . . . . . . . | 7 |
| **Figure 1.4** | The control hierarchies for the power converters . . | 9 |
| **Figure 1.5** | Security taxonomy for power electronic CPS . . . | 10 |
| **Figure 1.6** | Communication topologies for cyber structures: (a) star topology (b) bass topology (c) ring topology (d) Daisy chain topology . . . . . . . . . . . . . . | 11 |
| **Figure 1.7** | Voltage source converter with main types: (a) power supply network and (B) mains voltage source converter . . . . . . . . . . . . . . . . . . . . . . | 12 |
| **Figure 1.8** | Resource threats in CPS and their attack properties | 14 |
| **Figure 2.1** | The schematic diagram of the B2B HVDC transmission systems. . . . . . . . . . . . . . . . . . . | 23 |
| **Figure 2.2** | The schematic diagram of the P2P HVDC transmission systems. . . . . . . . . . . . . . . . . . . . | 24 |
| **Figure 2.3** | Two MT-HVDC grids: a) three-terminal b) eight-terminal. . . . . . . . . . . . . . . . . . . | 25 |
| **Figure 2.4** | The schematic diagram of the MT-HVDC grids: a) Radial MT-HVDC grid; b) Ring MT-HVDC grid; c) Lightly-Meshed MT-HVDC grid; and d) Densely-Meshed MT-HVDC grid. . . . . . . . . . . . . . | 26 |
| **Figure 2.5** | The schematic diagram of the Monopolar SWRT Configuration. . . . . . . . . . . . . . . . . . . . . | 28 |
| **Figure 2.6** | The schematic diagram of (a) Asymmetric Monopolar configuration (b) Symmetric Monopolar configuration. . . . . . . . . . . . . . . . . . . . . | 29 |

**Figure 2.7** Schematic diagram of the Bipolar HVDC configuration. . . . . . . . . . . . . . . . . . . . . . 29
**Figure 2.8** Two improved Bipolar HVDC configurations: a) Bipolar configuration with Metallic Return Path b) Bipolar configuration with Series-Connected Converters. . . . . . . . . . . . . . . . . . . . . . 30
**Figure 2.9** Schematic diagram of the Homopolar HVDC configuration. . . . . . . . . . . . . . . . . . . . . . 31
**Figure 2.10** Schematic diagrams of Hybrid HVDC configurations: a) Bipolar HVDC with Monopolar SWER; b) Homopolar HVDC with Symmetric Monopolar; and c) Homopolar HVDC with Bipolar HVDC. . . 32
**Figure 2.11** General categorization of AC/DC power converters. . . . . . . . . . . . . . . . . . . . . . 33
**Figure 2.12** The general schematic diagram of a B2B VSC. . . 34
**Figure 2.13** Three-phase two-level VSC. . . . . . . . . . . . 34
**Figure 2.14** Output waveform of a multilevel converter. . . . . 35
**Figure 2.15** Schematic diagram of the NPC three-level converters. . . . . . . . . . . . . . . . . . . . . . 36
**Figure 2.16** Schematic diagram of the FC three-level converters. . . . . . . . . . . . . . . . . . . . . . 38
**Figure 2.17** SMs of MMCs: a) Half-Bridge; b) Full-Bridge; c) Mixed; d) Asymmetrical; e) Cross-Connected and Parallel; f) Clamped; g) FC-Type; and h) NPC-Type. . . . . . . . . . . . . . . . . . . . 41
**Figure 2.18** The general structure of MMCs. . . . . . . . . . . 42
**Figure 2.19** Chain-links of power electronics SMs: a) Chain-links of Half-Bridges; b) Chain-links of Full-Bridges; c) Chain-links of Double Clamped; d) Chain-links of Mixed; e) Chain-links of Cross-Connected; f) Chain-links of Asymmetrical; g) Chain-links of Stacked FC-Type; h) Chain-links of Series FC-Type; and i) Chain-links of NPC-Type. . 43
**Figure 2.20** Schematic diagram of a complete three-phase MMC based on: a) Half-Bridge sub-modules; b) Full-Bridge sub-modules; and c) Comparison between a two-level VSC and a MMC . . . . . . . 45
**Figure 2.21** The general structure of AAMMCs . . . . . . . . . 46

**Figure 2.22** Hybrid MMCs with Monolithic Director Switches: a) Hybrid MMC with arm chain-link SMs; b) Hybrid two-level with AC-bus chain-link SMs; c) Hybrid two-level with midpoint chain-link SMs; d) Hybrid three-level FC with midpoint chain-link SMs; and e) Hybrid three-level NPC with arm chain-link SMs. . . . . . . . . . . . . . . . . . . . 48
**Figure 2.23** Hybrid MMCs with H-bridge Director Switches: a) H-bridge Hybrid MMC with parallel chain-link SMs; b) Three-phase Hybrid MMC with parallel chain-link SMs. . . . . . . . . . . . . . . . . . . . 49
**Figure 2.24** General schematic diagram of a B2B CSC. . . . . 49
**Figure 2.25** Schematic diagram of CSCs: a) Three-phase 6-pulse; and b) Three-phase 12-pulse. . . . . . . . . 50
**Figure 2.26** SMs of MCSCs: a) Half-Bridge; b) Full-Bridge; c) Three-phase; d) Mixed; and e) Clamped. . . . . . . 52
**Figure 2.27** Schematic diagram of the conventional MCSCs: a) Three-phase MCSC; and b) Single-phase MCSC. . 54
**Figure 2.28** The schematic diagram of the MMCSCs: a) CS arm; b) Single-phase MMCSC. . . . . . . . . . . . 55
**Figure 2.29** The schematic diagram of a three-phase MMCSC. . . . . . . . . . . . . . . . . . . . . . . 56
**Figure 2.30** Schematic diagrams of two Hybrid MCSCs: a) Single-phase; b) Three-phase. . . . . . . . . . . . 56
**Figure 2.31** Schematic diagrams of two new hybrid MCSCs: a) Series-connected CS arm; b) Parallel-connected CS arm. . . . . . . . . . . . . . . . . . . . . . . . . 57
**Figure 2.32** The schematic diagram of an HCVSC and its output waveforms. . . . . . . . . . . . . . . . . . . . . 57
**Figure 2.33** Schematic diagram of (a) Hybrid CS-VS arms; (b) Hybrid CS-VS SMs. . . . . . . . . . . . . . . . . 58
**Figure 2.34** General categorization of the DC/DC converters. . 59
**Figure 2.35** Schematic diagram of the flyback/forward-based DC/DC converters: a) Modular topology; and b) Centralized coupled inductor topology. . . . . . . . 60
**Figure 2.36** Schematic diagram of the two-level DAB DC/DC converter . . . . . . . . . . . . . . . . . . . . . . 61

**Figure 2.37** Schematic diagrams of the fourfold cascaded DAB multi converter as the DC/DC converters: a) ISOS; b) IPOS; c) ISOP; and d) IPOP. . . . . . . . . . . . 62
**Figure 2.38** Schematic diagram of the conventional DAB-MMC as the DC/DC converter . . . . . . . . . . . . . . . 63
**Figure 2.39** Schematic diagram of the DAB-MMC based on CTB. 64
**Figure 2.40** The schematic diagram of the DAB-MMC based on TAC. . . . . . . . . . . . . . . . . . . . . . . . 65
**Figure 2.41** Schematic diagram of the DAB-AAMMC . . . . . 66
**Figure 2.42** Schematic diagram of the DAB-MMC based on HCTC. . . . . . . . . . . . . . . . . . . . . . . 67
**Figure 2.43** Schematic diagram of the DC Autotransformer. . . 68
**Figure 2.44** Schematic diagram of the resonant DC/DC converters. . . . . . . . . . . . . . . . . . . . . . 69
**Figure 2.45** Schematic diagram of the multi stage resonant DC/DC converter. . . . . . . . . . . . . . . . . . . 70
**Figure 2.46** Schematic diagram of the DC-MMCs: a) equipped with filter; and b) equipped with control scheme. . 71
**Figure 2.47** Schematic diagram of both types of Choppers: a) Capacitive Accumulated Choppers; and b) Inductive Accumulation Choppers. . . . . . . . . . . . . 72
**Figure 2.48** General classification of the DC-PFCs. . . . . . . . 73
**Figure 2.49** Schematic diagram of the R-type SDC-PFCs: a) Discrete R-type SDC-PFC; b) Variable R-type SDC-PFC. . . . . . . . . . . . . . . . . . . . . . 74
**Figure 2.50** Schematic diagram of the V-type SDC-PFC. . . . . 75
**Figure 2.51** Conceptual structure of the CDC-PFC. . . . . . . . 75
**Figure 2.52** Schematic diagram of an IDC-PFC: a) Original Structure b) Simplified Structure. . . . . . . . . . . 76
**Figure 3.1** Daily load demand varriation . . . . . . . . . . . . 88
**Figure 3.2** Daily variation of PVDG output power. . . . . . . 89
**Figure 3.3** Daily variation of WTDG output power. . . . . . . 90
**Figure 3.4** Single diagram of test systems: a). IEEE 33-bus, b). IEEE 69-bus. . . . . . . . . . . . . . . . . . . . . 94
**Figure 3.5** Boxplot of MOI after applying the algorithms for the IEEE 33-bus: a). PVDG case, b). WTDG case. . 99
**Figure 3.6** Boxplot of MOI after applying the algorithms for the IEEE 69-bus: a). PVDG case, b). WTDG case. . 100

**Figure 3.7** Convergence curves of the applied algorithms for the IEEE 33-bus: a). PVDG Case, b). WTDG Case. . . . . . . . . . . . . . . . . . . . 101
**Figure 3.8** Convergence curves of the applied algorithms for the IEEE 69-bus: a). PVDG Case, b). WTDG Case. . . . . . . . . . . . . . . . . . . . 102
**Figure 3.9** Total voltage derivation variation for both test systems: a). IEEE 33-bus, b). IEEE 69-bus. . . . . . . 103
**Figure 3.10** Daily voltage profiles variation for the IEEE 33-bus: a). Basic Case, b). After PVDG, c). After WTDG. . . . . . . . . . . . . . . . . . . . . . . . . 105
**Figure 3.11** Daily voltage profiles variation for the IEEE 69-bus: a). Basic Case, b). After PVDG, c). After WTDG. . . . . . . . . . . . . . . . . . . . . . . . . 106
**Figure 3.12** Active power loss variation for the IEEE 33-bus: a). Basic Case, b). After PVDG, c). After WTDG. . . 107
**Figure 3.13** Active power loss variation for the IEEE 69-bus: a). Basic Case, b). After PVDG, c). After WTDG. c). After WTDG. . . . . . . . . . . . . . . . . . . . . . 108
**Figure 3.14** The total operation time of overcurrent relay variation for both test systems:a). IEEE 33-bus, b). IEEE 69-bus. . . . . . . . . . . . . . . . . . . . . . . . . 109
**Figure 5.1** Schematic of small DC microgrid . . . . . . . . . 140
**Figure 5.2** Classification of different types of cyber attacks . . 143
**Figure 5.3** Fundamental structure of a Blockchain . . . . . . . 145
**Figure 5.4** Enactment of the differential protection scheme proposed . . . . . . . . . . . . . . . . . . . . . . . 147
**Figure 5.5** Capacitor and freewheeling stages . . . . . . . . . 148
**Figure 5.6** A line's equivalent model during a fault . . . . . . 148
**Figure 5.7** Fault current without the use of a method of isolation. . . . . . . . . . . . . . . . . . . . . . . . 150
**Figure 5.8** Identifying and isolating the fault with 1Ohm fault resistance. . . . . . . . . . . . . . . . . . . . . . . . 150
**Figure 5.9** Identifying and isolating the fault with 10 Ohms fault resistance. . . . . . . . . . . . . . . . . . . . . 151
**Figure 5.10** A Cyber-Attack on the protection system . . . . . 152
**Figure 5.11** Identifying and isolating the fault during cyber-attack with fault resistance 1 ohm . . . . . . 152

**Figure 5.12** Identifying and isolating the fault during cyber-attack with fault resistance 10 ohms . . . . . . . . 152
**Figure 6.1** Four critical steps in microgrid planning. . . . . . . 159
**Figure 6.2** Modeling of uncertainties for PV, WT, and load in optimal microgrid planning. . . . . . . . . . . . . 167
**Figure 6.3** Automation equipment used in active distribution systems. . . . . . . . . . . . . . . . . . . . . . . 173
**Figure 6.4** The steps of the system restoration. . . . . . . . . 175
**Figure 7.1** Global capacity of the PV energy system and its annual additions, 2007–2017 . . . . . . . . . . . . 186
**Figure 7.2** A share of grid-connected, grid-connected decentralized and off-grid installations, 2006–2016 . . . 187
**Figure 7.3** Unscheduled maintenance occurrences by components . . . . . . . . . . . . . . . . . . . 188
**Figure 7.4** Failure distribution of components in PV inverter . 189
**Figure 7.5** PV system configurations . . . . . . . . . . . . . . 190
**Figure 7.6** Typical failure rate curve as a function of time . . . 192
**Figure 7.7** The internal structure of an IGBT module . . . . . 195
**Figure 7.8** Lifetime vs Cost . . . . . . . . . . . . . . . . . . 197
**Figure 7.9** Electrolytic capacitor failure modes, failure mechanism, and their causes . . . . . . . . . . . . . . . . 198
**Figure 8.1** Classification of EES energy storage systems. . . . 215
**Figure 8.2** Classification of types of supercapacitors . . . . . 217
**Figure 8.3** Capacitor cloud storage . . . . . . . . . . . . . . . 218
**Figure 8.4** How a superconducting magnetic storage system works. . . . . . . . . . . . . . . . . . . . . . . . 219
**Figure 8.5** How the storage pump system works . . . . . . . . 221
**Figure 8.6** Scheme of operation of a CAES storage system . . 222
**Figure 8.7** How the flywheel storage system works and . . . . 223
**Figure 8.9** How high temperature thermal energy storage system works. . . . . . . . . . . . . . . . . . . . . 226
**Figure 8.10** How high temperature thermal energy storage system works. . . . . . . . . . . . . . . . . . . . . 227
**Figure 8.11** How lithium-ion battery works . . . . . . . . . . . 230
**Figure 8.12** High temperature battery operation (sulfur-sodium). 231
**Figure 8.13** Key technologies according to initial investment needs and technology risk . . . . . . . 233
**Figure 8.14** Classification of storage usage in energy system process. . . . . . . . . . . . . . . . . . . . . . . 234

**Figure 9.1** Wireless network . . . . . . . . . . . . . . . . . . . . 241
**Figure 9.2** Sensor networks protocol stack . . . . . . . . . . . 242
**Figure 9.3** Classification of network lifetime maximization techniques . . . . . . . . . . . . . . . . . . . . . . . . 243
**Figure 10.1** Basic configuration of distribution networks with a SOP. . . . . . . . . . . . . . . . . . . . . . . . . . . . 255
**Figure 10.2** Main circuit topology of the back-to-back VSC-based SOP. . . . . . . . . . . . . . . . . . . . . . 255
**Figure 10.3** Basic functions of SOPs. . . . . . . . . . . . . . . . 256
**Figure 10.4** Single line diagram of a simple distribution system, where option A represents a NOP connection and option B represents a SOP connection. . . . . . . . 257
**Figure 10.5** Power electronic devices in distribution networks for control of voltage and power flow. . . . . . . . 258
**Figure 10.6** A distribution network with a SOP at the remote ends of two feeders. . . . . . . . . . . . . . . . . . . 260
**Figure 10.7** An example of a SOP's operating point: active and reactive power provided by a SOP with two voltage source converters with the same rating. . . . . . . . 261
**Figure 10.8** Flow chart for the supply restoration strategy using SOPs. . . . . . . . . . . . . . . . . . . . . . . . . . . 262
**Figure 10.9** Classification of existing SOP configurations. . . . 264
**Figure 10.10** Connection diagram of existing SOP configurations. . . . . . . . . . . . . . . . . . . . . 265
**Figure 10.11** Power flow of a three-terminal SOP under various operating conditions. . . . . . . . . . . . . . . . . 266
**Figure 10.12** Planning of SOP placement in active distribution networks. . . . . . . . . . . . . . . . . . . . . . . . 268
**Figure 10.13** Process flow diagram to select candidate locations of SOPs in a distribution system based on PFBI and VDI. . . . . . . . . . . . . . . . . . . . . . . . . . . 269
**Figure 10.14** Locations of SOPs in Taiwan power company distribution system. . . . . . . . . . . . . . . . . . 270
**Figure 10.15** The framework for a two-stage coordinated optimization. . . . . . . . . . . . . . . . . . . . . . 271
**Figure 10.16** Block diagram of the power control loop of the SOP in the power flow control mode. . . . . . . . . . . 273
**Figure 10.17** Current control loop block diagram of the SOP in the power flow control mode. . . . . . . . . . . . . 274

**Figure 10.18** PLL block diagram of the SOP in the power flow control mode. . . . . . . . . . . . . . . . . . . . . . . 275
**Figure 10.19** The control block diagram of the interface voltage source converter for the supply restoration control mode. . . . . . . . . . . . . . . . . . . . . . . . . . . 276
**Figure 10.20** The control mode transition system. . . . . . . . . 277
**Figure 11.1** Different types of AWE systems. . . . . . . . . . . 289
**Figure 11.2** Extraction of energy from wind. . . . . . . . . . . 290
**Figure 11.3** Different types of floating wind structures. . . . . . 291
**Figure 11.4** Fast rotating rotor. . . . . . . . . . . . . . . . . . . . 296
**Figure 11.5** Multiple rotor setup (MRS). . . . . . . . . . . . . . 298
**Figure 11.6** Multi-rotor DAWT. . . . . . . . . . . . . . . . . . . . 300
**Figure 11.7** Different small wind turbines. . . . . . . . . . . . . 301
**Figure 11.8** Wind induced energy harvesting aeroelastic phenomena. . . . . . . . . . . . . . . . . . . . . . . . . . 303
**Figure 11.9** VIV and galloping device. . . . . . . . . . . . . . . 304

# List of Tables

**Table 1.1** Comparison between conventional charging stations and SST-based intelligent photovoltaic charging stations . . . 4
**Table 1.2** Cyber-attacks on smart grids . . . 13
**Table 2.1** The comparison of various voltage source SMs. . . . 42
**Table 2.2** The comparison of various current source SMs. . . . 53
**Table 3.1** The daily main characteristics of the investigated SDG test systems. . . . 94
**Table 3.2** The obtained results after optimization for IEEE 33-bus. . . . 95
**Table 3.3** The obtained results after optimization for IEEE 69-bus. . . . 96
**Table 5.1** The important DC microgrid's energy resources . . . 140
**Table 5.2** Different energy storage system technologies . . . 141
**Table 6.1** Summary of objectives used for optimal microgrid planning in the literature. . . . 162
**Table 6.2** Summary of different optimization methodologies used in literature. . . . 171
**Table 6.3** A comparison between remotely controlled switches and manual switches. . . . 174
**Table 6.4** Summary of objectives used for optimal placement of switches in the literature. . . . 176
**Table 7.1** Critical stressors for power devices (High to the low level of importance $\rightarrow$***$\rightarrow$**$\rightarrow$*) . . . 194
**Table 7.2** Model parameters of the Semikorn model . . . 201
**Table 7.3** Bayerers model parameters . . . 202
**Table 7.4** Summary and comparative analysis of different lifetime models . . . 202
**Table 8.1** Some energy storage technologies and examples of related projects . . . 213

**Table 8.2** Types of advanced electrical energy storage technologies. . . . . . . . . . . . . . . . . . . . . . . . . 216
**Table 10.1** Comparison of emulation and nomograms . . . . . . 257
**Table 10.2** Power electronics devices for network compensation in distribution systems. . . . . . . . . . . . . . . . . . 259
**Table 10.3** Control modes of the back-to-back voltage source converter–based SOP. . . . . . . . . . . . . . . . . . . 261
**Table 10.4** Contol modes of DC-SOPs . . . . . . . . . . . . . . . 267
**Table 10.5** The location and capacity of SOPs. . . . . . . . . . . 269
**Table 10.6** Optimization objectives of SOPs under normal conditions . . . . . . . . . . . . . . . . . . . . . . . . . 275
**Table 10.7** The summary of objectives for service restoration using SOPs in distribution networks . . . . . . . . . . 278

# List of Contributors

**Abbasipour, M.,** *The University of Saskatchewan, Saskatoon, SK S7N 5A9, Canada; E-mail: mehdi.abbasipour@usask.ca*

**Agarwal, A.K.,** *Chitkara University Institute of Engineering and Technology, Chitkara University, Punjab, India; E-mail: ambuj4u@gmail.com*

**Alok Misra, A.,** *Institute of Engineering and Technology, Lucknow; E-mail: alokalokmm@gmail.com*

**Appadurai, M.,** *Department of Mechanical Engineering, Dr. Sivanthi Aditanar College of Engineering, Tamilnadu, India; E-mail: appadurai86@gmail.com*

**Belbachir, N.,** *Department of Electrical Engineering, University of Mostaganem, Mostaganem, Algeria; E-mail:*

**El-Bayeh, C.Z.,** *Canada Excellence Research Chairs Team, Concordia University, Montréal, Algeria; E-mail:*

**Eskandari, M.,** *The School of Electrical Engineering and Telecommunications, University of New South Wales, Sydney, NSW 2052, Australia; E-mail: m.eskandari@unsw.edu.au*

**Faried, S.O.,** *The University of Saskatchewan, Saskatoon, SK S7N 5A9, Canada; E-mail: sherif.faried@usask.ca*

**Gatla, R.K.,** *Department of Electrical and Electronics Engineering, Institute of Aeronautical Engineering, Hyderabad, India; E-mail: ranjith.gatla@gmail.com*

**Haque Shawon, S.M.R.,** *The University of Saskatchewan, Saskatoon, SK S7N 5A9, Canada; E-mail: shs054@mail.usask.ca*

**Irudaya Raj, E.F.,** *Department of Electrical and Electronics Engineering, Dr. Sivanthi Aditanar College of Engineering, Tamilnadu, India; E-mail: fantinraj@gmail.com*

**Jahangiri, A.,** *Hamedan Branch, Islamic Azad University, Hamedan, Iran; E-mail: Jahangiri87@gmail.com*

**Khullar, V.,** *Chitkara University Institute of Engineering and Technology, Chitkara University, Punjab, India; E-mail: vikas.khullar@gmail.com*

**kumar, D.G.,** *Department of Electrical and Electronics Engineering, B.V Raju Institute of Technology, Narsapur, Telangana, India; E-mail:*

**Liang, X.,** *The University of Saskatchewan, Saskatoon, SK S7N 5A9, Canada; E-mail: xil659@mail.uasak.ca*

**Lu, J.,** *School of Information Science and Engineering, Wuhan University of Science and Technology, Wuhan, P R Chinal; E-mail:*

**Manimala, K.,** *Department of Electrical and Electronics Engineering, Dr. Sivanthi Aditanar College of Engineering, Tamilnadu, India; E-mail: smonimala@gmail.com*

**Mitolo, M.,** *Irvine Valley College, Irvine, CA 92618, USA; E-mail: mmitolo@ivc.edu*

*Dr. Sivanthi Aditanar College of Engineering, Tamilnadu, India; E-mail: fantinraj@gmail.com*

**Mohapatra, B.,** *Einstein Academy of Technology and Management. Odisha; E-mail:*

**Moradi, M.H.,** *Hamedan Branch, Islamic Azad University, Hamedan, Iran*

**Nasab, M.A.,** *CTIF Global Capsule, Department of Business Development and Technology, Herning 7400, Denmark; E-mail: m.aziminasab@iaub.ac.ir*

**Nithya, S.,** *Department of EEE, SRMIST, Ramapuram, Chennai; E-mail: nithisavidhina@gmail.com*

**Nourizadeh, H.,** *Department of Electrical Engineering, Shahid Beheshti University of Tehran, Iran; E-mail: H_nourizadeh@sbu.ac.ir*

**Padmanaban, S.,** *CTIF Global Capsule, Department of Business Development and Technology, Herning 7400, Denmark; E-mail: Sanjeev@btech.au.dk*

**Parimala, D.M.,** *Department of ECE, Velalar College of Engineering and Technology, Erode, Tamil Nadu; E-mail: parimaladevi.vlsi@gmail.com*

**Prasad, K.N.V.,** *School of Electrical and Automation Engineering, Hefei University of Technology, Hefei, PR China*

**Saaklayen, Md.A.,** *The University of Saskatchewan, Saskatoon, SK S7N 5A9, Canada; E-mail: vpu975@mail.usask.ca*

**Samavat, T.,** *CTIF Global Capsule, Department of Business Development and Technology, Herning 7400, Denmark; E-mail:*

**Sanjeevikumar, P.,** *CTIF Global Capsule, Department of Business Development and Technology, Herning 7400, Denmark; E-mail: sanjeev@btech.au.dk*

**Settoul, S.,** *Department of Electrotechnic, University of Constantine 1, Constantine, Algeria; E-mail:*

Shahbazi, M., *CTIF Global Capsule, Department of Business Development and Technology, Herning 7400, Denmark; E-mail: M.shahbazi@basu.ac.ir*

**Sridhar, P.,** *Department of Electrical and Electronics Engineering, Institute of Aeronautical Engineering, Hyderabad, India*

**Tiwari, R.G.,** *Chitkara University Institute of Engineering and Technology, Chitkara University, Punjab, India; E-mail: rajgaurang@gmail.com*

**Vijayalakshmi, K.,** *Department of EEE, SRMIST, Ramapuram, Chennai; E-mail: vijayalk1@srmist.edu.in*

**Zand, M.,** *CTIF Global Capsule, Department of Business Development and Technology, Herning 7400, Denmark; E-mail: mo.zand@iaub.ac.ir*

**Zellagui, M.,** *Département de Génie* Electrique, *École de Technologie Supérieure, Montréal, Canada, Department of Electrical Engineering, University of Batna 2, Batna, Algeria; E-mail: m.zellagui@univ-batan2.dz*

# List of Abbreviations

| | |
|---|---|
| $\alpha$ and $\beta$ | Shape parameters of the beta distribution |
| $\Delta V$ | Voltage drops of the distribution line |
| $\Delta V_{max}$ | Maximum voltage drops |
| $\delta_i$, $\delta_j$ | Angles at buses $i, j$ |
| $\eta_{\mathrm{MT}}$ | Electrical efficiency of MT |
| $\imath$Prated | Rated output power |
| $\lambda$ | Parameter of load power demand |
| $\mu$ and $\sigma$ | Mean and standard deviation of historical solar data |
| $\mu$, $\sigma$ | Mean and standard deviation |
| $\mu_{\mathrm{P}}$ and $\mu_{\mathrm{Q}}$ | Means of normal PDF for active and reactive power |
| $\sigma_{\mathrm{P}}$ and $\sigma_Q$ | Variances of normal PDF for active and reactive power |
| $\mathrm{C_{MT}}$ | Cost function of the generated power of MT |
| $\mathrm{C_{ng}}$ | Price of natural gas |
| f(s) | Beta PDF of the solar irradiation |
| F | Fuel consumption of diesel engine |
| $\mathrm{F_0}$ and $\mathrm{F_1}$ | Fuel consumption curve fitting coefficients |
| $f\,(v)$ | Weibull PDF of the wind $\mathrm{v_c}$, $\mathrm{v_r}$ and $\mathrm{v_f}$ Cut-in, rated and cut-out wind speed |
| FF | Fill factor $\mathrm{V_{MPP}}$ and $\mathrm{I_{MPP}}$ Voltage and current at the maximum power point |
| $\mathrm{I_{SC}}$ and $\mathrm{V_{OC}}$ | Short circuit current and open circuit voltage |
| $\mathrm{K_i}$ and $\mathrm{K_V}$ | Temperature coefficients of current and voltage |
| $\mathrm{K_P}$ | Temperature coefficient of power |
| $\mathrm{L_h}$ | Low-hot value of natural gas |
| NOT | Nominal operating temperature |
| $\mathrm{P_{MT}}$ | Generated power by each MT unit |
| $\mathrm{P_S}$ | Probabilistic generation of each PV module |
| $\mathrm{P_{STC}}$ | Output of the PV module under standard test conditions |

| | |
|---|---|
| Pgenerated | Actual output power |
| S(t) | SOC at the time $t$ |
| s | Solar irradiation |
| $S_{AV}$ | Average solar irradiation |
| $S_{STC}$ | Solar irradiation under standard test conditions |
| $T_C$ | Working temperature of solar panels |
| $T_C$ and $T_A$ | Cell and the ambient temperature of PV module |
| $V_m$ | Mean value of the wind speed |
| $A, B$ | Parameters of $f_b(s)$ |
| $A, B$ | Constants equal to 0.14 and 0.02 |
| $APL_{Before/AfterRDG}$ | Active power losses before and after RDG |
| $c$ | Scale index |
| $CTI_{Before/AfterRDG}$ | Coordination time interval before and after RDG |
| $f(P_L)$ and $f(Q_L)$ | Normal PDF of active and reactive power |
| $f_b(s)$ | Beta distribution function of $s$ |
| $I_F$ | Fault current |
| $I_P$ | Pickup current |
| $I_{sc}$ | Short-circuit current $(A)$ |
| $K_v, K_i$ | Current/voltage temperature coefficients ($A/°$C and $V/°$C) |
| $N_{bus}$ | Bus number |
| $N_{OT}$ | Nominal operating temperature of cell (°C) |
| $n_{RDG,i}$ | Location of RDG units at bus $i$ |
| $N_{RDG.max}$ | Maximum number of RDG units |
| $N_{RDG}$ | Number of RDG units |
| $N_{PR}, N_{BR}$ | Number of primary and the backup overcurrent relays |
| $OT_{Bachup}$ | Backup relay's operation time |
| $OT_{Before/AfterRDG}$ | Operation time before and after RDG |
| $OT_{Primary}$ | Primary relay's operation time |
| $P_{SB}(t)$ | Charging or discharging power at the time $t\eta_c$ and $\eta_d$ Charge and discharge efficiency |
| $P_{ij}, Q_{ij}$ | Active and reactive powers of the line |
| $P_L$ and $Q_L$ | Active and reactive power |
| $P_{out}$ and $P_{rated}$ | Output and rated power of each WT |
| $P_{RDG}^{min}, P_{RDG}^{max}$ | Active power output limits of RDG |
| $P_{WT}$ | WT generator output power |
| $P_D, Q_D$ | Total active and reactive of demand load |
| $P_G, Q_G$ | Total active and reactive powers of generator |

| | |
|---|---|
| $P_{RDG}$ | Total active power injected from RDG |
| $Q_k$, $P_k$ | Reactive and active powers injected at bus $k$ |
| $Q_{ok}$, $P_{ok}$ | Reactive and active load powers at rated voltage at bus $k$ |
| $Q_{RDG}^{min}, Q_{RDG}^{max}$ | Reactive power output limits of RDG |
| $Q_{RDG}$ | Total reactive power injected from RDG |
| $R_{ij}$, $X_{ij}$ | Resistance and reactance of the line |
| $RDG_{Position}$ | Position of RDG units |
| $RPL_{Before/AfterRDG}$ | Reactive power losses before and after RDG |
| $S$ | Solar irradiance's random variable ($kW/m^2$) |
| $s_1, s_2$ | Solar irradiance limits of state $s$ |
| $S_{ij}$ | Apparent power in branch |
| $S_{max}$ | Maximum apparent power |
| $T_{cy}$, $T_A$ | Cell and ambient temperatures (°C) |
| $V$ | Average wind speed over each period ($hour$) |
| $v$ | Wind speed |
| $v_{ci}, v_{ci}$, and $v_{co}$ | Cut-in speed, rated speed, and cut-off speed of the WT |
| $V_i$, $V_j$ | Voltage at bus $i$, and $j$ |
| $V_{MPP}$, $I_{MPP}$ | Maximum power point's voltage and current |
| $V_{oc}$ | Open-circuit voltage ($V$) |
| $V_{min}, V_{max}$ | Specified voltages limits |
| $VD_{Before/AfterRDG}$ | Voltage deviation before and after RDG |
| AAMMCs | Alternative arm modular multi-level converter |
| AC | Alternating current |
| ADN | Active distribution network |
| AGSO | Adaptive group search optimization |
| AHP | Analytic hierarchical process |
| ALO | Ant lion optimisation |
| APL | Active power loss |
| APLI | Active power loss index |
| B2B | Back-to-Back |
| B-t-B-VSC | Back-to-back voltage source converter |
| BTSO | Backtracking search optimization |
| CCG | Column-and-constraint generation |
| CDC-PFC | Cascaded dc power flow controller |
| CS | Current source |
| CSC | Current source converters |
| CTB | Controlled transition bridge |

| | |
|---|---|
| CTI | Coordination time interval |
| CTII | Coordination time interval index |
| CVR | Conservation voltage reduction |
| DAB | Dual active bridge |
| DAB-AAMMC | Dual active bridge-alternative arm modular multi-level |
| DC | Direct current |
| DC-CFCs | Dc current flow controllers |
| DC-PFCs | Dc power flow controllers |
| DG | Distributed generation |
| DG | Distributed generation |
| DR | Demand response |
| DRS | Distributed reactive source |
| DSO | Distribution system operator |
| DSO | Distribution system operator |
| D-STATCOM | Distribution static synchronous compensator |
| D-UPFC | Distribution unified power flow controller |
| EENS | Expected energy not supplied |
| EIR | Energy index of reliability |
| EMI | Electromagnetic interference |
| ENS | Energy not supplied |
| ES | Energy storage |
| ESS | Energy storage systems |
| FACTS | Flexible alternating current transmission system |
| FACTS | Flexible ac transmission systems |
| FC | Flying capacitor |
| FF | Fill factor |
| GA | Genetic algorithm |
| GOA | Grasshopper optimization algorithm |
| GSO | Group search optimization |
| GWO | Grey wolf optimizer |
| HCTC | Hybrid cascaded tow-level converters |
| HCVSCs | Hybrid current & voltage source converters |
| HCVSCs | Hybrid current and voltage source converters |
| HV | High voltage |
| HVDC | High voltage direct current |
| IDC-PFC | Interline dc power flow controller |
| IGBT | Insulated gate bipolar transistor |
| IPFC | Interline power flow controller |

| | |
|---|---|
| IPOP | Input-parallel output-parallel |
| IPOS | Input-parallel output-series |
| ISOP | Input-series output-parallel |
| ISOS | Input-series output-series |
| KCL | Kirchhoff's current law |
| KVL | Kirchhoff's voltage law |
| LF | Load factor |
| LSF | Loss sensitivity factor |
| LVRT | Low-voltage ride-through |
| MAIFIe | Momentary average interruption event frequency index |
| MCSCs | Modular current source converters |
| MF | Medium frequency |
| MFO | Moth flame optimizer |
| MG | Microgrid |
| MILP | Mixed integer linear programming |
| MINLP | Mixed integer non-linear program |
| MINLP | Mixed integer non-linear programming |
| MISOCP | Mixed integer second order cone programming |
| MMCs | Modular multi-level Converters |
| MMCSCs | Modular multi-level current source converters |
| MOI | Multi-objective indices |
| MPA | Marine predators algorithm |
| MT | Micro turbine |
| MT-HVDC | Multi-terminal HVDC |
| N | Modules' number |
| NMG | Networked microgrid |
| NOPs | Normally open points |
| NPC | Neutral point clamped |
| NSGA | Non-dominated sorting genetic algorithm |
| OLTC | On-load tap changer |
| OT | Operation time |
| OTI | Operation time index |
| P2P | Point-to-Point |
| PDF | Probability density function |
| PDF | Probability distribution function |
| PFBI | Power flow betweenness index |
| PLL | Phase-locked loop |
| PSO | Particle swarm optimization |

| | |
|---|---|
| PSO | Particle swarm optimization |
| PV | Photovoltaic |
| PV | Photovoltaic |
| PVDG | Photovoltaic distributed generator |
| PWM | Pulse-width-modulation |
| PWM | Pulse width modulation |
| RB | Reverse blocking |
| RC | Reverse conducting |
| RDG | Renewable distributed generator |
| RESs | Renewable energy sources |
| RO | Robust optimization |
| RPL | Reactive power loss |
| RPLI | Reactive power loss index |
| R-type | Resistance-type |
| SAIDI | System average interruption duration index |
| SAIFI | System average interruption frequency index |
| SDC-PFC | Series dc power flow controller |
| SDG | Smart distribution grid |
| SHE | Selective harmonic elimination |
| SMs | Sub-modules |
| SOC | State of charge |
| SOP | Soft open point |
| SPPs | Solar power plants |
| SQP | Sequential quadratic programming |
| SSSC | Synchronous series compensator |
| SST | Solid state transformer |
| ST | Smart transformer |
| SWRT | Single-wire earth return |
| TAC | Transition arm converter |
| TDS | Time dial setting |
| THD | Total harmonic distortion |
| VD | Voltage Deviation |
| VDI | Voltage Deviation Index |
| VDI | Voltage deviation index |
| VS | Voltage source |
| VSC | Voltage source converters |
| VSC | Voltage source converters |
| VSF | Voltage sensitivity factor |
| V-type | Voltage-type |

| | |
|---|---|
| WPPs | Wind power plants |
| WT | Wind turbine |
| WT | Wind turbines |
| WTDG | Wind turbine distributed generator |
| ZVS | Zero-voltage switching |

# 1

# Power Electronics in Smart Grid

**Padmanaban Sanjeevikumar[1,*], Mohammad Zand[1], Morteza Azimi Nasab[1], Mohsen Eskandari[2], Tina Samavat[1], Alireza Jahangiri[3], and Mohammad H. Moradi[3]**

[1]CTIF Global Capsule, Department of Business Development and Technology, Herning 7400, Denmark
[2]The School of Electrical Engineering and Telecommunications, University of New South Wales, Sydney, NSW 2052, Australia
[3]Hamedan Branch, Islamic Azad University, Hamedan, Iran
E-mail: m.aziminasab@iaub.ac.ir; mo.zand@iaub.ac.ir; sanjeev@btech.au.dk; Jahangiri87@gmail.com; m.eskandari@unsw.edu.au
*Corresponding Author

## Abstract

The high penetration of emerging power electronics-based distributed energy resources into electric grids has raised control, management, and stability issues. Besides, the electrical grids have been converted to multiinput multi output cyber-physical systems (CPSs) that need advanced communication infrastructure to be monitored and optimally controlled. The smart grid (SGs) concept has been introduced as the platform that facilitates dealing with the challenges of modern electric grids. Monitor and control capacities, automatic grid control, and active consumer participation in the energy industry expand the SGs influence. Mentioned emphatic features of SGs have been achieved by combining power system technologies with emerging power electronics and digital telecommunications technologies. Since almost all the distributed energy resources (DERs) technologies, such as renewable energy resources, micro sources, and energy storage systems, use the power electronic interfaces, it is vital to match the characteristics of the DERs with the grid requirements. On the other hand, power electronics provide more flexibility, controllability, and better power quality, which are essential factors in SGs. However, with the rapid progression of physical systems in power electronics applications to interface DERs combined with cyber frameworks, the threats of cyber have a significant impression on SG security and performance. A comprehensive study of the vital power electronic role in SGs as CPSs to provide sustainable/reliable energy resources is provided in this chapter. Specifically, cyber-attacks in power electronics systems are studied briefly.

## 1.1 Introduction

Nowadays, with the increase in the amount of demanded electricity, sustainable energy has become a challenging issue in the electric power industry. Considering the environmental effects of traditional plans production, the joining of green sources signifies a brighter future for the next generation. Therefore, SGs are an essential factor in facilitating sustainable clean electricity that provides new capacities for DESs [1].

In order to convert the energy from micro/renewable resources to the grid, various electromechanical and power converters are used [2]. DERs in SGs have received more attention owing to providing opportunities to take advantage of cooperation of heat and power, demand reduction and peak shaving, backup power, better power quality, and ancillary services. Power electronics is a critical component in DERs technologies. Also, transforming the produced energy into usable energy that can be provided directly to the grid may require up to 40% of a distributed energy cost. Accordingly, improving DERs can dramatically reduce the costs of this sector [3].

Nevertheless, more analyses are required to guarantee security and dependability, different protections, and designs for various DERs [3].

In this chapter, the application of power electronics in the SGs is reviewed. The smart grid and its benefits are reviewed in the next section, and in the third section, power electronics in the SG are introduced. The fourth section describes the application of power electronics devices (power converters, FACTS, and HVDC devices) in SGs. In the fifth section, approaches to reduce cyber-attacks in electronic power systems in SGs are reviewed.

## 1.2 Smart Grid, Components and Advantages

Consumers' demand patterns and generators' productivity can be intelligently scrutinized through a smart grid network. Also, by power transmitters, the network will be under proper supervision and reach and persist in a steady state.

Another target is to change the power flow pattern. By now, there is only one flow from supplier to consumers; but the ideal layout is a two-way connection between them. In addition, solid-state transformer (SST) has been used recently for integration and intelligent energy management, which has the ability of coordination and intelligent planning.

### 1.2.1 Structure of Photovoltaic Intelligent Charging Station (PVCS) Based on SST Solid State Transformer

The structure of a photovoltaic charging station based on a solid-state transformer is shown in Figure 1.1. Includes AC/DC bidirectional converter module, DC/DC isolated bidirectional converter, DC/DC converter connected to a photovoltaic (PV) system, and DC/DC bidirectional converter for electric vehicles. The SST controller is designed to integrate the energy management components, the central controller, the PV controller, and the charger controller; therefore, the SST controller can perform almost all the PV charging station structure functions. Compared to a conventional PV charging station, the advantages of an SST-based PV charging station are as follows [14]:

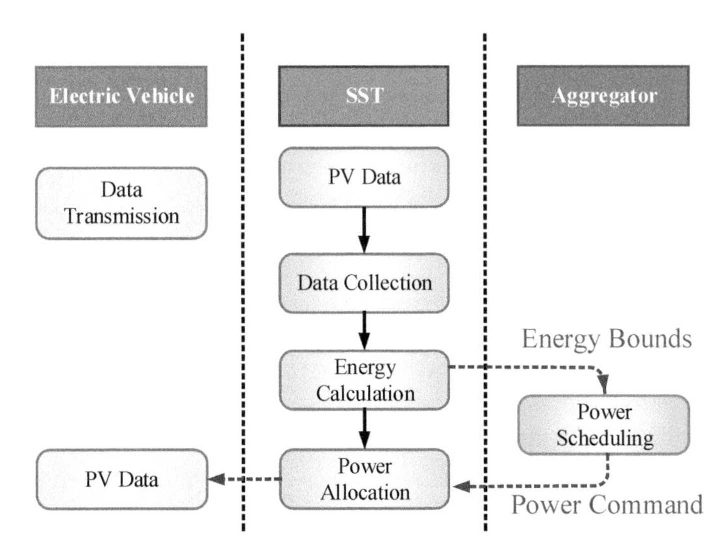


**Figure 1.1** SST-based scheduling for an example of energy-intensive smart grid with integrated parking charge and photovoltaic f

**Table 1.1** Comparison between conventional charging stations and SST-based intelligent photovoltaic charging stations

| Device Name | SST Based Charging Station | Typical charging Station |
|---|---|---|
| Monitoring system | No | Yes |
| Central controller | Yes | Yes |
| Independent photovoltaic controller | No | Yes |
| Independent chargers | No | Yes |
| Independent one-way DC/DC modules | No | Yes |
| Integrated DC/DC chargers | Yes | No |
| Communication system | Simple | Complicated |
| Two-way AC/DC converter | Yes | Yes |
| Active power filter | No | Yes |
| Industrial frequency transformers | No | Yes |
| Medium frequency transformer | Yes | No |

1. PV charging station can be considered as a controlled system that functions as an active charge.
2. Medium frequency transformers are utilized to transmit power, making the volume of the PV charging station much smaller.
3. Making energy management and controlling the power of electric vehicles easier, the DC / DC converter charging controllers are integrated with SST controllers.
4. Energy management is becoming more efficient in real-time. Because of: a) the information sampling process is completed in a shorter period. b) control functions and energy management are integrated.
5. As a fully integrated construction, the complexity and costs of the communication system have been reduced.

Table 1.1 compares the relevant devices between the SST-based photovoltaic charging station and the standard photovoltaic charging station. The comparison results show that some primary devices are integrated, and some other devices are required. Therefore, the cost of an integrated system can be lessened [1].

The main advantage of a smart grid is defined as follows [4]:

1. Reliability

The most important feature of a smart grid is reliability, which results from different sources coordinated beside absolute control over demand response and energy management. Therefore, widespread blackout's possibility decreases.

2. Economy

These networks provide low-cost electricity owing to emerge of intelligent and high-tech devices. Integrating several alternative energy sources can lead to a reduction in the generating, delivering, and consuming costs. Also, new jobs will be available for local experts and talented researchers to take economic advantage of proposed job opportunities.

3. Performance

Issues such as the current climate change around the world have increased the tendency to green resources. This effort can reduce greenhouse gas emissions, employ several renewable sources, and improve electricity generation, delivery, and consumption. Also, automatic control and the presence of artificial intelligence increase efficiency and performance.

4. Security

It is obtained by reducing the probability and consequences of cyber-attacks and natural disasters.

5. Safety

Due to great observation over the function of the grid, error and needed time to respond to faults decrease, and mostly the final decisions are made by controlling systems that themselves have minor errors compared to manually applied changes while facing fails.

## 1.3 Power Electronic Converters in Smart Grids

Power converters are vital components in microgrids and smart grids. To improve the control system, most parts are envisaged to be joined to buses via power converters. Thus, the power stability and dynamic response in the micro/smart grid affect the performance of the power converters and control loops. The DC microgrid architecture is shown in Figure 1.1, containing renewable sources such as PV and wind turbine, joined to the High voltage DC bus which is considered to be about 400 volts. Additionally, local storage complements can be connected to high or low-voltage buses (48V, 24V, 12V). Among these, for safety reasons, loads are connected to LV buses, all sources, storage units, and often to different buses via an electronic power converter (EPC). Figure 1.2 represents the micro grid architecture [5].

Figure 1.3 shows the applications of electronic power converters for RESs and ESS in smart grids. As can be seen, a DC-AC electronic power converter

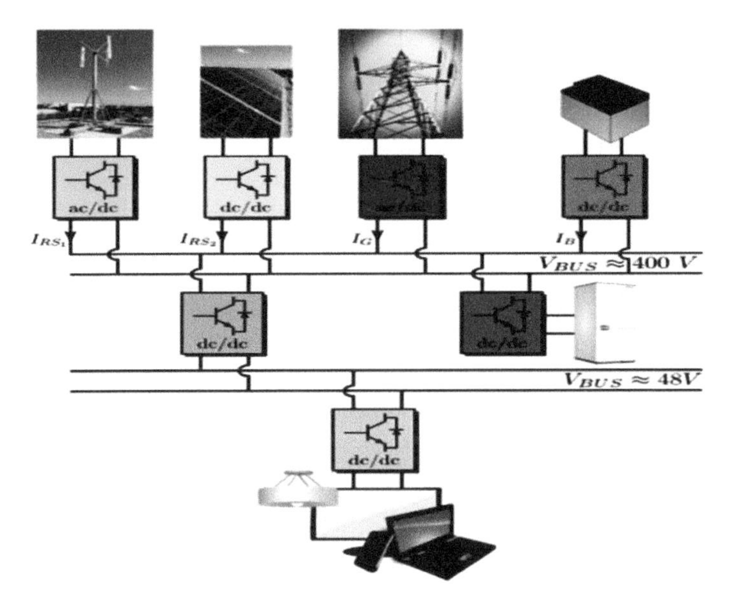


**Figure 1.2** Micro grid architecture design [5]

for the following three levels in the power system electrical is required in smart grids [6]:

1. Power distribution level
2. Industrial and service level
3. Residential level

## 1.4 Application of Power Electronic Technology in Smart Grid

Emerge of power electronics led to an evolution in the world. Particularly in SGs, the role of power electronics and its relevant technologies cannot be neglected. In such networks employing electronic power devices let experts control and transforms electrical energy [7]. Power circuits including HVDC controllers, FACTS controllers, static synchronous compensation (STATCOM), inverter-based DERs, and power converters help control and facilitate the transmission and distribution of generated power [8].

In addition, controllable electronic devices have been developed to fine-tune voltage, current, and frequency ranges while increasing the quality of energy supplied to consumers. [8].

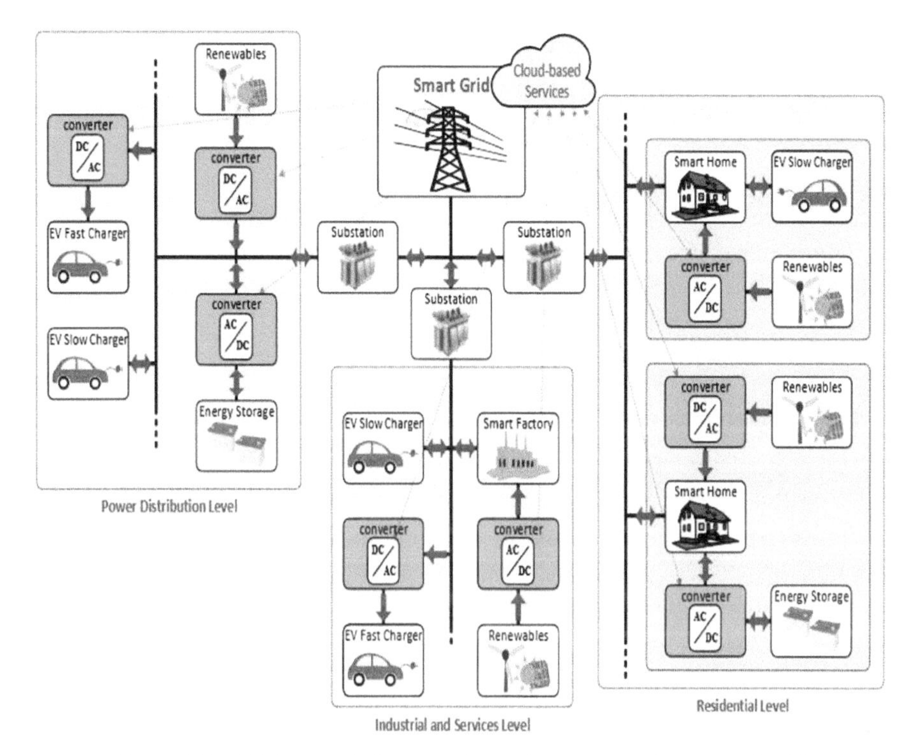


Figure 1.3 Grid-connected power electronic converters in smart grids [6]

### 1.4.1 The Application of DC-AC and DC-AC Converters in SG

As mentioned before, the use of SGs to meet the consumers' needs demands power converters. Specifically, these technologies communicate with the power grid using separate equipment (by a combination of ac-dc to dc-dc converters), which helps to boost performance [1].

Developing these power converters with modular design, lower cost, and improved reliability will improve smart grids' overall cost and durability of integrated renewable energy systems. In using a smart network, a two-way converter must have the following conditions [9]:

1. Transfer energy from/to the storage device.
2. Must implement charging algorithms for various storage elements: lead-acid, Ni-Cd, lithium-ion batteries, or large capacitors
3. Easily controllable.
4. Must be able to receive commands from a power management controller [9].

There are several distributed sources in a smart grid, and renewable sources are the most favorable type. An increase in resources and generators means facing the harmonics problem in a network. While the reduction of harmonics at the point of standard coupling (PCC) is of great importance and the opposite leads to technical problems. Multi-level inverters (MLIs) reduce harmonics in distributed sources and thus reduce the need for costly PCC filters. One of the most critical issues facing energy conservation from PV panels is the small DC voltage available. Therefore, increasing this voltage is essential. So, an appropriate amplifier converter stands before converting the DC voltage to AC [10].

### 1.4.2 The Application of HVDC Technology in Smart Grid

HVDC, at present, the direct current transmission system is mainly in the transmission link, while the power generation and power system used in the electric power system is still ongoing. The alternating current in the transmission line is transmitted to the rectifier equipment through the transformer equipment on the power supply side, and the AC current is converted to a direct high voltage current and then to the DC transmission line. Then, through the DC/DC transmission line from the inverter to return to the terminal equipment at the converter station, the high voltage direct current is converted to alternating current, and the current of the converter transformer is transferred to the AC system [11].

### 1.4.3 The Application of FACTS Technology in Smart Grid

The purpose of flexible AC transmission technology is to complement the new FACTS devices that are extensively utilized in smart grids [12].

FACTS technology combines electronic power equipment and the renovated control technology to obtain the transmission system parameters and quick and adaptable control over the network, which dramatically improves the capacity of transmission lines [12].

Figure 1.4 shows an ideal interconnected transmission network. Permissible spots for the installation of the FACTS are identified. The name implies that FACTS is a flexible intermittent transmission system, while the separate devices that assist the overall FACTS system are called FACTS devices. [13].

For making vital connections between renewable energy resources and transmission networks, FACTS has a crucial role. Also, it allows RES to meet network needs and is helpful to rise penetration while the quality of power and system stability are compromising [13].

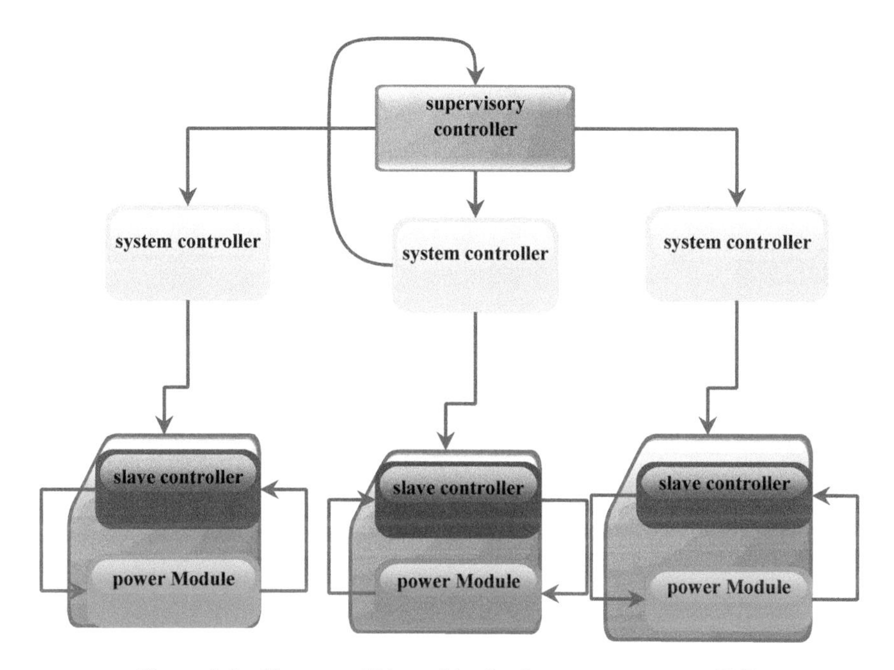


**Figure 1.4** The control hierarchies for the power converters [14]

## 1.5 CPS Attacks Mitigation Approaches on Power Electronic Systems for Smart Grid Applications

Physical systems have evolved rapidly during the last decades, particularly in the application of power electronics for utilizing renewable energy combined with the cyber framework, the performance of smart grid affected by cyber threats. Hackers may aim at communication networks where electronic devices are located. In case of such a cyber-attack digital controller can be disconnected through the physical control loop. Accordingly, specific approaches must be considered to protect the smart grid against an attacker.

Cyber-physical (CP) refers to devices that are directly connected to both the physical and cyber worlds. Also, distribution plants have CP aspects due to their direct connection with the physical aspects of smart grids [15].

Network power converters can be used in renewable energy systems, smart grids, telecommunications, smart homes, car drives, battery management systems [16, 17].

Figure 1.4 shows the power converters with hierarchical control variables. Three rows are the most widely used hierarchy [14].

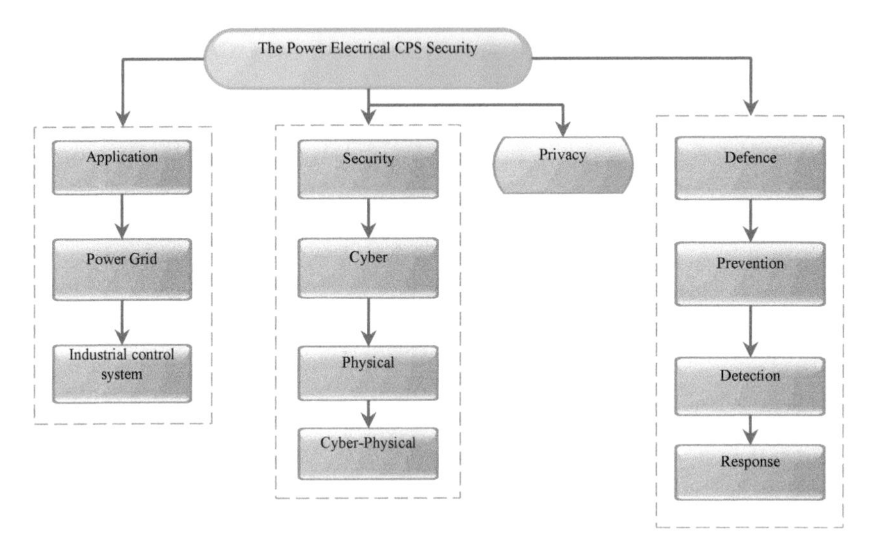


**Figure 1.5** Security taxonomy for power electronic CPS [14]

The monitoring controller may be a programmable logic controller (PLC) or other controllers that can turn off dynamic level commands of the subsystem and dynamic controllers such as speed/torque with emergency commands [18].

The concept of security in intelligent systems has significant importance to users. This issue is even more important in the electricity and smart grid industry. Because in general, energy supply and management groundworks are considered as sensitive sectors due to the dependence of all divisions of an association on electricity. There are countless complex connections in the smart grid, and vital data is recorded. In the case of cyber-attacks, the control over generators, users, and devices can be lost, making consequential trouble for the target region. Along these, the pattern of consumers and related information to each consumer is recorded, threatening their privacy. Therefore in CPSs, vital protection tools must be employed [19].

Figure 1.5 shows that the CPS electronic power system focuses on CPS electronic power systems in the smart grid and industrial control system.

### 1.5.1 Architecture of Digitally-Controlled Power Electronic CPSs

This structural design is commonly used to form cooperation of renewable resources (such as photovoltaic, wind), energy storage devices, and electrical charge arrangement through a smart grid [11].

Voltage source converter systems are connected to a comprehensive intelligent CP network via communications. In order to protect the network, for precise control, the steps mentioned below should be considered [14]:

1. Physical stage
2. Cyber stage
3. Voltage source converters role

1. Physical stage

Source Converter From inside the mains input, a mains voltage converter is connected as an interface between the two DC phases and the mains. In Figure 1.6, the input works through an internal interface to a micro AC network, stand-alone AC loads, or an AC network. Due to the connection between the different AC phases, there are countless suitable standards [12].

2. Cyber stage

There are several voltage source converters in each smart grid. Simultaneously the network is controlled with the help of synchronous generators, and also all these units are considered a standard factor for a smart grid with interconnected voltage source converters. In networks, connection topologies are represented as physical layout and generally form the links with cables. Some of the frequently used communication topologies [13] are tree, star, bass, mesh, ring, binary, Daisy chain, and combination, shown in Figure [13].

3. Voltage source converters role

In microgrids and renewable power systems, the principal role of voltage source converters as grid power supply units is network support. These roles can be described as follows [24]:

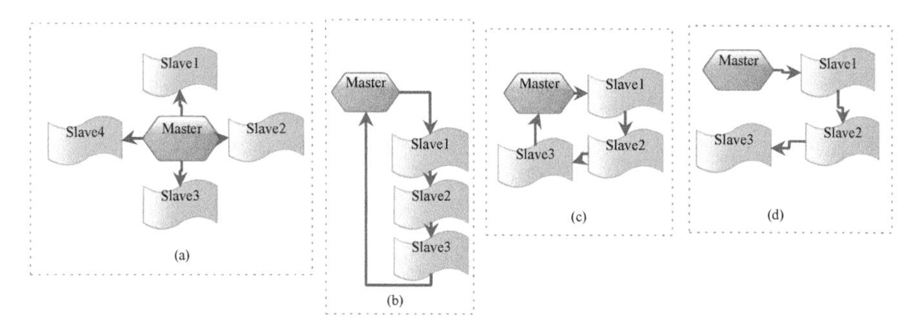


**Figure 1.6** Communication topologies for cyber structures: (a) star topology (b) bass topology (c) ring topology (d) Daisy chain topology [14]

1) Network power supply voltage converter unit

The primary function of this unit is to inject indeterminate current into the network.

2) Network converter voltage source converter unit

This unit is used to adjust the local voltage [20].

This graphic image shows both cyber structures, as dotted lines indicate information about it. For power converters, most control topologies include star, bus, ring, and the Daisy chain [14].

A centralized communication can be considered to connect PMUs (Phasor measurement units) and local controllers. SCADA is considered the most efficient coordination technique between international devices to facilitate monitoring in smart grids. For several factors, this method not only requires significant communication resources but is also sensitive to possible cyber-attacks. Decentralized control uses a design only as a local measurement. At the same time, the distributed control model is justifiable because computational resources are constantly allocated to achieve coordination. Therefore, it is low [15].

Figure 1.5 shows the outline of the control of voltage source converters linked to the AC network according to their time interval.

In both microgrids and renewable power systems, the primary roles of voltage source converters are considered as network power supply and network support units. These roles can be described as follows [14]:

1) Mains supply voltage converter unit:

The primary function of this unit is to inject unknown current into the network. Therefore, those are shown as current sources, as shown in

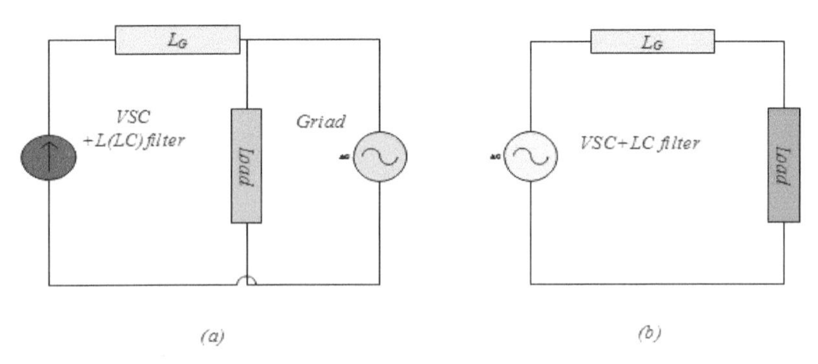


**Figure 1.7** Voltage source converter with main types: (a) power supply network and (B) mains voltage source converter [14]

Figure 1.7(a). For real-time operation, they include a dedicated synchronization unit, an external DC voltage control.

2) Network converter voltage source converter unit:

The task of this unit is to adjust the local voltage. Therefore, the ideal voltage source is chosen to simulate this section in Figure 1.7(b). Due to the voltage regulation, this becomes the main point of the system, which draws the local AC network.

Due to the parallel voltage source converters in the independent microgrid, the control law can be applied to both reactors to match the frequency $\omega^*$ with the voltage $V^*$ for synchronization, respectively [14].

$$V^* = V_{ref} - N_Q(Q - Q^*) \tag{1.1}$$

$$\omega^* = \omega_{ref} - M_P(P - P^*) \tag{1.2}$$

## 5-2. SCENARIOS OF CYBER-ATTACKS

In general, cyberattacks can affect four main aspects of power systems [26]:

1. Energy market
2. Status assessment
3. Voltage control
4. Automatic production control

FDI attacks can deceive system operators into believing that operating conditions are economically and physically viable if they do not exist [16].

An attacker could affect a communication network, for instance, by trying to connect to smart electronic devices and can lead to spy on telecommunications. Telecommunications providers, Connected SCADA, EMS, ISPs, and IT systems can be possible targets for Attackers.

**Table 1.2** Cyber-attacks on smart grids [17]

| | Transmission System | Distribution System | Device | System | Attack Type |
|---|---|---|---|---|---|
| Data Concentrator (DC) | ✓ | ✓ | | ✓ | FDIA/Delay/Jamming |
| SCADA | ✓ | ✓ | | ✓ | FDIA/DOS |
| Control System | ✓ | | ✓ | ✓ | FDIA/DOS |
| State Estimator | ✓ | | | ✓ | FDIA |
| Communication Channel | ✓ | ✓ | | ✓ | DOS/Jamming/Delay |
| Power Market | ✓ | | | ✓ | FDIA/DOS/Delay |
| Remote Termial Unit (RTU) | ✓ | ✓ | ✓ | | FDIA/DOS/Delay/Jamming |
| Phasor Measurement Unit(PMU) | ✓ | ✓ | ✓ | | FDIA/Delay/Jamming |
| Programmable logic controller (PLC) | ✓ | ✓ | ✓ | | FDIA/Delay/Jamming |
| Advanced Meter Infrastructure (AMI) | | ✓ | ✓ | | FDIA/Jamming |
| Intelligent Electronic Device (IED) | | ✓ | ✓ | | FDIA/Jamming |

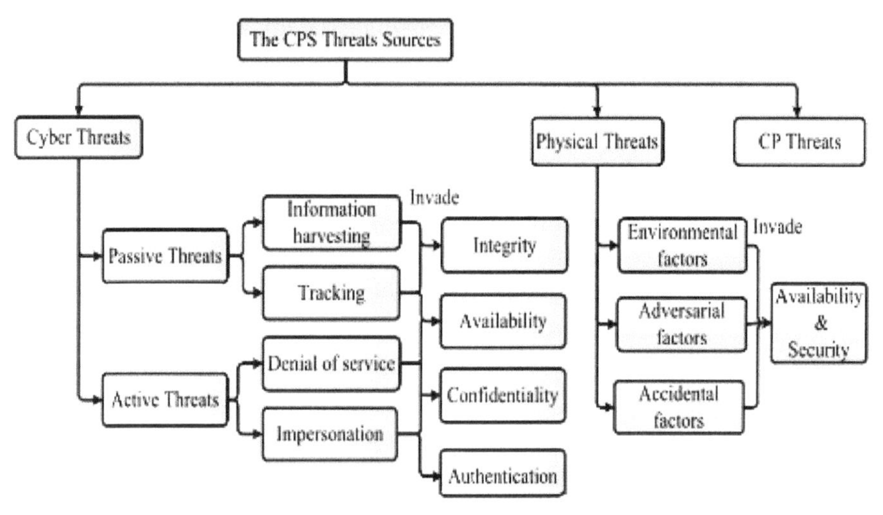


**Figure 1.8** Resource threats in CPS and their attack properties [14]

Their SCADA-connected systems can significantly increase the vulnerability of the smart grid [18].

### 1.5.2 CPS Protection Vulnerabilities and the Applications of Power Electronics

This part reports the principal causes for CPS security vulnerabilities, including power electronics in applications. These can be classified into three types [29]:
Vulnerabilities in cyberspace

1. CP vulnerabilities
2. Physical vulnerabilities

Figure 1.8 shows the sources of cyber threats (active and passive threats included).

## 1.6 Conclusion

In general, in this chapter, the basics of smart networks and the use of power converters were studied. Then the vulnerability analysis of cyber-attacks in the control of the power converter can be summarized at the end of the two cyber security discussions and the vulnerability analysis of cyber-attacks in the control of voltage source converters. As follows:

ANALYSIS AND IMPACT OF CYBER-ATTACKS VULNERABILITY ON CONTROLLING THE VOLTAGE SOURCE CONVERTERS

### 1.6.1 Topology Detection and Cyber Attack

Smart grid monitoring and challenges ahead:

D-PMU uses GPS technology to measure real-time mains voltage, current, and other data, and real-time scheduling [5]. The data is sent to an intelligent control station, which is used to protect and dynamically monitor the power system, fault detection system, and other areas. This is an important piece of equipment to ensure the safe operation of the smart grid.

The IoT can enable accurate and timely use of smart grid power information. It can also provide reliable and various data and technical support for power applications such as fault detection and location in the distribution network. With the integration of telecommunication satellites, 4G and 5G wide-area wireless communication and wireless private network telecommunication technologies, different types of error detection, and analysis are provided [63]. The core technology of intelligent performance management and control directed towards the IoT can provide error location and analysis, smart alerts, and timely monitoring. Complete decision-making, performance planning, service process, and complete information exchange process create an effective platform that is shared to remove individual islands, collect data and perform a complete analysis. Interactive and intelligent computing technology based on automated information and service provides high-precision distribution network error detection methods.

PMUs have been widely and successfully used as key monitoring technology in power transmission systems (T-PMUs). PMU reporting times are much shorter than older Supervisory Control and Data Acquisition (SCADA) systems. They can provide synchronized phase measurements that can generate voltage phases for different network nodes. PMUs measure voltage and current phases, frequency, and frequency change rate (ROCOF) with a common time source [1]. The time source is generated by GPS signals. Each report has its own time tag when it comes to the GPS signal. Different phase phasors will be comparable because of this time tag.

The need to use PMUs in Distribution - Phasor Measurement Unit Smort (SD-PMU) is increasing as renewable energy generators, energy storage equipment, and more loads are connected to the grid. Renewable energy sources are connected to distribution networks through electronic power converters and have little or no inertia, and if a high percentage of these converters are connected to the grid, the stability of the grid may be disrupted. D-PMUs can provide real-time monitoring of such networks.

But T-PMUs are not suitable for distribution networks due to many factors. First, distribution networks have different phase differences and smaller sizes between different nodes due to shorter distances. This limitation requires D-PMUs to be more accurate. Second, the intermittent behavior of generators, loads, and storage equipment requires short delays. Third, higher distortion requires a redefinition of harmonic and subharmonic constraints.

Today, the use of phasor measuring units in distribution networks is very widespread. These applications mainly include error type and location detection [2], microgrid frequency control in island mode [3], and mode estimation [4]. With the development of smart grids, power system requirements require more information, type, and accuracy. The Internet of Things (IoT) is a technology that makes all kinds of equipment installed in power stations intelligent for the exchange and transmission of information, which improves the depth and breadth of information understanding in many aspects of the power source and improves analysis. , Warning, self-healing, and disaster prevention capabilities in the power system help [5]. The IoT has the characteristics of comprehensive understanding, efficient information processing, and flexible and convenient application. It also provides a supportive role in energy production, consumption, transmission, service, and other areas of application requirements of several types of distribution networks.

It is necessary for the developed technology and complete role-playing to improve the performance level of the power system [6]. But the development of the IoT has faced many problems and difficulties in various aspects.

1. The electricity industry has its own practical characteristics and the position, practical goals and expectations of each power network are different. Different electrical applications include different electrical equipment, structures, and electrical systems.
2. The IoT depends on different sensors and their corresponding smart meters and terminals. There are many major differences between different IoT functions and methods of use, with different purposes.
3. IoT data has various characteristics, large volume, and continuous production, so it is necessary to solve the problems of interference, aggregation, sharing, and collaboration of IoT information [7].

Cyber protection solutions in smart energy networks:

With the rapid spread of communication technologies, confusion in cyber components is becoming a reality. These disturbances can significantly affect the performance of the smart grid. In fact, due to the increasing penetration of voltage converters in smart grids, their profound impact on the system cannot be ignored [20].

Cyber-attacks can be carried out on load collectors, smart meters, and sensors in an active distribution network to disrupt power management, voltage stability, frequency settings based on ancillary services. [17].

If resources are changed, system dynamics can be changed to activate the protection layer or create instability. This can be described by showing the voltage state space using mathematics [21].

Measurements can be used to confirm the topology of a distribution network, such as open/closed switches. Topology detection involves situations where SCADA data at remote terminals is either unavailable or unreliable for any reason [50]. This is especially important in the case of real topologies to prevent consumers from leaving and violating restrictions. A particular case in point is the detection of a cyber-attack that intentionally obscures or distorts information, or masks a physical attack, or tricks the operator into making decisions that inadvertently damage the network. . Such attacks can be detected by checking the compatibility of SCADA data with data that is located independently in the cyber network. Therefore, in order to escape being detected, an attacker has to attack both networks simultaneously in full coordination. Cyberattacks will also be detected by detecting unexpected operations or topological changes through measurements.

The topology detection algorithm is generally proposed in three ways:

Residual state estimation error: This method, which depends on the absolute measurement accuracy, relates the residual state estimation error to the difference between the estimated and actual topology. An appropriate algorithm can then identify which set of possible topologies minimizes the residual error and is therefore the most plausible real topology [21].

Time series signals of topological changes: This method focuses on detecting and inferring transitions between modes as the keys are opened or closed.

Source impedance method: This method also relies on time series and examines the effective impedance from the network point of view through step changes in voltage and current and reflects the recovery of the circuit topology.

### 1.6.2 Vulnerability Analysis of Cyber-attacks in the Control of Voltage Source Converters

1) Networking control for voltage source converters:

Network generator voltage source converters can adjust the frequency as well as the voltage locally. To synchronize with other AC sources, the initial control must be localized using existing measurements. This structure is entirely secure in terms of cyberspace because attackers cannot access the physical layer. In addition, beam formation is widely used as security for suitable physical layers. This problem has been remedied by using a secondary controller using information from other voltage source converters. Centralized or distributed secondary control strategies can be imposed on the primary control law to control compensation. However, this is a compassionate and significant space for hackers to find the attacked data or in the controller, communication link, or sensors [12].

Regular penetration techniques for manipulating each component include [14]:

1. Sensors
2. Communication links
3. Controller

2) Power supply and network support for voltage source converters: For voltage source converters, mains power control is used to inject reactive/active power into network generating units. This strategy is widely used in grid-connected applications for the integration of renewable energy sources [33]. As a result, the CP security smart grid must consider both physical securities along with cyber security.

## References

[1] M. Zand, M. A. Nasab, A. Hatami, M. Kargar and H. R. Chamorro, “Using Adaptive Fuzzy Logic for Intelligent Energy Management in Hybrid Vehicles, ” 2020 28th ICEE, pp. 1-7, doi: 10.1109/ICEE50131.2020.9260941. IEEE Index

[2] Hamed Ahmadi-Nezamabad, et al.. “Multi-objective optimization based robust scheduling of electric vehicles aggregator.” Sustainable Cities and Society vol. 47,101494, 2019.

[3] Zand, Mohammad; Nasab, Morteza Azimi; Sanjeevikumar, Padmanaban; Maroti, Pandav Kiran; Holm-Nielsen, Jens Bo: ’Energy management strategy for solid-state transformer-based solar charging

station for electric vehicles in smart grids', IET Renewable Power Generation, 2020, DOI: 10.1049/iet-rpg.2020.0399 IET Digital Library, https://digital-library.theiet.org/content/journals/10.1049/iet-rpg.2020.0399

[4] Ghasemi M, et al. (2020). An Efficient Modified HPSO-TVAC-Based Dynamic Economic Dispatch of Generating Units, Electric Power Components and Systems doi.org/10.1080/15325008.2020.1731876

[5] Nasri, Shohreh, et al, Maximum Power Point Tracking of Photovoltaic Renewable Energy System Using a New Method Based on Turbulent Flow of Water-based Optimization (TFWO) Under Partial Shading Conditions. 978-981-336-456-1

[6] Rohani A, et al, "Three-phase amplitude adaptive notch filter control design of DSTATCOM under unbalanced/distorted utility voltage conditions," Journal of Intelligent & Fuzzy Systems, 2020, 10.3233/JIFS-201667

[7] M. Zand, M. A. Nasab, O. Neghabi, M. Khalili and A. Goli, "Fault locating transmission lines with thyristor-controlled series capacitors By fuzzy logic method," 2020 14th International Conference on Protection and Automation of Power Systems (IPAPS), Tehran, Iran, 2019, pp. 62-70, doi: 10.1109/IPAPS49326.2019.9069389.

[8] Z. Zand, M. Hayati and G. Karimi, "Short-Channel Effects Improvement of Carbon Nanotube Field Effect Transistors," 2020 28th Iranian Conference on Electrical Engineering (ICEE), Tabriz, Iran, 2020, pp. 1-6, doi: 10.1109/ICEE50131.2020.9260850.

[9] Lilia Tightiz et all, An intelligent system based on optimized ANFIS and association rules for power transformer fault diagnosis, ISA Transactions, Volume 103, 2020, Pages 63-74, ISSN 0019-0578, https://doi.org/10.1016/j.isatra.2020.03.022.

[10] Zand M, , et al "A Hybrid Scheme for Fault Locating in Transmission Lines Compensated by the TCSC," 2020 15th International Conference on Protection and Automation of Power Systems (IPAPS), 2020, pp. 130-135, doi: 10.1109/IPAPS52181.2020.9375626

[11] M. Zand, M. Azimi Nasab, M. Khoobani, A. Jahangiri, S. Hossein Hosseinian and A. Hossein Kimiai, "Robust Speed Control for Induction Motor Drives Using STSM Control," 2021 12th (PEDSTC), 2021, pp. 1-6, doi: 10.1109/PEDSTC52094.2021.9405912.

[12] P. Sanjeevikumar, M. Zand, M. A. Nasab, M. A. Hanif and M. S. Bhaskar, "Spider Community Optimization Algorithm to Determine UPFC Optimal Size and Location for Improve

Dynamic Stability," 2021 IEEE 12th Energy Conversion Congress & Exposition - Asia (ECCE-Asia), 2021, pp. 2318-2323, doi: 10.1109/ECCE-Asia49820.2021.9479149

[13] Azimi Nasab, M.; Zand, M.; Eskandari, M.; Sanjeevikumar, P.; Siano, P. Optimal Planning of Electrical Appliance of Residential Units in a Smart Home Network Using Cloud Services. Smart Cities 2021, 4, 1173–1195. https://doi.org/10.3390/smartcities4030063

[14] Azimi Nasab, M.; Zand, M. Sanjeevikumar, P. et, al. An efficient robust optimization model for the unit commitment considering of renewables uncertainty and Pumped-Storage Hydropower. Computers and Electrical Engineering 2022,

[15] Azimi Nasab, Mortez, Zand, Mohammad, Padmanaban, Sanjeevikumar,Dragicevic, Tomislav, Khan, Baseem, "Simultaneous Long-Term Planning of Flexible Electric Vehicle Photovoltaic Charging Stations in Terms of Load Response and Technical and Economic Indicators" World Electric Vehicle Journal ,2021, P 190, doi:10.3390/wevj12040190

[16] Zand, Mohammad, et al. "Big Data for SMART Sensor and Intelligent Electronic Devices–Building Application." Smart Buildings Digitalization. CRC Press 11-28. https://doi.org/10.1201/9781003201069, 2022.

[17] B. Zhang, P. Dehghanian, and M. Kezunovic, "Optimal Allocation of PV Generation and Battery Storage for Enhanced Resilience," *IEEE Trans. Smart Grid*, vol. 10, no. 1, pp. 535–545, Jan. 2019.

[18] M. Milton, C. D. La O, H. L. Ginn, and A. Benigni, "Controller-Embeddable Probabilistic Real-Time Digital Twins for Power Electronic Converter Diagnostics," *IEEE Trans. Power Electron.*, vol. 35, no. 9, pp. 9850–9864, Sep. 2020.

[19] R. V. Yohanandhan, R. M. Elavarasan, P. Manoharan, and L. Mihet-Popa, "Cyber-Physical Power System (CPPS): A Review on Modeling, Simulation, and Analysis With Cyber Security Applications," *IEEE Access*, vol. 8, pp. 151019–151064, 2020.

[20] S. Sahoo, T. Dragicevic, and F. Blaabjerg, "Cyber Security in Control of Grid-Tied Power Electronic Converters–Challenges and Vulnerabilities," *IEEE Journal of Emerging and Selected Topics in Power Electronics*. pp. 1–1, 2019.

[21] D. Perez-Estevez, J. Doval-Gandoy, and J. M. Guerrero, "AC-Voltage Harmonic Control for Stand-Alone and Weak-Grid-Tied Converter," *IEEE Trans. Ind. Appl.*, vol. 56, no. 1, pp. 403–421, Jan. 2020.

# 2

# Power Electronics in HVDC Transmission Systems

**Mehdi Abbasipour[1], Xiaodong Liang[1], and Massimo Mitolo[2]**

[1]The University of Saskatchewan, Saskatoon, SK S7N 5A9, Canada
[2]Irvine Valley College, Irvine, CA 92618, USA
E-mail: mehdi.abbasipour@usask.ca; xil659@mail.uasak.ca
mmitolo@ivc.edu

## Abstract

High Voltage Direct Current (HVDC) transmission is widely used for bulk power transmission, asynchronous connection, and marine power transmission. The principle of HVDC transmission relies on AC to DC power conversion and vice versa. Generally, power electronics converters in HVDC transmission systems can be classified into three major groups: 1. power converters, 2. DC/DC converters, and 3. DC power flow controllers. Power converters convert AC power to DC power and vice versa. DC/DC converters (or named DC transformers) provide DC voltage matching; they can also be used to divide large Multi-Terminal HVDC (MT-HVDC) grids into several smaller protection zones, regulate DC voltage, isolate faults, and connect Bipolar/Monopolar configurations. DC power flow controllers (DC-PFCs) or DC current flow controller (DC-CFCs) control power flow in HVDC transmission systems, especially in MT-HVDC grids. In this chapter, power electronic converters and inverters, commonly used in HVDC transmission systems are summarized through extensive literature review, and their various topologies are introduced.

**Keywords:** DC/DC Converters, DC Power Flow Controllers, HVDC Transmission Systems, Power Converters, Power Electronics.

## 2.1 Introduction

Direct Current (DC) technology was firstly introduced for power transmission and electrification in the late 1880s and early 1890s by Thomas Edison's company. Meanwhile, George Westinghouse's company proposed the Alternating Current (AC) technology. The competition between the technologies was later known as the War of the Currents. Eventually, Westinghouse's AC technology became dominant for electrification [1]. Advantages of DC and HVDC systems over the AC transmission on certain applications, such as bulk power transmission, asynchronous connection, and marine power transmission, attracted attention again in the 1930s. The first commercialized HVDC transmission systems were constructed in the Soviet Union and Sweden in 1951 and 1954, respectively. Since then, several HVDC transmission systems have been installed [2]. Since 2008, many countries have changed their energy policies due to the rapid growth of electricity demand and environmental concerns related to fossil fuel-based power generation. Renewable energy sources (RESs) play nowadays an increasing role in modern power systems. HVDC transmission has gained significant interest in the arena of renewable energy integration [3, 4].

Power electronic converters and inverters are important in HVDC transmission systems. Considerable developments in the design of novel and sophisticated high-voltage and high-power switches have enabled the advancement of power electronics converters and inverters in HVDC applications.

In this chapter, we focus on existing power electronics converters and inverters for HVDC transmission systems. In Section two, HVDC transmission systems are examined. In Sections three through five, power converters, DC/DC converters, and DC power flow controllers in HVDC transmission systems are analyzed. A summary is presented in Section six.

## 2.2 HVDC Transmission Systems

We will start by introducing existing HVDC transmission systems, their configurations, and the importance of power electronics converters.

### 2.2.1 Brief Overview on HVDC Transmission Technologies

The operation of HVDC transmission systems requires the conversion of AC power to DC power and vice versa. Typically, rectifiers perform the conversion of AC to DC power at generation stations, or at wind power

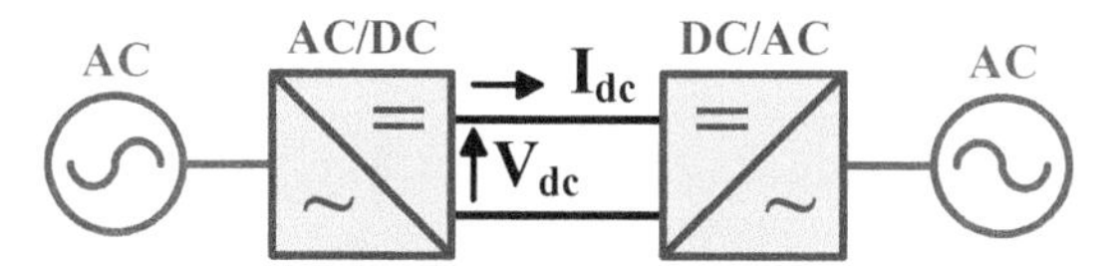


**Figure 2.1** The schematic diagram of the B2B HVDC transmission systems [5].

plants (WPPs), solar power plants (SPPs), and hydropower plants. Inverters are used at consumption stations or receiving ends, to convert from DC back to AC. HVDC converting stations can provide both rectification and inversion, enabling bidirectional power flow. Similar to the AC transmission technology, HVDC stations include switchgear, converters, and transformers, which facilitate the AC/DC and DC/AC power conversion and transmission.

Power electronics converters in HVDC transmission systems perform power conversion, harmonic reduction, voltage regulation, fault protection, and reliability enhancement. HVDC transmission technologies include two types of converters: Current Source Converters (CSC) and Voltage Source Converters (VSC).

HVDC transmission systems can also be classified into three major groups: 1. Back-to-Back HVDC Transmission Systems; 2. Point-to-Point HVDC Transmission Systems; and 3. Multi-Terminal HVDC Transmission Systems [3].

#### 2.2.1.1 Back-to-Back HVDC transmission

In the Back-to-Back (B2B) configuration, no transmission medium (or short of transmission medium) is present between transmitting and receiving ends of an HVDC system (Figure 2.1). B2B systems are primarily used to provide an asynchronous connection between two AC networks with different operating frequencies. An example of B2B connection is the power exchange and energy trade between the U.S. (60 Hz operating frequency) and Mexico (50 Hz operating frequency) [5]. Both CSC-HVDC and VSC-HVDC technologies can be used to provide asynchronous connections.

#### 2.2.1.2 Point-to-Point HVDC transmission

In the Point-to-Point (P2P) configuration, a transmission medium between transmitting and receiving ends exists (Figure 2.2). HVDC transmission can transfer bulk power to remote load centers. Remarkable examples are the submarine HVDC transmission system in Gotland, Sweden, and the overhead-line HVDC transmission system in Manitoba, Canada.

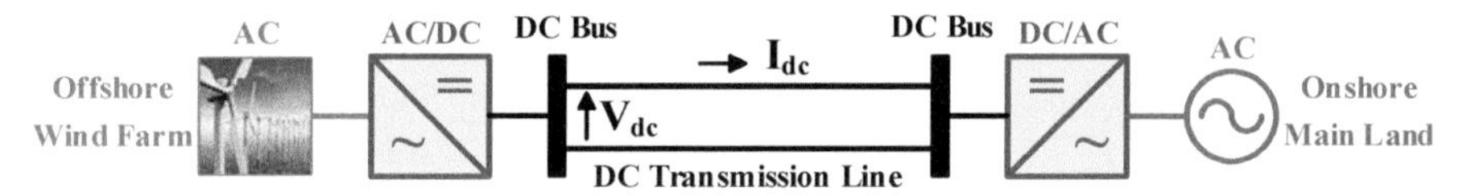


**Figure 2.2** The schematic diagram of the P2P HVDC transmission systems [3].

Most P2P systems are built based on the CSC-HVDC technology, which is more mature than the VSC-HVDC in ultra-bulk power transmission. However, the rapid development of the VSC-HVDC technology may soon allow P2P systems to be built based on both systems [6–8].

#### 2.2.1.3 Multi-Terminal HVDC grids

Generally, an HVDC grid is composed of at least three HVDC stations (i.e., a three-terminal HVDC grid) (Figure 2.3(a)). Multi-Terminal HVDC (MT-HVDC) grids have been developed with multiple HVDC terminals and lines. MT-HVDC grids offer many advantages, such as the efficient transfer of bulk power across long distances (i.e., mainland overhead lines) and short distances (i.e., marine cables), the connection of various types of RESs, which improves the capability of RESs of meeting the power demand and with lower losses.

An eight-terminal MT-HVDC grid is shown in Figure 2.3(b) [9, 10]. Generally, MT-HVDC grids are classified into three types: 1.) Radial MT-HVDC grids; 2. Ring MT-HVDC grids, and; 3. Meshed MT-HVDC grids.

Radial MT-HVDC grids are very similar to their AC counterpart, resembling a star configuration with no loops. Radial MT-HVDC grids are relatively simple and require low initial investment. HVDC converters are connected to only one HVDC line/cable and only one HVDC line/cable connects the generation HVDC station to a specific mainland HVDC station. Drawbacks of this configuration include low reliability since a DC fault can put the HVDC station offline. The schematic diagram of a Radial MT-HVDC grid is shown in Figure 2.4(a).

MT-HVDC grids in a Ring configuration also have a simple structure and are similar to their AC counterpart. They resemble a ring with one large loop. In this configuration, all HVDC stations are connected to two HVDC lines/cables. A DC fault cannot entirely interrupt an HVDC station because of the redundancy offered by the second HVDC line/cable. Thus, the reliability of this configuration is higher than that of the Radial MT-HVDC grid. The schematic diagram of a Ring MT-HVDC grid is shown in Figure 2.4(b).

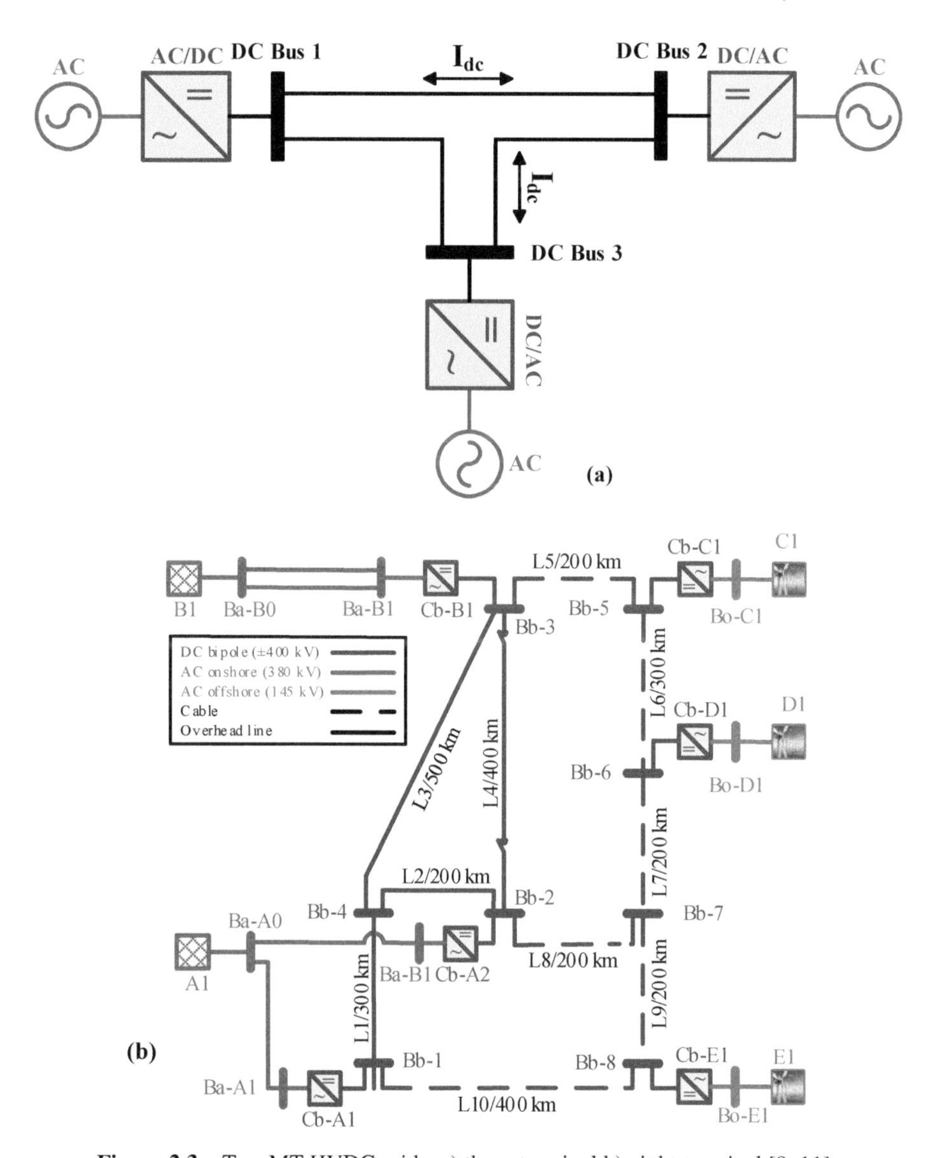


**Figure 2.3** Two MT-HVDC grids: a) three-terminal b) eight-terminal [9, 11].

Meshed MT-HVDC grids are the most complex configuration and are very similar to their AC counterpart. This configuration is a combination of Ring and Radial MT-HVDC grids, with at least one loop, and requires more investment than the other cases. In this arrangement, all HVDC stations are

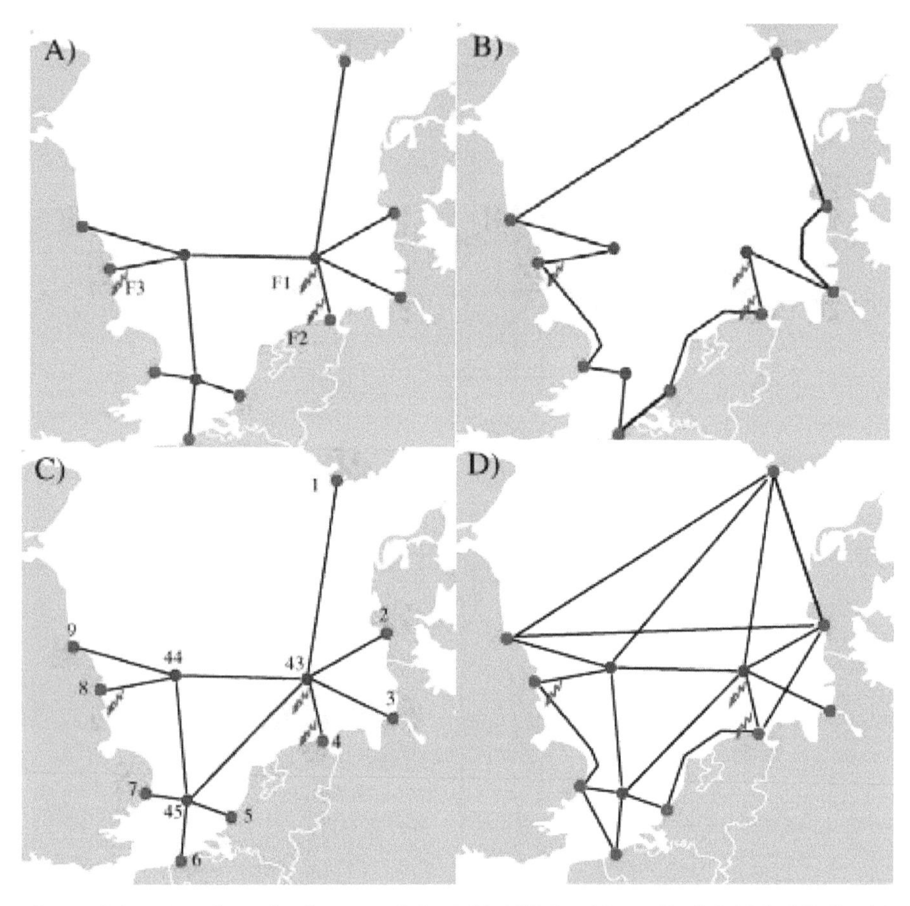


**Figure 2.4** The schematic diagram of the MT-HVDC grids: a) Radial MT-HVDC grid; b) Ring MT-HVDC grid; c) Lightly-Meshed MT-HVDC grid; and d) Densely-Meshed MT-HVDC grid [10].

connected to two or more HVDC lines/cables. A DC fault cannot disconnect the HVDC station thanks to the multiple HVDC lines/cables, thus the reliability of this configuration is higher than that of Radial or Ring MT-HVDC grids.

Generally, meshed MT-HVDC grids can be Lightly-Meshed and Densely-Meshed. Lightly-Meshed MT-HVDC grids are simpler, with fewer HVDC lines/cables in their structure, whereas Densely-Meshed MT-HVDC grids feature numerous HVDC lines/cables with several loops in their structure. Densely-Meshed MT-HVDC grids have the highest reliability among all existing MT-HVDC transmission systems. The schematic diagrams of the Lightly- and Densely-Meshed MT-HVDC grids are shown in Figure 2.4(c)–(d).

### 2.2.2 HVDC Configurations

Among various types of HVDC transmission systems, Meshed MT-HVDC grids are expected to play an essential role in tomorrow's power systems. It is anticipated that future Meshed MT-HVDC grids will be based on VSC-HVDC technology, as they can transmit power at a constant voltage.

Generally, HVDC configurations can be categorized into four major groups based on their electrical circuits and the number of poles: 1. Monopolar; 2. Bipolar; 3. Homopolar; and d) Hybrid [12–15].

#### 2.2.2.1 Monopolar

The electrical circuit of a Monopolar HVDC configuration is completed through one or two metallic conductor(s). Generally, Monopolar HVDC configurations can be separated into two subgroups based on the voltage's symmetry and its polarity: 1. Asymmetric Monopolar; and 2. Symmetric Monopolar [12, 13].

In the Asymmetric Monopolar configuration, one metallic path operates at the system's nominal voltage ($+U_n$), and the return path operates at zero volts (or very low voltage). The term Monopolar indicates that one converter is present at each HVDC terminal, and it operates at the system's rated voltage. The term Asymmetric indicates that one path operates at the nominal voltage, whereas the operating voltage of the return path is zero: there is no symmetry between the two voltages. In symmetric configurations, the voltages of the electrical paths are the same but with opposite polarity.

The operating polarity of the conductor can be either positive or negative; however, it is preferred to assign the negative polarity ($-U_n$) to the conductor to prevent the Corona effect. We can have two configurations for the Asymmetric Monopolar arrangement: 1. Asymmetric Monopolar configuration with the Earth Return Path; and 2. Asymmetric Monopolar configuration with Metallic Return Path [12, 13].

The asymmetric Monopolar configuration with the Earth Return Path is also known as the Monopolar Single-Wire Earth Return (SWRT), where just one conductor is used at the system rated voltage ($+U_n$ or $-U_n$), and the return path is the actual earth. Two ground electrodes are installed at the HVDC station to drive the operating current into the earth. Because the power

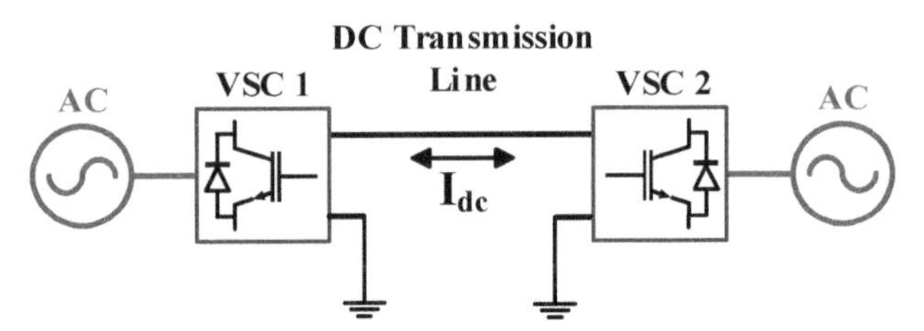


**Figure 2.5** The schematic diagram of the Monopolar SWRT Configuration [12, 13].

can only flow in one direction, one HVDC station is the rectifier, and the other one is the inverter. A capacitor is used in both HVDC terminals to maintain the voltage at the desired value. The schematic diagram of the Monopolar SWRT configuration is shown in Figure 2.5 [12, 13].

Asymmetric Monopolar configuration with a Metallic Return Path is concisely named "Asymmetric Monopolar", and unlike the SWER, two metallic conductors are employed. One conductor is used at the system rated voltage ($+U_n$ or $-U_n$), and the other is used as the return path at zero voltage (or close to zero voltage) to complete the circuit. Thanks to the metallic return path, the Asymmetric Monopolar configuration can be used for bidirectional power flow. The schematic diagram of the Asymmetric Monopolar configuration is shown in Figure 2.6(a) [12, 13].

In the Symmetric Monopolar configuration, two conductors operate at half of the system's rated voltage with different polarities ($\pm U_n/2$). The term Symmetric indicates that the voltages of the two paths have the same magnitude but opposite polarities. The capacitance is divided into two parts with its midpoint grounded. The schematic diagram of the Symmetric Monopolar configuration is shown in Figure 2.6(b) [12, 13].

### 2.2.2.2 Bipolar

The Bipolar HVDC transmission is the most popular configuration, which generally employs two conductors to complete the circuit. The term bipolar indicates the presence of two converters in each HVDC terminal. This configuration is symmetric, as conductors and converters operate at half of the system rated voltage with opposite polarities ($\pm U_n/2$). Bidirectional power flow is possible due to the two conductors. The schematic diagram of the Bipolar HVDC configuration is shown in Figure 2.7 [12, 13].

As shown in Figure 2.8, the midpoint between the two converters may be grounded. Therefore, a Bipolar HVDC configuration has the unique capability of being able to operate in the Monopolar SWER mode if one of the

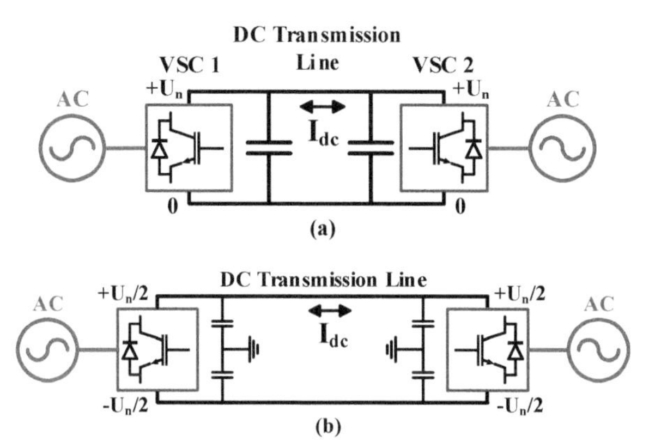


**Figure 2.6** The schematic diagram of (a) Asymmetric Monopolar configuration (b) Symmetric Monopolar configuration [12, 13].

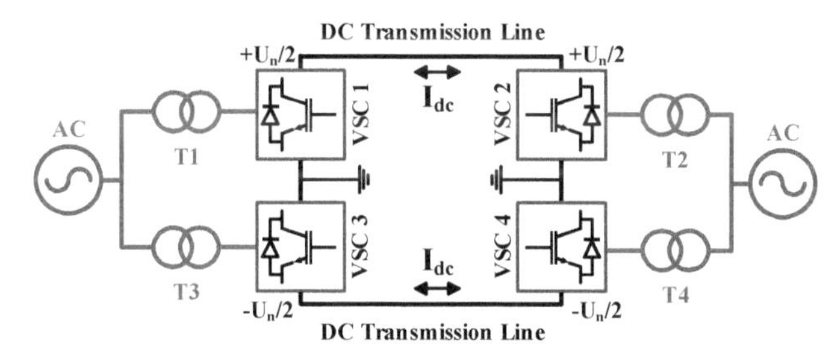


**Figure 2.7** Schematic diagram of the Bipolar HVDC configuration [12, 13].

converters at HVDC terminals has issues. Power transmission can still be assured at half of its capacity, which is a significant feature.

Several Bipolar configurations are proposed in the literature and two are shown in Figure 2.8 [14].

The Bipolar configuration with Metallic Return Path in Figure 2.8(a) can operate in the Asymmetric Monopolar configuration during contingencies, such as a fault at each converter, or temporary outages for the converter maintenance. The difference between monopolar operations and the basic bipolar configuration is the metallic return path, which provides advantages similar to the Asymmetric Monopolar configuration. The Bipolar configuration with Series-Connected Converters of Figure 2.8(b) is another improved arrangement, where each converter can be bypassed during maintenance, without disrupting operations, thanks to the bipolar or monopolar SWER configurations [12–14].

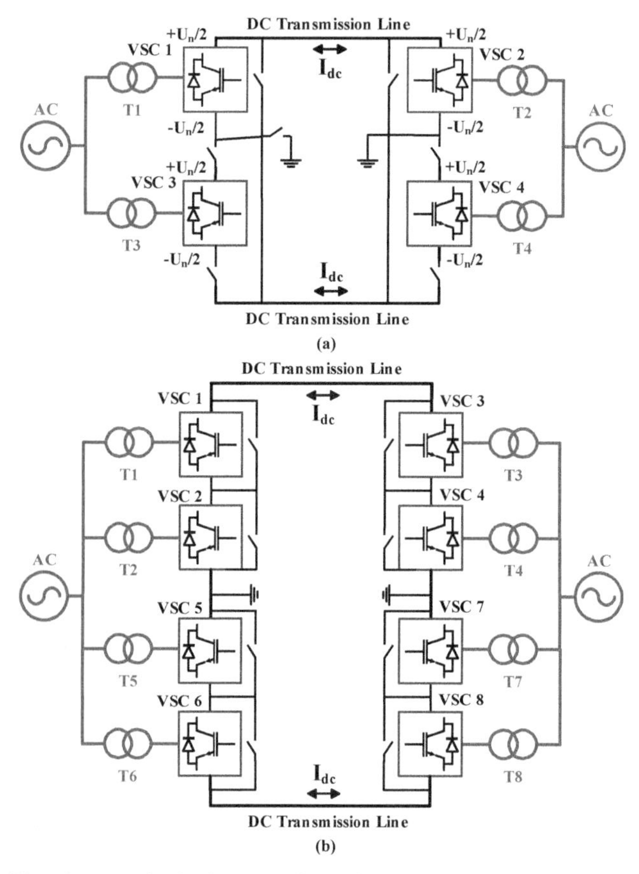


**Figure 2.8** Two improved Bipolar HVDC configurations: a) Bipolar configuration with Metallic Return Path b) Bipolar configuration with Series-Connected Converters [14].

### 2.2.2.3 Homopolar

The Homopolar HVDC configuration features two conductors with the same voltage magnitude and polarity and employs the earth as the return path. In this configuration, the conductors' polarity can be negative or positive ($+U_n$ or $-U_n$), but the negative polarity is preferred as it reduces Corona's effects. These features are the main differences between Homopolar and Bipolar HVDC configurations. Since the earth return path is used, the Homopolar HVDC solution can only transfer power unidirectionally. Although this configuration is very similar to the arrangement of the Bipolar, it has similar performance to the Monopolar SWER configuration, as if two Monopolar SWER configurations were merged to create a Homopolar configuration.

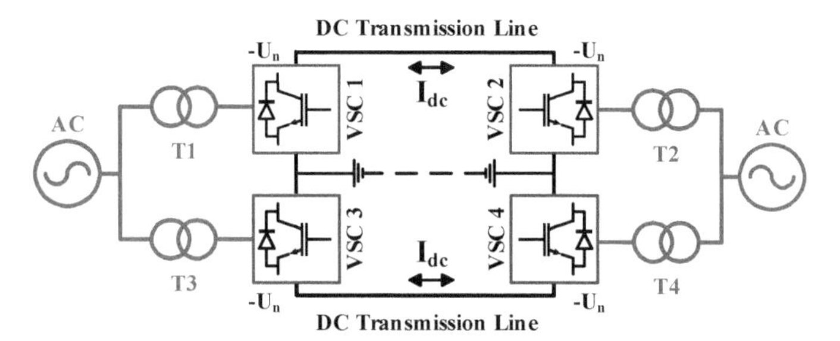


**Figure 2.9** Schematic diagram of the Homopolar HVDC configuration [15].

The schematic diagram of the Homopolar HVDC configuration is shown in Figure 2.9 [15].

#### 2.2.2.4 Hybrid

The hybrid HVDC configuration is the combination of HVDC configurations. Monopolar, Bipolar, and Homopolar HVDC configurations can be combined to satisfy the requirements of HVDC systems. Schematic diagrams of Hybrid HVDC configurations are shown in Figure 2.10 [13].

### 2.2.3 Power Electronics Converters in HVDC Transmission Systems

HVDC transmission systems include various components, such as DC cables, power conversion units, filters, DC circuit breakers, and control systems, where power electronics converters play a critical role. Recently, HVDC transmission systems received a renewed interest due to renewable energy integration, and thus, experienced rapid developments. New technologies, such as VSC-HVDC, and MT-HVDC grids, and new HVDC configurations, such as Bipolar and Hybrid HVDC, became available. Current research on HVDC transmission includes protection, power flow, harmonics, control, and reliability. Rapid developments in semiconductor materials and switches enable the advancement of new power electronic converters with novel and unique features.

This chapter reviews existing power electronics converters in HVDC transmission systems [16]. Generally, power electronics converters in HVDC transmission systems can be classified into three major groups based on their structure and duties: 1. Power Converters; 2. DC/DC Converters; and 3. DC Power Flow Controllers.

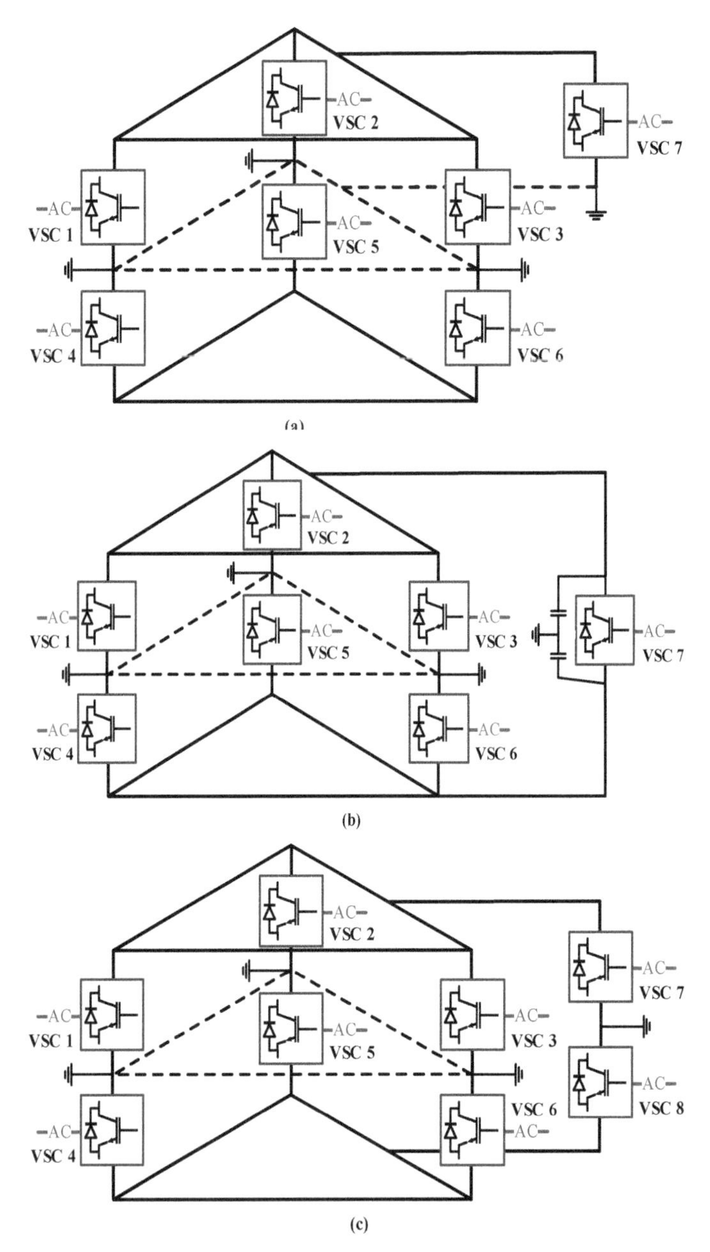


**Figure 2.10** Schematic diagrams of Hybrid HVDC configurations: a) Bipolar HVDC with Monopolar SWER; b) Homopolar HVDC with Symmetric Monopolar; and c) Homopolar HVDC with Bipolar HVDC [13].

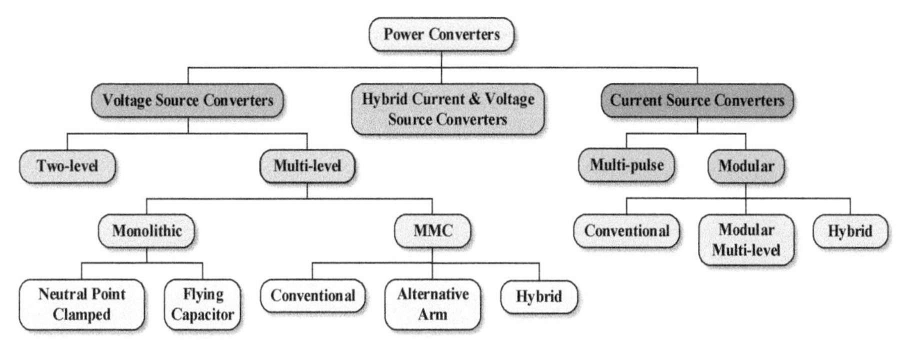


**Figure 2.11** General categorization of AC/DC power converters.

## 2.3 Power Converters

The primary duty and purpose of power converters are to convert power from AC to DC and vice versa. Studies have been conducted to develop novel topologies of power converters and improve the performance of existing power converters in HVDC transmission systems. We will first introduce existing AC/DC power converters [12, 17–23].

All AC/DC power converters are generally categorized into three major groups: 1. VSCs; 2. CSCs; and 3. Hybrid Current & Voltage Source Converters (HCVSCs). Each type of these AC/DC power converters can be divided into other subgroups based on their structure (Figure 2.11) [12, 17–23].

### 2.3.1 Voltage Source Converters

VSCs are AC/DC power converters that maintain their output voltage at a predetermined value, regardless of the magnitude and direction of the current flowing through them. VSCs can be controlled to perform as rectifiers or inverters by controlling the magnitude and phase angle of the AC output voltage (when they are connected to an active DC system). Such control is realized by leading or lagging power reactive generation/consumption in the system. Power reversal can be realized by switches that have the bidirectional current-conducting capability (also known as Reverse Conducting (RC)). The direction of $I_{dc}$in the system is controlled while $V_{dc}$is fixed at a constant polarity. A VSC can be connected to AC systems through a coupling reactance. The general schematic diagram of a B2B VSC is shown in Figure 2.12 [19].

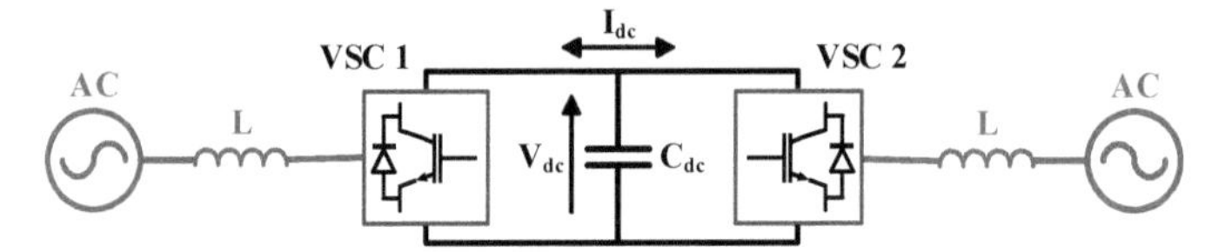


**Figure 2.12** The general schematic diagram of a B2B VSC [19].

### 2.3.1.1 Two-level VSCs

The two-level VSCs are the first generation of AC/DC power converters in HVDC transmission systems. This topology has been firstly commercialized by ABB and is known as the "HVDC Light." The schematic diagram and output voltage of a two-level VSC are shown in Figure 2.13 [24].

This topology is called two-level because it provides two output voltage levels ($+U_{dc}$ and $-U_{dc}$). Only one dc voltage is present, and it is used to generate the reference AC waveform. To generate a suitable AC waveform at the output of the VSC in the inverting mode, the Pulse Width Modulation (PWM) scheme is employed. IGBT switches are commonly used in this topology; however, their inherent low power loss (about 3% of the total power) is still considered very high. The total harmonic distortion (THD)

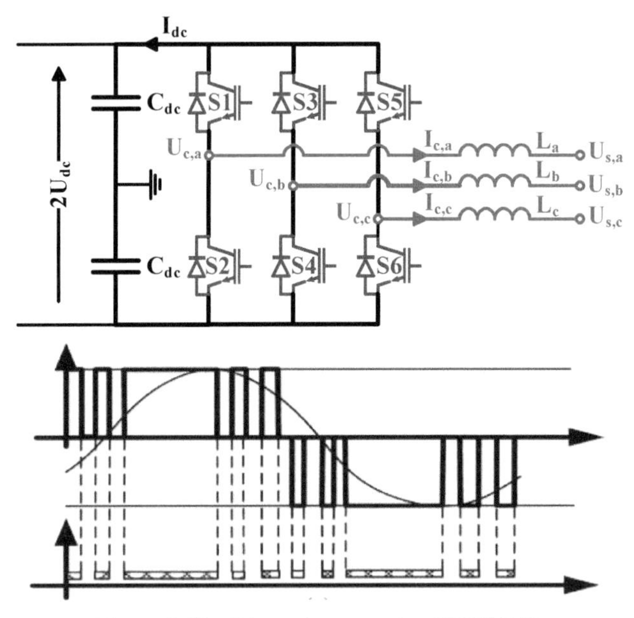


**Figure 2.13** Three-phase two-level VSC [19].

is also very high, and thus, filters are required. Generally, this topology was exploited in early HVDC transmission systems and currently is no longer employed [12].

#### 2.3.1.2 Multilevel converters

Multi-level Converters are the advanced generation of AC/DC power converters in HVDC transmission systems. The output waveform of a multilevel converter is shown in Figure 2.14.

Multilevel converters produce output voltage waveforms with different voltage levels. As shown in Figure 2.14, multilevel converters have the output waveform very similar to a "staircase", which is much closer to the sinusoidal AC waveform, when compared to an output of two-level VSCs. Suitable switching schemes can be used to reduce the output THD by eliminating or reducing low-frequency harmonics. This topology requires much smaller filters, which is the significant advantage of multilevel converters over the two-level configuration. Also, the voltage stress (*dv/dt*) is considerably lower than that of two-level VSCs, which significantly reduces the electromagnetic interference in converters. Furthermore, multilevel converters' switching can be conducted at a lower frequency. Lower switching frequency and voltage stress level are two remarkable factors, which result in a significant reduction in switching power loss. However, multi-level converters have one major drawback: the total energy stored in passive components is significantly higher than that in two-level VSCs for the same rated power. In general, multilevel converters are superior to their two-level counterparts in terms of power quality [12, 17, 19, 20, 23].

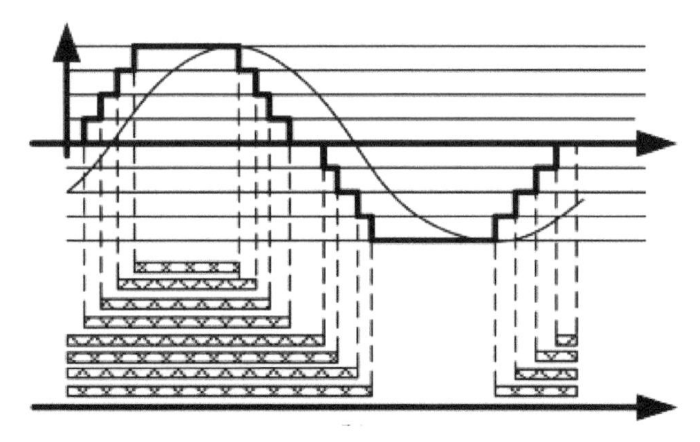

**Figure 2.14** Output waveform of a multilevel converter [19].

Multilevel converters are generally categorized into two major subgroups based on the modularity or solidarity of their structure: 1. Monolithic Multilevel Converters; and 2. Modular Multilevel Converters (MMCs) [19].

#### 2.3.1.2.1 Monolithic multilevel converters

Monolithic Multilevel Converters have no modular or extendible features. Many series-connected switches are employed to withstand the high operating voltage of HVDC transmission systems. Two well-known topologies are present in the monolithic multi-level converter's family: Neutral Point Clamped (NPC) Multilevel Converters and Flying Capacitor (FC) Multilevel Converters [12, 17].

##### 2.3.1.2.1.1 Neutral point clamped multilevel converter

NPC Multilevel Converters are also known as Diode Clamped Multilevel Converters. The schematic diagram of the NPC three-level converters is shown in Figure 2.15.

NPC multilevel converters generate a multilevel output voltage. The output voltage of an NPC three-level converter may have three magnitudes: $+V_{dc}/2$, *0*, and $-V_{dc}/2$. NPC multilevel converters can be optimally utilized because only a few capacitors are needed with a single DC source. However, maintaining the neutral point voltage at the half, or full, a voltage of the DC

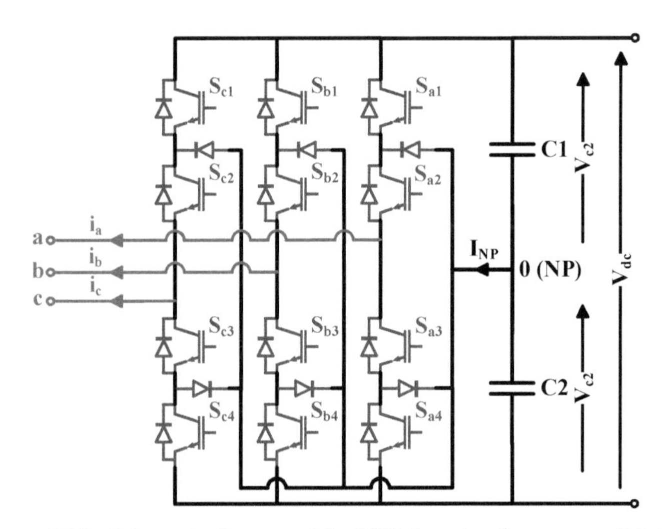


**Figure 2.15** Schematic diagram of the NPC three-level converters [12, 17].

input source may be a challenge. The main advantages and disadvantages of this topology are summarized below [12, 17].

Advantages:

1. Only one isolated DC source is required.
2. Small harmonic filters may be used to eliminate undesired harmonics.
3. Reactive power regulation can be achieved.

Disadvantages:

1. More diodes are needed to reach more voltage levels.
2. Some switching patterns cannot maintain the capacitor voltage at the desired value, which limits the choice of patterns.
3. Additional complex control schemes are needed to balance neutral point voltage in topologies with more than three voltage levels.
4. Complex active power control of individual NPC converters is needed due to improper balancing of the capacitor.
5. Decreased performance may occur in redundancy applications of HVDC systems.

#### 2.3.1.2.1.2 Flying capacitor multilevel converter

FC Multilevel Converters are also known as Capacitor Clamped Multilevel Converters. The schematic diagram of this converter is shown in Figure 2.16. This topology is very similar to the NPC multilevel converter, the only difference is that capacitors are clamped at the midpoint of switches to share the voltage. A three-level FC converter can produce a three-level voltage: $+V_{dc}/2$, 0, and $-V_{dc}/2$. The main advantages and disadvantages of this topology are summarized below [12, 17].

Advantages:

1. A single DC source is required.
2. Active power and reactive power regulation can be achieved.
3. No filters are needed for harmonic reduction.
4. The THD level is low in high-level FC multilevel converters.

Disadvantages:

1. More capacitors are needed to reach more voltage levels, which makes the system bulky.
2. More complex control schemes are required for high-level FC multilevel converters.
3. Switching losses are high because of the high-frequency operations.

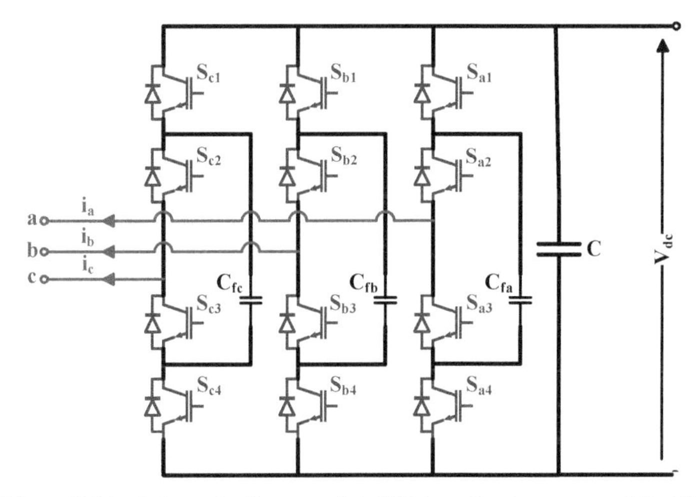


**Figure 2.16** Schematic diagram of the FC three-level converters [12, 17].

### 2.3.1.2.2 Modular multilevel converters

Monolithic multilevel converters have no modularity features, which complicates their maintenance. Their structure also poses limitations for their use in high-voltage and high-power applications, especially in HVDC transmission systems. In this regard, to improve the performance of the monolithic multilevel converters, the concept of modularity has been proposed. Modularity indicates the combination of several subsystems to form a larger system, which, in power converters, is realized through the cascaded connection (the chain-links) of several small converter cells. The chain-links of converter cells allow high-voltage and high-power applications of monolithic multilevel converters, with high-quality output waveforms. In this configuration, converter cells and building blocks/sub modules (SMs) are connected together, forming the intermediate stage of integrated passive elements and power-switching devices. To increase voltage-withstanding and current-conducting capabilities, the SMs are connected in series and in parallel.

Monolithic multilevel converters can also be used in Modular Multilevel Converters as 1. chain links of power electronic SMs; and 2. the combination of SMs and monolithic multi-level converters [19, 20].

#### 2.3.1.2.2.1 Power electronics voltage source sub modules

Power electronics building blocks are the fundamental bricks of MMCs. The SM of an MMC is DC/DC or DC/AC converters, which can be connected in parallel (to increase the current-conducting capability) or in series (to increase the voltage-withstanding capability). The combined series and parallel connections can be used to meet the required specifications in various applications. The popular SMs of MMCs, and their related output voltage waveform, are summarized in Figure 2.17. The important features of each SM are explained below [19, 20].

1. **Half-Bridge** SM*:* It is the simplest and fundamental SM, capable of providing bidirectional current flow and unidirectional voltage blocking. It generates a unipolar output voltage by chopping the DC link voltage. Its output voltage waveform allows the operation in two quadrants by generating a two-level output voltage ($U_C$ and *0*) (Figure 2.17(a)) [19].
2. **Full-Bridge** SM**:** It is implemented by connecting two half-bridges in parallel and operates in four quadrants by generating both positive and negative dc output voltages ($+U_C$, *0*, and $-U_C$) (Figure 2.17(b)). Because of the two half-bridges, the number of switches of this SM is twice that of half-bridge SMs, but it enables bidirectional current flow [19].
3. **Mixed** SM**:** It is implemented by connecting the half-bridge and the fullbridge SMs. This configuration can offer both bipolar and unipolar benefits. This SM produces asymmetric four-level output voltages ($+2U_C$, $+U_C$, *0*, and $-U_C$) (Figure 2.17(c)) [19].
4. **Asymmetrical double** SM**:** it consists of two half-bridges, similar to the mixed SM, and generates asymmetric four-level dc output voltages ($+2U_C$, $+U_C$, *0*, and $-U_C$) (Figure 2.17(d)). This is another alternative for asymmetric four-level SMs [19].
5. **Cross-Connected and parallel** SMs**:** These two building blocks are very similar and obtained by connecting two full-bridges. With similar performance, they both generate a symmetric bipolar five-level DC output voltage ($+2U_C$, $+U_C$, *0*, $-U_C$, and $-2U_C$) (Figure 2.17(e)). There is a structural difference between these two SMs: in the cross-connected SM, it is possible to generate more voltage levels by cross-connecting more intermediate capacitors; in the parallel SM, the parallel connection of the capacitors helps the SM reduce voltage ripples in capacitors [19].
6. **Clamped** SM*:* this SM consists of two full-bridges, and generates an asymmetric four-level output voltage ($+2U_C$, $+U_C$, *0*, and $-U_C$), (Figure 2.17(f)). By employing suitable switching schemes, capacitors can be connected in series or parallel. A precise design and control schemes are required when this SM operates in the full-bridge mode, because of the danger of the parallel connection of capacitors with two different

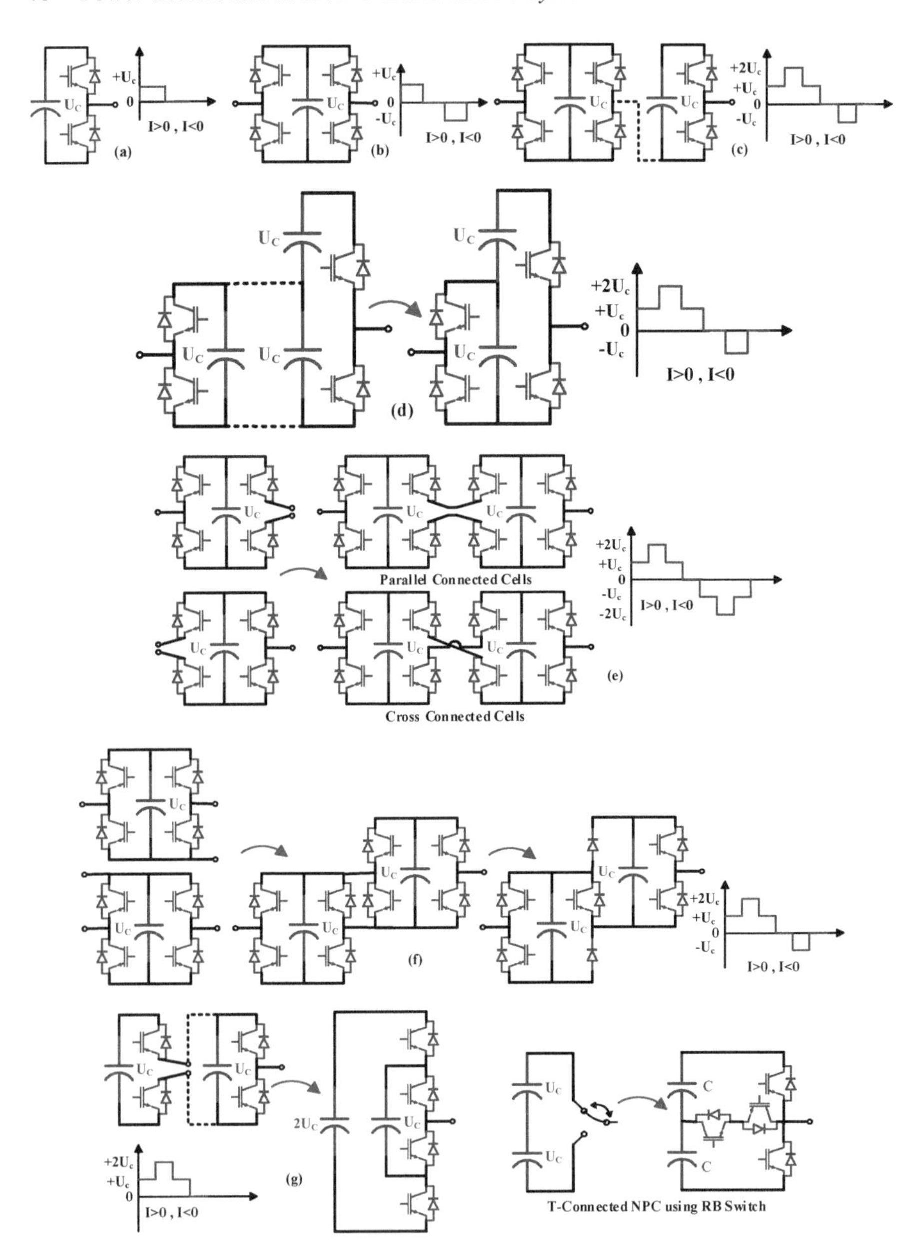


**Figure 2.17** Continued.

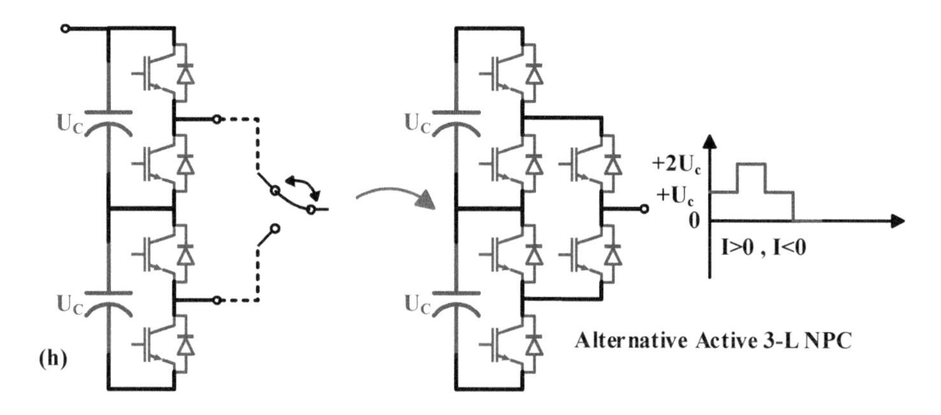


**Figure 2.17** SMs of MMCs: a) Half-Bridge; b) Full-Bridge; c) Mixed; d) Asymmetrical; e) Cross-Connected and Parallel; f) Clamped; g) FC-Type; and h) NPC-Type [19].

voltages. One way to avoid this issue is to replace active switches with diodes (Figure 2.17(f)), but this approach limits the full-bridge mode to a three-quadrant operation [19].

7. **FC-Type** SM**:** This SM is based on connecting half-bridges in a nested configuration and generates an asymmetric three-level output DC voltage ($+2U_c$, $+U_c$, and $0$) (Figure 2.17(g)). With a three-level voltage, its intermediate capacitor's voltage is half of the DC voltage [19].
8. **NPC-Type** SM**:** This SM can be obtained with two different approaches: 1) connecting two half-bridges in series; and 2) connecting switches in a T-shape, where the midpoint of the switch package must block both polarities, ("T-connected NPC sub-module"). This SM generates an asymmetric three-level output DC voltage ($+2U_c$, $+U_c$, and $0$) (Figure 2.17(h)). The switches of the first NPC SM block the same voltages, since one of the capacitors, is present in each switching process; the upper and lower arm's switches in the T-connection NPC SM block the double DC link voltage, as both capacitors are present in the switching process [19].

A comparison among various SMs is provided in Table 3.1 in terms of the output voltage levels, voltage blocking level, number of switches, and the complexity level of control and design.

It is usually preferred to design a SM following specific technical features, such as high voltage blocking, symmetrical voltage levels, bipolar operation, and the least cost. Other factors, including cell mechanical design, protection

**Table 2.1** The comparison of various voltage source SMs [19].

| Characteristics \ Sub-module Type | a | b | c | d | e | f | g | h |
|---|---|---|---|---|---|---|---|---|
| **Output Voltage Levels** | *2* | *3* | *4* | *4* | *5* | *4* | *3* | *3* |
| **Voltage Blocking Level** | *Uc* | *Uc* | *2Uc* | *2Uc* | *2Uc* | *2Uc* | *2Uc* | *2Uc* |
| **No. of Switches (Normalized by Uc)** | *2* | *4* | *6* | *6* | *8* | *7* | *4* | *6* |
| **Max no. Switches (for conducting)** | *1* | *2* | *3* | *3* | *4* | *3* | *2* | *2* |
| **Bipolar Output Voltage** | *No* | *Yes* | *Yes* | *Yes* | *Yes* | *Yes* | *No* | *No* |
| **Design Complexity Level** | *Low* | *Low* | *Low* | *High* | *High* | *High* | *High* | *High* |
| **Control Complexity Level** | *Low* | *Low* | *Low* | *High* | *Low* | *Low* | *High* | *High* |

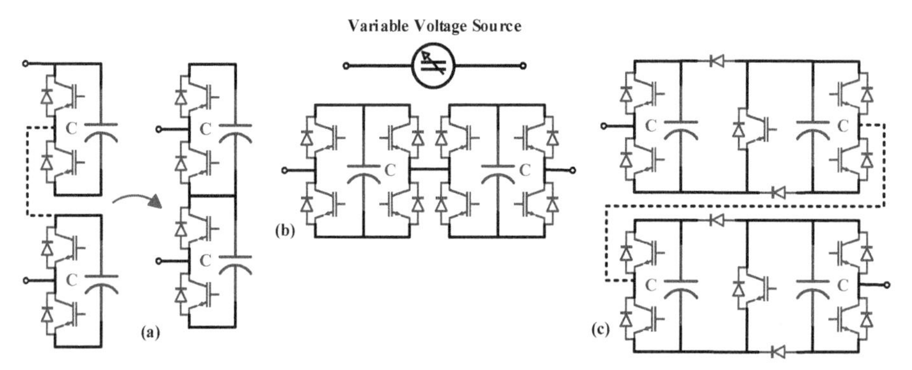


**Figure 2.18** The general structure of MMCs [12, 17, 19].

schemes (for internal faults), and the control complexity, should also be considered. There is always a trade-off between the cell complexity and its functionality/reliability for an optimal SM design [19].

#### 2.3.1.2.2.2 Conventional MMCs

MMCs are composed of various SMs, and the general structure based on chain-links is shown in Figures 2.18 and 2.19; it can be seen that the main structure of MMCs is similar to that of a two-level VSC. There is, however, a remarkable difference: the conventional RC devices in each arm of two-level VSCs are replaced by the chain-links of SMs. Thus, the energy storage is distributed in the MMCs' arms [12, 17, 19].

Advantages:

1. A superior performance.
2. The *dv/dt* challenge is eliminated by multilevel steps on the AC side, and consequently, the transformer insulation requirement in high-voltage applications is minimized.

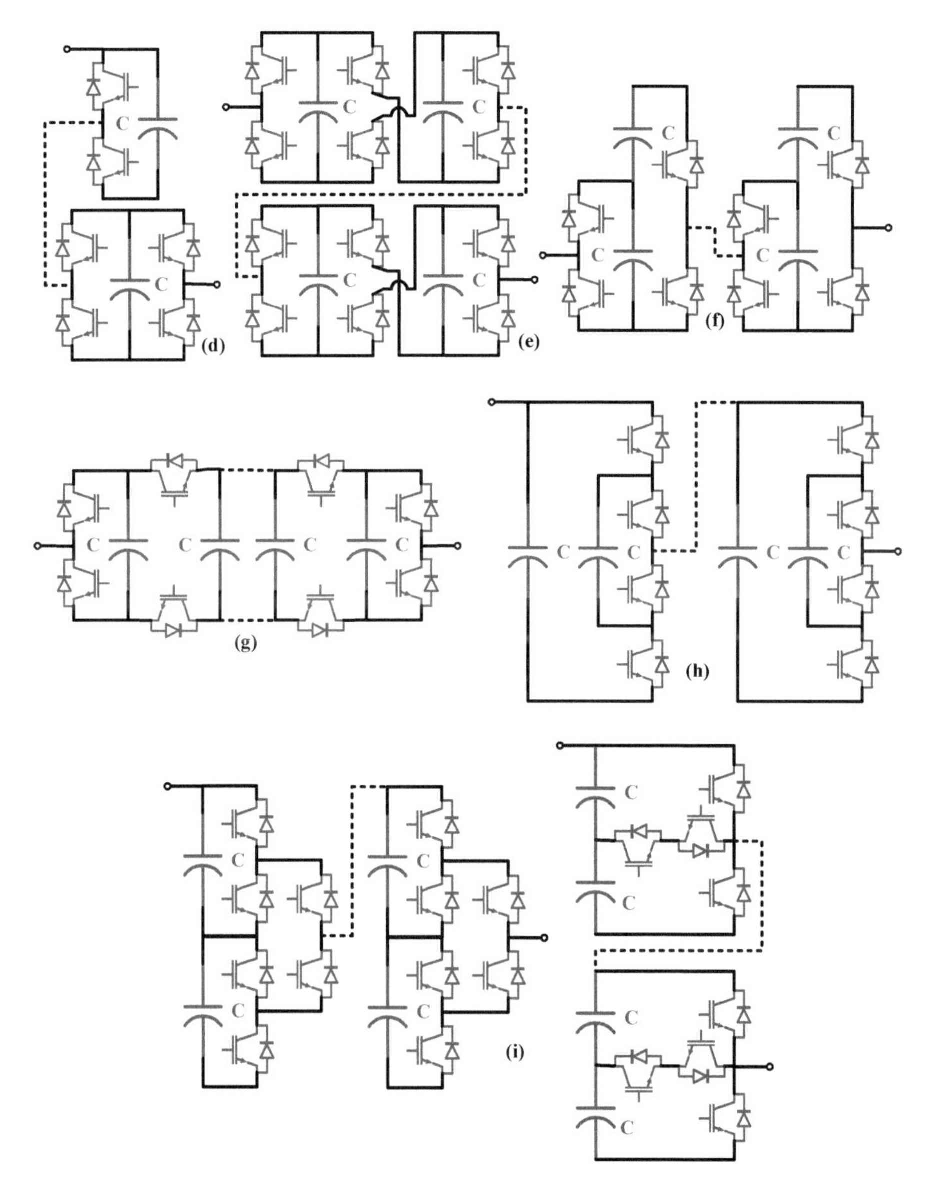


**Figure 2.19** Chain-links of power electronics SMs: a) Chain-links of Half-Bridges; b) Chain-links of Full-Bridges; c) Chain-links of Double Clamped; d) Chain-links of Mixed; e) Chain-links of Cross-Connected; f) Chain-links of Asymmetrical; g) Chain-links of Stacked FC-Type; h) Chain-links of Series FC-Type; and i) Chain-links of NPC-Type [19].

3. Fault blocking capability, which suppresses fault currents.
4. Low distortion for medium voltage motor drives.
5. A compatible structure with economical IGBT switches, with cost reductions.
6. The required switching frequency is reduced considerably, because of the high number of SMs in each converter's arm.
7. Harmonics are reduced significantly, and the filter can be reduced in size or eliminated.

Disadvantages:

1. A complex control scheme is required for high-level converters.

Most MMCs are based on half-bridge and full-bridge sub-modules. Figure 2.20(a)-(b) shows MMCs based on half-bridge and full-bridge, respectively. Figure 2.20(c) shows the comparison of output currents and voltages between a two-level VSC and an MMC [12, 19].

Due to the high power quality at the output, MMCs are considered promising power converter technology for HVDC transmission systems, considering the potential development of semiconductors and power electronics. However, MMCs have some issues in high and medium power applications. One critical challenge is the current circulation in their circuitry caused by energy differences in the converter's arms. Protection against internal and external faults is another challenge. Solving these issues is an important research goal [12, 17, 19, 20].

#### 2.3.1.2.2.3 Alternative arm modular multilevel converters

The Alternative Arm Modular Multilevel Converter (AAMMCs) is another variant of MMCs, and is designed by changing the SM structure or chain-link configuration; its general structure is shown in Figure 2.21 [17, 19].

The alternative chain-link configuration in AAMMCs provides several interesting features. The capacitors in sub-modules can be used in either polarity in the chain-link configurations of Figure 2.19(b)-(g). This unique feature provides a fault blocking capability for AAMMCs in the case of a short-circuit fault between positive and negative DC terminals. This characteristic offers extra controllability in VSCs in the case of a temporary fault along overhead lines. Bipolar chain-link provides a decoupling between AC and DC voltages, which allows VSCs to be utilized in either polarity, similar to LCC-HVDC systems; thus, they can be employed to connect CSC-HVDC and VSC-HVDC technologies to create a Hybrid-HVDC system [19].

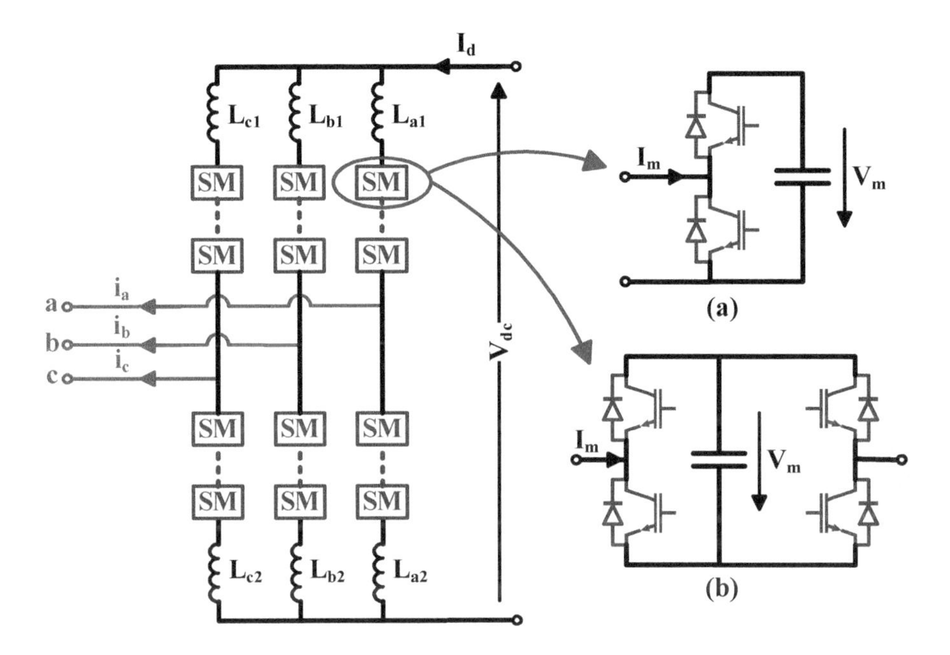


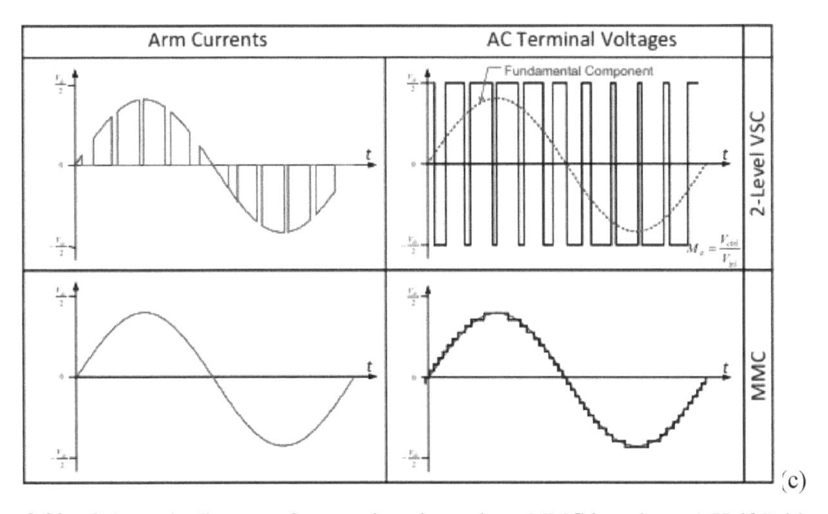


**Figure 2.20** Schematic diagram of a complete three-phase MMC based on: a) Half-Bridge sub-modules; b) Full-Bridge sub-modules; and c) Comparison between a two-level VSC and a MMC [12, 19].

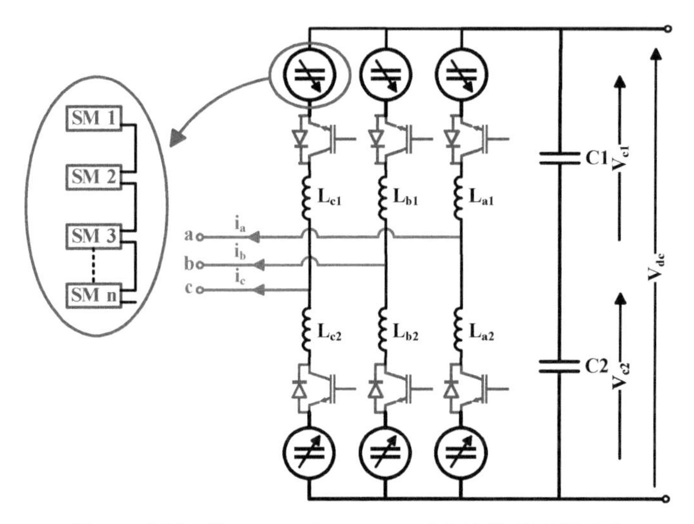


**Figure 2.21** The general structure of AAMMCs [17, 19].

However, these prominent features come with issues, such as a high number of power electronics devices and high power losses, compared to unipolar configurations. The design of AAMMCs must also consider extra protection, measurement, and cooling equipment for the bipolar SMs. The main advantages and disadvantages of AAMMCs are summarized below [17].

Advantages:

1. A modular and scalable structure.
2. A DC fault capability and management that can omit the need for large AC/DC breakers.
3. The need for large AC filters is eliminated, and consequently, the converter size is reduced.
4. A sinusoidal AC waveform at the output terminal with negligible harmonic contents or harmonic-free.

Disadvantages:

1. A complex control scheme is required for high-level converters.
2. A higher number of power electronics components than that of other MMC topologies.

#### 2.3.1.2.2.4 Hybrid MMCs

Hybrid MMCs of different topologies are designed by combining chain-link SMs and monolithic MMCs, and thus, have advantages and disadvantages of both chain-link SMs and monolithic MMCs. The low number of power electronics elements, high level of modularity, and high power quality are the major advantages of Hybrid MMCs. High-voltage stress and the series connection of devices are major disadvantages. Generally, hybrid MMCs can be designed in two ways: 1. Hybrid MMCs with Monolithic Director Switches; and 2. Hybrid MMCs with H-bridge Director Switches. The series-connected devices (also known as director switches) are responsible for generating output waveforms [19, 20].

Hybrid MMCs with Monolithic Director Switches are designed to connect series-connected chain-link SMs to different points (i.e., converter leg, AC side, or DC midpoint) of monolithic two-level converters. The primary purpose of using chain-link SMs at the DC or AC side of monolithic MMCs is to mitigate harmonics of the square-shaped output waveforms generated by director switches (monolithic converters). Switching is basically performed according to the zero-voltage switching (ZVS) approach, to minimize switching losses and voltage stress. The chain-link SMs can generate bipolar multilevel voltage waveforms and produce a near-sinusoidal waveform at the output terminals. Thus, just bipolar chain-link SMs can be used (Figure 2.19(b)-(g)). The higher number of SMs in chain-links leads to a higher quality of the output voltage. Various topologies of such hybrid MMCs are shown in Figure 2.22 [19].

This type of Hybrid MMCs is expected to reduce the number of SMs and power electronics components, as the DC voltage is shared between director switches and chain-link SMs. However, the ZVS approach introduces a specific relationship between AC and DC voltages, which is based on maintaining power balance (zero average power) in chain-link SMs. This AC and DC voltage correlation affects active and reactive power controls. One way to facilitate the decoupling of AC and DC voltages is to use three-level monolithic FC- and NPC-type director switches. However, this scheme may increase the complexity of the design and the control. Suitable schemes will have to be introduced, which makes this issue an interesting research topic [19].

Hybrid MMCs with H-bridge Director Switches are designed by combining H-bridge or a three-phase monolithic converter with chain-link SMs. Its operation principle consists of generating a bipolar multilevel waveform by synthesizing a rectified half-wave multilevel voltage (generated by chain-link sub-modules), and by the polarity adjustment through director switches (H-bridge or three-phase monolithic converters). The chain-link SMs are connected to H-bridge or three-phase monolithic converters in parallel.

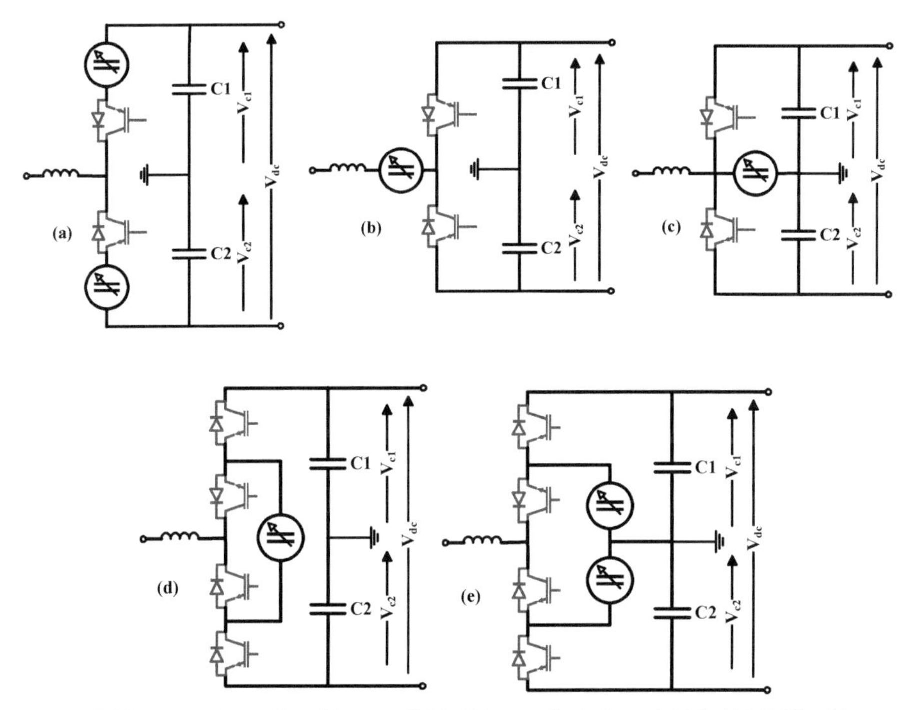


**Figure 2.22** Hybrid MMCs with Monolithic Director Switches: a) Hybrid MMC with arm chain-link SMs; b) Hybrid two-level with AC-bus chain-link SMs; c) Hybrid two-level with midpoint chain-link SMs; d) Hybrid three-level FC with midpoint chain-link SMs; and e) Hybrid three-level NPC with arm chain-link SMs [19].

Also, their switching is performed under the ZVS approach to minimize switching losses and voltage stresses.

The various topologies of this type of Hybrid MMCs are shown in Figure 2.23 [19].

### 2.3.2 Current Source Converters

CSCs are AC/DC power converters that maintain their output current in their predetermined value, regardless of the polarity and magnitude of the voltage at the poles. CSCs are controlled to act as rectifiers or inverters by controlling the magnitude and phase angle of the AC output current. The performance of CSCs relies on large inductors at the DC side. Unlike VSCs, the power reversal is achieved by reversing the voltage polarity in CSCs, and the current can flow in one direction only. Power reversal is realized through

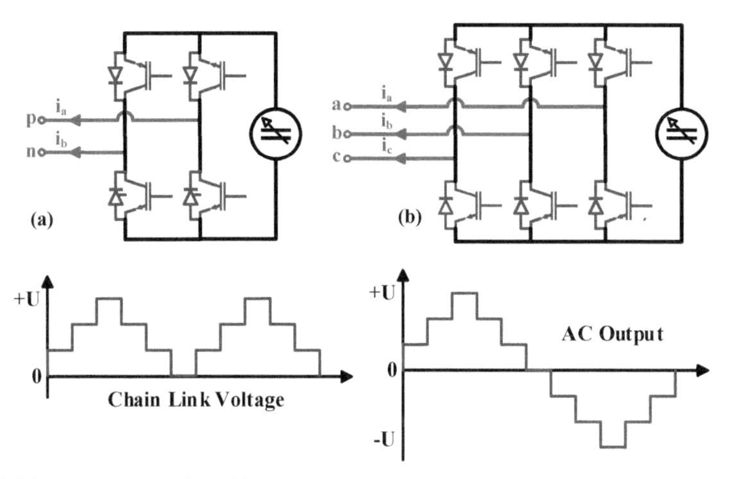


**Figure 2.23** Hybrid MMCs with H-bridge Director Switches: a) H-bridge Hybrid MMC with parallel chain-link SMs; b) Three-phase Hybrid MMC with parallel chain-link SMs [19].

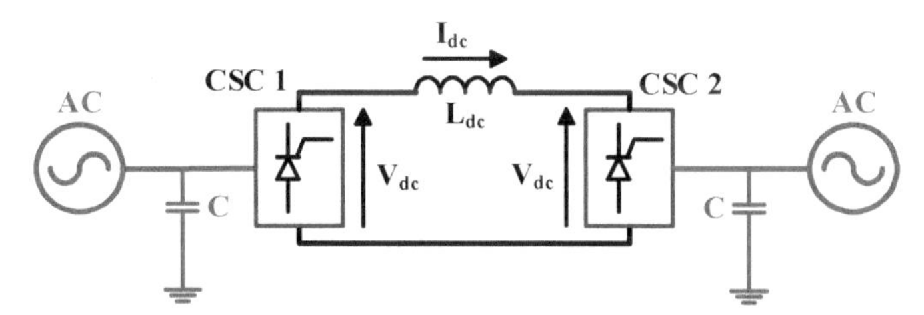


**Figure 2.24** General schematic diagram of a B2B CSC [19].

existing switches with a bidirectional voltage blocking capability (also known as Reverse Blocking (RB)). The polarity of DC link voltage $V_{dc}$ is controlled, but the DC link current $I_{dc}$ is controlled in a fixed direction. Because the AC system has a considerable inductance in its structure, a shunt intermediary capacitor is used at the AC side of the converter. The general schematic diagram of a back-to-back CSC is shown in Figure 2.24. Relevant CSCs, present in Figure 2.11, are examined as follows [19].

#### 2.3.2.1 Multipulse CSCs

Multipulse CSCs are the first generation and simplest current source AC/DC for power converters in HVDC transmission systems. Mercury-arc valves were firstly used until they became obsolete, then thyristors were used in multipulse CSCs as rectifiers/inverters. Basically, six thyristor switches must be used to form the simplest CSC. Other multipulse CSCs must be made by

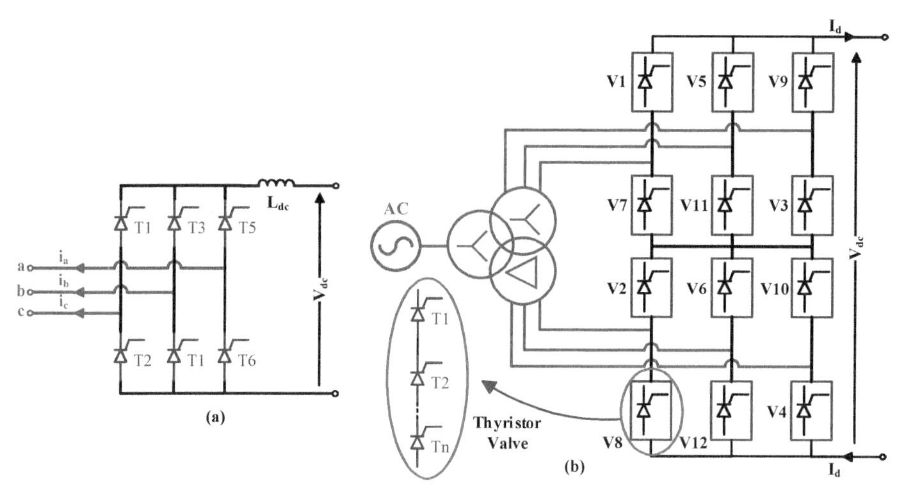


**Figure 2.25** Schematic diagram of CSCs: a) Three-phase 6-pulse; and b) Three-phase 12-pulse [12].

using a proper number of thyristor switches, in multiples of six. Accordingly, various topologies, such as 6-, 12-, 18-, and 24-pulse CSCs have been proposed, with 6- and 12-pulse CSCs being the prominent types. The schematic diagrams of 6- and 12-pulse CSCs are shown in Figure 2.25. These topologies are known as multipulse because they chop the voltage based on the number of thyristors being used. As shown in Figure 2.25(a), a 6-pulse CSC chops the input voltage into six periods based on the order of switching. The 6-pulse CSC is also known as the Graetz bridge, named after Leo Graetz, who first proposed a similar structure in the 1890s [12].

Thyristors can withstand voltages in either polarity when they are turned off and keep this condition until they receive a current signal to their gate. They are then turned on and keep this condition until the current flowing through them drops to zero, or an external circuit imposes a negative voltage at their terminals. These topologies cannot control the voltage at the AC side, because it is not possible to turn off thyristors without interfering with the AC voltage when external circuits are not used. The switching approach is based on firing thyristors, at a suitable angle and forward condition, to conduct the current; the thyristors are then turned off when the current flowing through them drops to zero. In these topologies, it is possible to control the DC side voltage by controlling the DC links at both ends of the HVDC transmission systems. The DC side current is kept at its predetermined value by using large inductors and the appropriate switching approach; the current direction cannot be reversed due to the presence of thyristors, but the power and polarity of the DC side voltage can [12].

#### 2.3.2.2 Modular current source converters

Similar to VSCs, Modular Current Source Converters (MCSCs) can be derived considering the existing duality concept between the VSC and the CSC (i.e., most MCSCs can be designed based on the duality principle in circuit topologies of VSCs). However, the VSC must have a dual topology to be used under the current source approach. The majority of SMs and single-phase structures have a dual circuit, but this is not necessarily true for all other three-phase topologies [19].

Multipulse CSCs lack modularity, like the case of monolithic multilevel VSCs, which makes their maintenance difficult. The definition of modularity in the current source approach is the same as the previous definition for the voltage source concept; other concepts, such as chain-link, building blocks, or SMs, are also similar. MCSCs consist of chain-links of sub-modules, which can be connected in series or parallel, like in the case of modular VSCs. MCSCs can be designed via 1. chain links of power electronics SMs, and 2. A combination of SMs and monolithic CSCs.

##### 2.3.2.2.1 Power electronics current source SMs

Power electronics current source sub-modules are the fundamental building blocks of any MCSCs. The SM of an MCSC is a DC/DC or DC/AC converter, which can be connected in parallel (to increase the current-conducting capability) or in series (to increase the voltage-withstanding capability). The combined series and parallel connections can be used to meet the specifications of a variety of applications. The popular SMs of MCSCs, and their output current waveforms, are shown in Figure 2.26. The features of each SM are explained as follows [19].

*a)* **Half-Bridge** SM*:* It is the simplest and fundamental sub-module, and has a dual half-bridge SM (Figure 2.17(a)). In this SM, the switches must be able to block voltage in any polarity and conduct current in one or in two directions (bidirectional). Since the duality concept can be applied, the output current of this current source sub-module is similar to the output voltage of its voltage source SM counterpart. The Half-Bridge SM generates a unidirectional output current, which operates in two quadrants by generating a two-level output current ($I_c$ and *0*) (Figure 2.26(a)) [19].

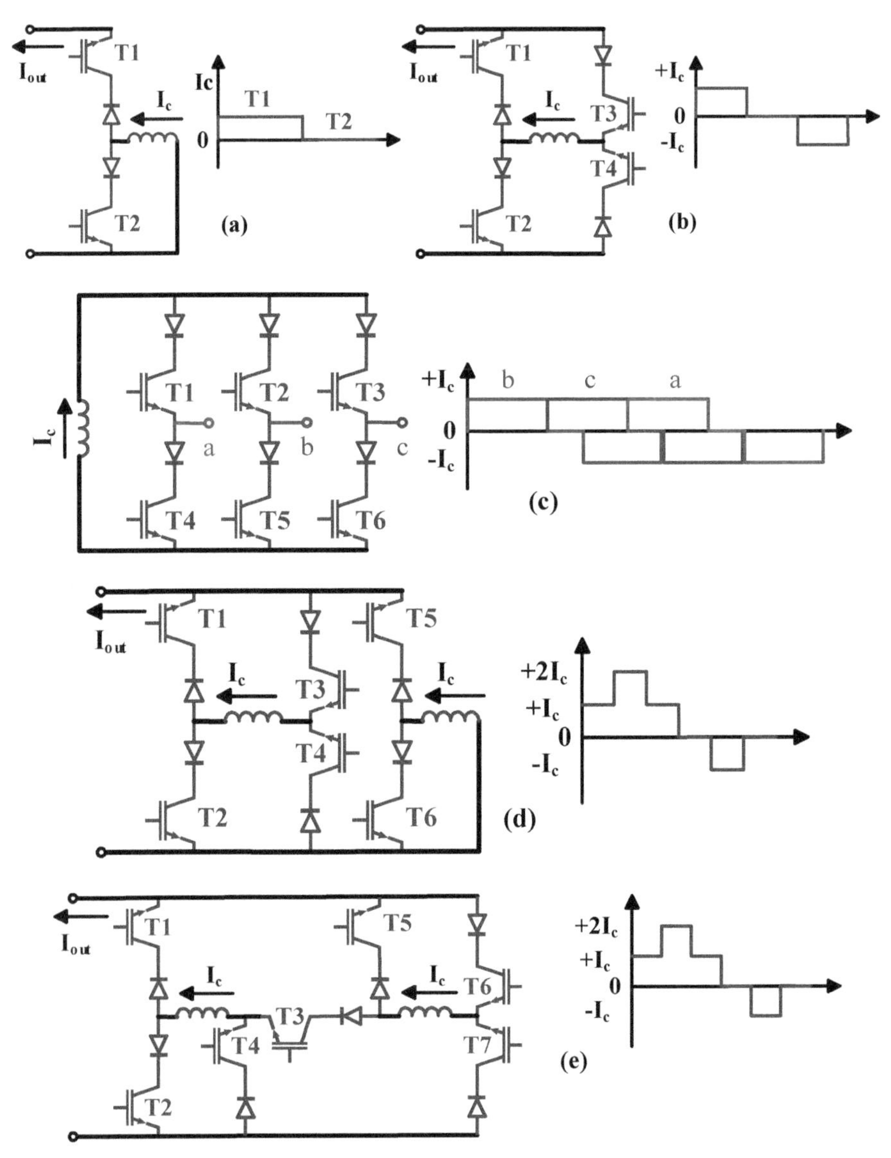


**Figure 2.26** SMs of MCSCs: a) Half-Bridge; b) Full-Bridge; c) Three-phase; d) Mixed; and e) Clamped [19].

*b)* **Full-Bridge** SM**:** It is designed by connecting two half-bridges in parallel (Figure 2.17(b)). This SM can conduct current in bidirectional paths and can block the voltage of either polarity. It operates in four

Table 2.2 The comparison of various current source SMs [19].

| Characteristics \ Sub-module Type | *a* | *b* | *c* | *d* | *e* |
|---|---|---|---|---|---|
| **Output Current Levels** | *2* | *3* | *3* | *4* | *4* |
| **Max. Output Current** | *Ic* | *Ic* | *Ic* | *2Ic* | *2Ic* |
| **No. of RB Switches** | *2* | *4* | *6* | *6* | *7* |
| **No. of Switches (for Conducting)** | *1* | *2* | *2* | *3* | *4* |
| **Bidirectional Output Current** | *No* | *Yes* | *Yes* | *Yes* | *Yes* |
| **Design Complexity Level** | *Low* | *Low* | *Low* | *Low* | *High* |
| **Control Complexity Level** | *Low* | *Low* | *High* | *Low* | *Low* |

quadrants and generates three-level output currents ($+I_C$, $0$, and $-I_C$ (Figure 2.26(b) [19].

*c)* **Three-phase** SM**:** This SM generates a symmetrical three-level output DC current ($+I_C$, $0$, and $-I_C$) (Figure 2.26(c)) [19].

*d)* **Mixed** SM**:** This SM is designed by connecting half-bridge and full-bridge SMs. The output current waveform is an asymmetric four-level current ($+2I_C$, $+I_C$, $0$, and $-I_C$) (Figure 2.26(d)) [19].

*e)* **Clamped** SM**:** It consists of two full-bridge SMs, and its simplified version is shown in Figure 2.26(e). This SMs is equivalent to two parallel-connected half-bridge SMs under a specific switching (T3 off, T4, and T5 on). It is equivalent to a full-bridge SMswith two series-connected inductors for a different specific switching (T3 on, T4, and T5 off). The presence of two series-connected inductors limits the *di/dt*, providing an interesting feature in the full-bridge mode of operation. Precise design and control schemes are required to prevent the possible connection of the two inductors with different currents. This SM generates an asymmetrical four-level output current ($+2I_C$, $+I_C$, $0$, and $-I_C$) (Figure 2.26(e)) [19].

A comparison among various current source SMs, in terms of output current levels, maximum output current, number of switches, and complexity of control and design, is shown in Table 2.2.

#### 2.3.2.2.2 Conventional MCSCs

Conventional MCSCs consist of parallel-connected full-bridge and three-phase current source SMs. The reason for the parallel connection is the nature of the CSCs. The schematic diagram of the conventional MCSCs is shown in Figure 2.27 [19].

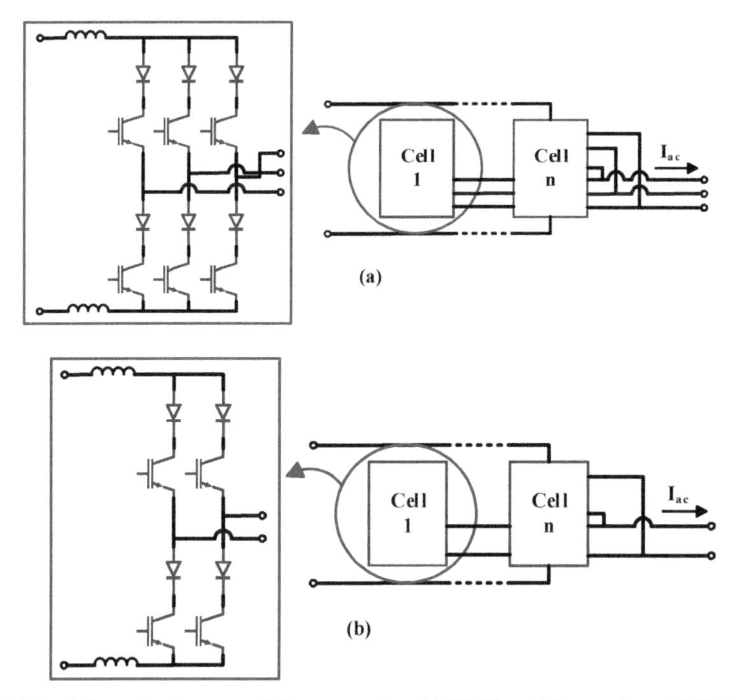


**Figure 2.27** Schematic diagram of the conventional MCSCs: a) Three-phase MCSC; and b) Single-phase MCSC [19].

The DC side of each current source cell is connected through inductors and is opened to minimize circulating currents between current source cells. It is possible to switch each current source at the fundamental frequency, providing that a sufficient number of SMs are used in the structure. Each valve should be able to withstand a full AC line-to-line voltage. For high-voltage applications, a large number of SMs should be connected in series [19].

#### 2.3.2.2.3 Modular multilevel current source converters

Modular Multilevel Current Source Converters (MMCSCs) can be designed by applying the duality concept. The current source arm and single-phase MMCSC can be obtained from their voltage source counterparts. The schematic diagrams of the current source (CS) arm and single-phase MMCSC are shown in Figure 2.28 [19].

In single-phase MMCSCs, the arms are complementary, i.e., the sum of currents flowing through "CS Arm 1" and "CS Arm 3" is constant and are both equal to $I_{dc}$. The AC output current of this topology is equal to the

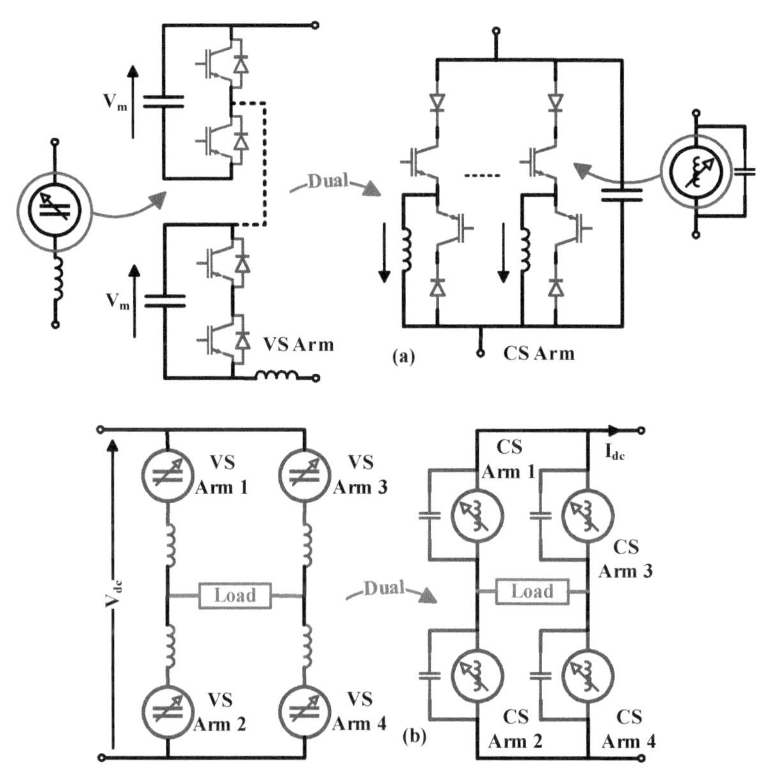


**Figure 2.28** The schematic diagram of the MMCSCs: a) CS arm; b) Single-phase MMCSC [19].

current difference of "CS Arm 1" and "CS Arm 2." In the case of half-bridge sub-modules, the CS arm must be operated under unidirectional condition, and consequently, the CS arm voltage must be bipolar to realize zero average power.

The duality concept cannot be applied for three-phase MMCSCs, because their voltage source counterpart is not planar. To eliminate this problem, the higher-level circuit topology of the voltage source counterpart can be copied. The schematic diagram of a three-phase MMCSC is shown in Figure 2.29 [19].

#### 2.3.2.2.4 Hybrid MCSCs

Hybrid MCSCs are designed by combining CS arms and monolithic CSCs. In this configuration, the CS arm is used as a current shaper to modify the output current of monolithic CSCs. Two hybrid MCSCs are shown in Figure 2.30,

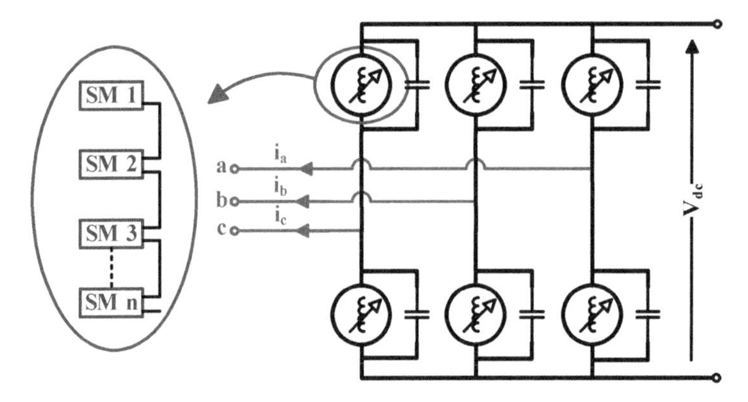


**Figure 2.29** The schematic diagram of a three-phase MMCSC [19].

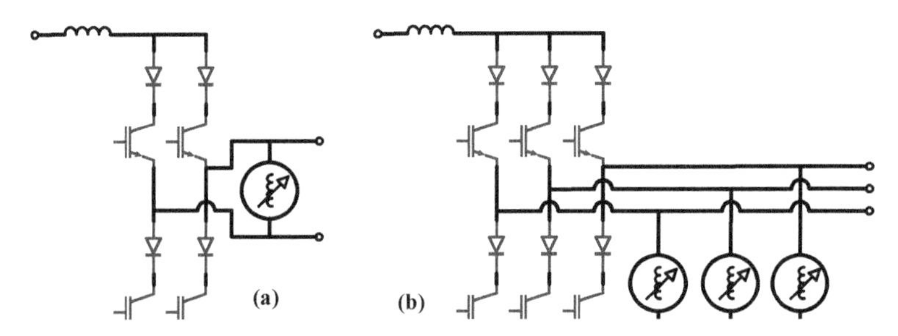


**Figure 2.30** Schematic diagrams of two Hybrid MCSCs: a) Single-phase; b) Three-phase [19].

where the current shaper can only synthesize quadrature and/or harmonic currents, to preserve zero average power in the topology [19].

An alternative design may use CS arms as current shapers at the DC side. The schematic diagrams of two new hybrid MCSCs based on monolithic full-bridge current source SMare shown in Figure 2.31. In series-connected CS arm hybrid MCSCs, the CS arm is responsible for synthesizing the rectified current, and the monolithic full-bridge SM CSC is used to invert the rectified current into a sinusoidal AC. In series-connected CS arm hybrid MCSCs, the CS arm has a different role, including synthesizing the current difference between a rectified current and the phase DC [19].

### 2.3.3 Hybrid Current and Voltage Source Converters

Hybrid Current and Voltage Source Converters (HCVSCs) combine both current source and voltage source (VS) concepts. Basically, the CS/VS arm

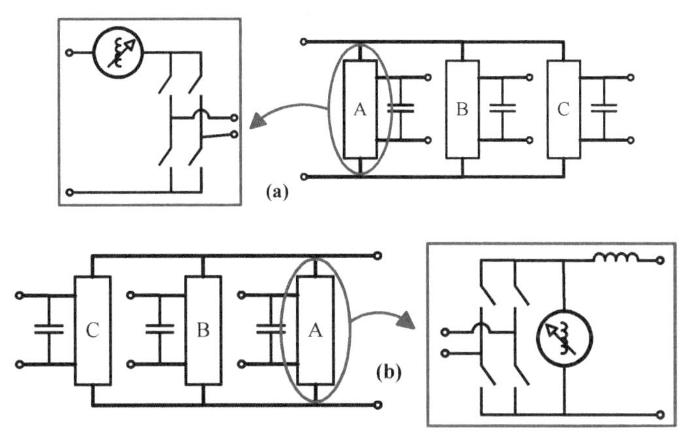


**Figure 2.31** Schematic diagrams of two new hybrid MCSCs: a) Series-connected CS arm; b) Parallel-connected CS arm [19].

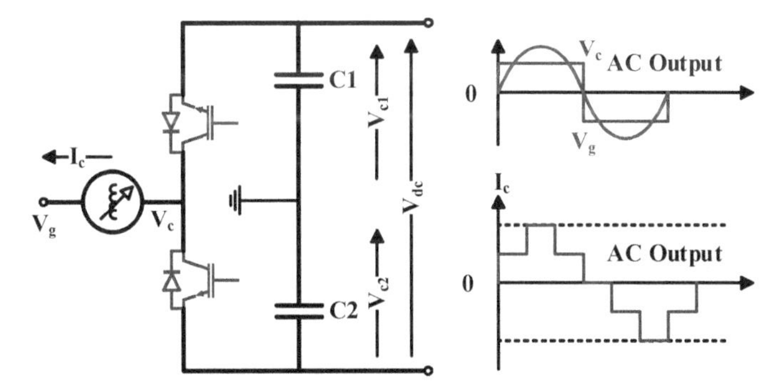


**Figure 2.32** The schematic diagram of an HCVSC and its output waveforms [19].

can be connected to a VSC/CSC. The general schematic diagram of an HCVSC and its output waveforms are shown in Figure 2.32.

In this configuration, the CS arm is used to synthesize the desired fundamental current, and the two-level VSC is used to maintain a zero average power. To do so, the two-level VSC is responsible for implementing a zero/90° phase displacement between the CS arm's voltage and it's current.

HCVSCs can also be constructed by using hybrid CS-VS arms consisting of hybrid CS-VS SMs. The schematic diagram of hybrid CS-VS arms and SMs is shown in Figure 2.33. In this topology, the role of the inductor is to synthesize the desired current waveform, and the role of the capacitor is to modify the voltage of CS arms. More studies should be performed on HCVSCs [19].

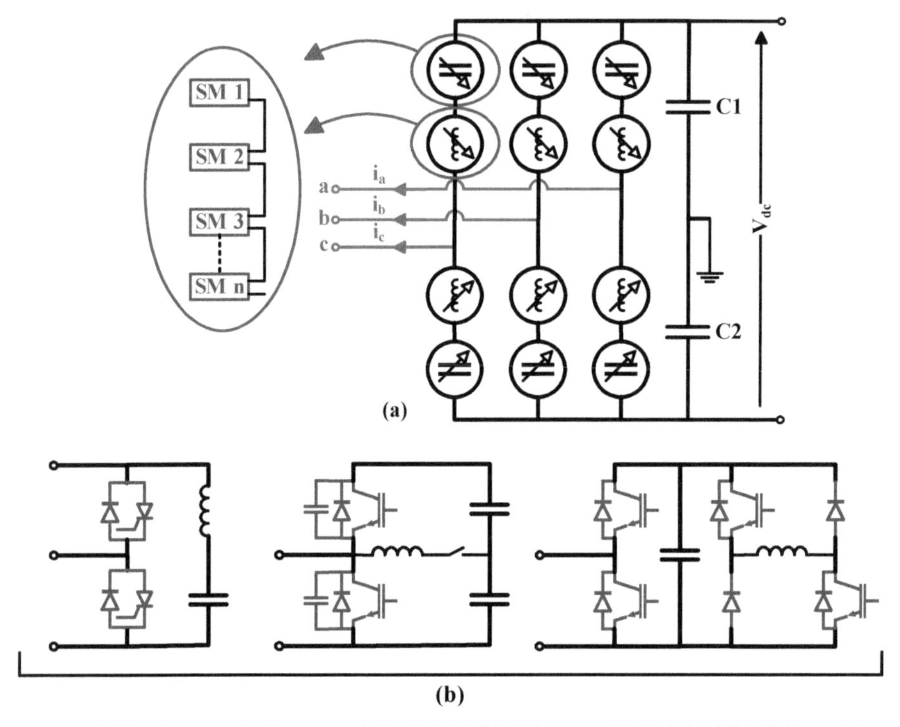


**Figure 2.33** Schematic diagram of (a) Hybrid CS-VS arms; (b) Hybrid CS-VS SMs [19].

## 2.4 DC/DC Converters

DC/DC converters, also known as DC Transformers, provide DC voltage matching. Besides this primary purpose, they can also be used to subdivide large MT-HVDC grids into several smaller protection zones, for DC voltage regulation, fault isolation, and to connect bipolar/monopolar configurations. CSC/VSC technologies have been considered for DC/DC converters. DC/DC converters play a pivotal role in HVDC transmission systems, especially large-meshed MT-HVDC grids. In this regard, various studies are conducted to develop novel practical topologies and improve the performance of existing DC/DC converters in HVDC transmission systems [25–29].

Generally, DC/DC power converters can be designed by considering two major approaches: 1. Isolated DC/DC converters; and 2. Non Isolated DC/DC converters. In the Isolated approach, an AC link exists in the heart of DC/DC converters, unlike the case of the Non Isolated converters, where the AC link is not employed. Each of these two configurations can be divided into

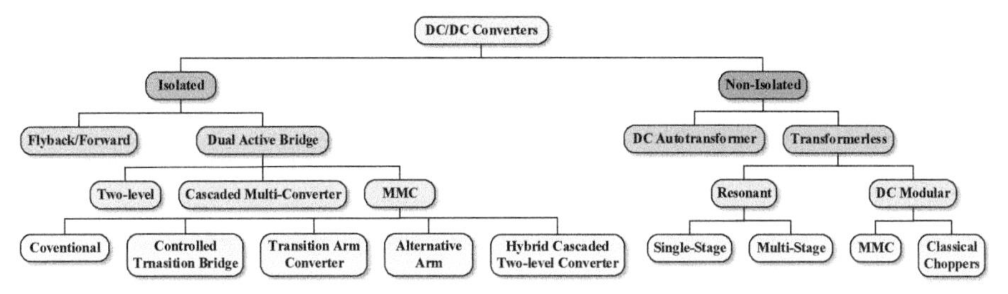


**Figure 2.34** General categorization of the DC/DC converters.

subgroups based on their structures. The general categorization of DC/DC converters is shown in Figure 2.34 [25–29].

## 2.4.1 Isolated DC/DC Converters

Isolated DC/DC converters belong to a large family, and their performance is based on the DC/AC/DC conversion. To design DC/DC converters, different AC/DC structures and topologies can be used. In isolated DC/DC converters, the galvanic separation is done deliberately by means of an intermediate AC link. A conventional AC transformer or coupled inductors allows the DC fault blocking capability because the DC fault is a controllable AC overcurrent at the healthy side of the converter. The isolation feature is used to enable different grounding schemes in HVDC transmission systems. Although grounding schemes can be implemented by other means, isolation schemes offer design simplicity and safety assessment [27].

### 2.4.1.1 Flyback/Forward-based

Generally, the most promising DC/DC converters for HVDC applications are developed under the front-to-front dual active bridge (DAB) concept. Other DC/DC converters that cannot be categorized into the DAB group are based on flyback/forward concepts. In some topologies of this group, a central coupled inductor and MMC-SMs are utilized for high-voltage purposes. Other modular structures based on coupled inductors have also been proposed, which are used for high step-up voltage ratios. High insulation requirements in coupled inductors-based DC/DC converters are a considerable challenge, and the high current requirement is a challenge in centralized inductor topologies. These issues have limited flyback/forward-based DC/DC converters in low-power applications. The schematic diagrams of the flyback/forward-based DC/DC converters are shown in Figure 2.35 [27].

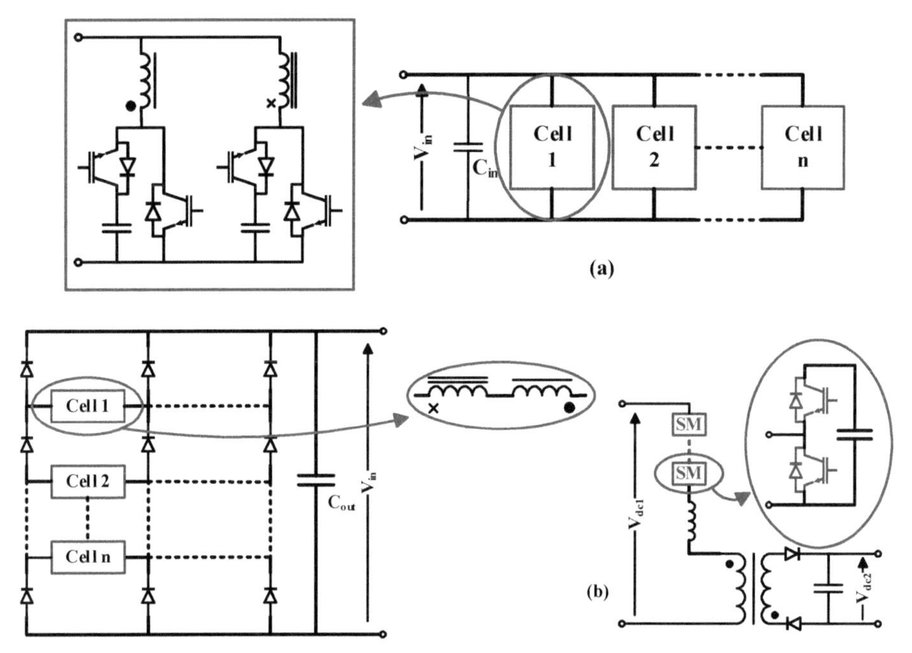


**Figure 2.35** Schematic diagram of the flyback/forward-based DC/DC converters: a) Modular topology; and b) Centralized coupled inductor topology [27].

### 2.4.1.2 DAB

Since most DC/DC converters are based on the DAB concept, we herein focus on the most important three subgroups in DAB DC/DC converters, as shown in Figure 2.34.

#### 2.4.1.2.1 Two-level DAB

Two-level DAB DC/DC converters can be built with two-level VSCs, with series-connected IGBTs connected via an AC transformer. The schematic diagram of the two-level DAB DC/DC converter is shown in Figure 2.36.

The operation of this configuration relies on the phase shift modulation at a fixed duty cycle. Typically, this topology operates in the square wave mode at the fundamental frequency, and each arm is used to conduct 180°. Various modulation schemes, such as PWM, may be used in this topology. Selective Harmonic Elimination (SHE) schemes are also employed to eliminate undesired harmonics. The two-level DAB DC/DC converter can be used for power flow control between two VSCs (i.e., VSC1 and VSC2). However, the main issue of this topology is its high *dv/dt* (which limits its

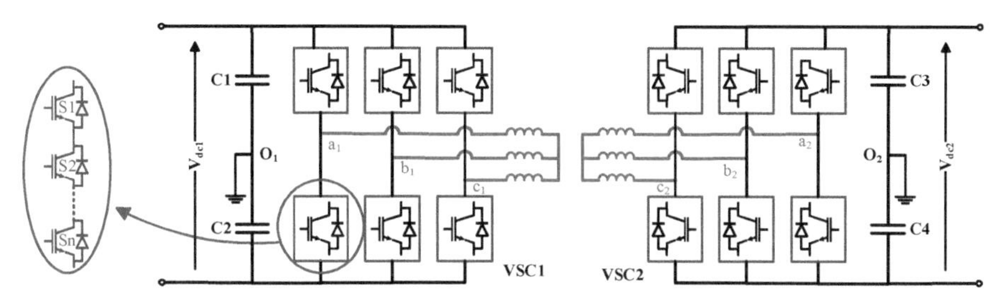


**Figure 2.36** Schematic diagram of the two-level DAB DC/DC converter [27, 28].

application to low power, and voltages up to ±200 kV), insulation issues, electromagnetic interference (EMI), unbalanced static and dynamic voltage sharing on transistor valves, and high switching losses. These issues make this topology not efficient for HV applications [27, 28].

#### 2.4.1.2.2 Cascaded DAB multilevel converter

Cascaded DAB Multi Converter topologies can be designed by using the series/parallel connection of several small DAB converters, which operate as elementary cells. In this topology, there are no series-connected switches since just a portion of the rated voltage should be withstood by small DAB converters. DAB converters can be connected in series or parallel at both input and output terminals, leading to four configurations: 1) Input-Series Output-Series (ISOS); 2) Input-Parallel Output-Series (IPOS); 3) Input-Series Output-Parallel (ISOP); and 4) Input-Parallel Output-Parallel (IPOP). The schematic diagrams of the fourfold cascaded DAB multi converter as DC/DC converters are shown in Figure 2.37 [27, 28].

Generally, the parallel connection is used to increase the current capability, and the series connection is used to increase the voltage capability. Therefore, these fourfold configurations of cascaded DAB multi converters can be used to meet the requirements of HVDC applications. Each DAB converter operates under the conventional standards of DAB converters but with an additional limitation: a balanced distribution of voltages and currents in the cells. It is possible to use this topology in medium frequency (MF, tens of kHz), leading to a considerable reduction in the size and weight of passive elements and the transformer. Soft switching can reduce switching losses, provided that this configuration is properly sized.

The prominent features of this family of DC/DC converters are their scalability and modularity, which allow different operating voltages and power levels. Other variations of this topology (especially IPOS) can be used

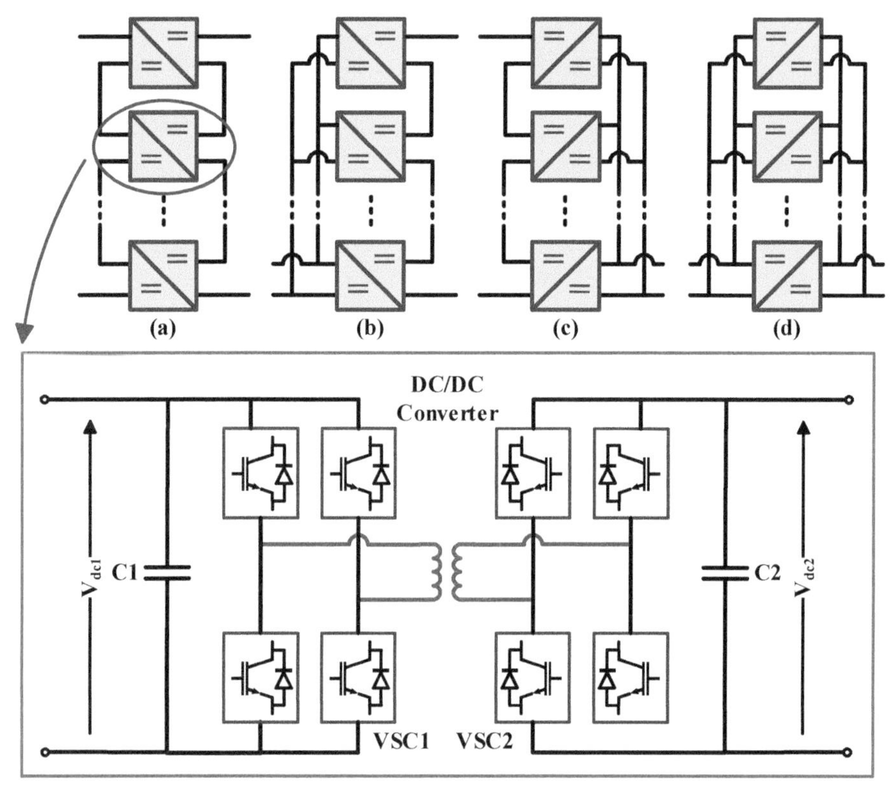


**Figure 2.37** Schematic diagrams of the fourfold cascaded DAB multi converter as the DC/DC converters: a) ISOS; b) IPOS; c) ISOP; and d) IPOP [27].

for high transformation ratio applications. Despite advantages, the main challenge of this family is the high insulation level requirement of the transformer, which limits its application to medium voltage [27].

#### 2.4.1.2.3 DAB-MMC

DAB-MMC is considered an appropriate topology for DC/DC converters for HV topologies. DAB-MMC is designed by connecting two MMCs via an MF transformer, where the AC link uses the front-to-front concept. The main reason to use the MF AC link is its capability to provide a compact transformer design and reduce the size of passive elements, including cell capacitance and arm inductance. The AC voltage is achieved by SMs used in the heart of MMCs. They employ sub-modules that can be of any type, but half-bridge and full-bridge are the most well-known types in this family. Half-bridge can provide unipolar voltages, and full-bridge can generate bipolar voltages at their terminals. The full-bridge converter is the preferred sub-module for DAB-MMC [27, 28].

#### 2.4.1.2.3.1 Conventional DAB-MMC

Conventional DAB-MMCs are DC/DC converters in which conventional MMCs, such as those used as power converters in VSC-HVDC transmission systems, are employed. The MMCs in this family rely on half-bridge and full-bridge SMs. The schematic diagram of the conventional DAB-MMC is shown in Figure 2.38 [27, 28].

Various modulation schemes/modes can be used for this topology, but there are two important modes of operation: 1. full multilevel modulation with sinusoidal voltages; and 2. quasi-two-level mode with trapezoidal voltages. In the first mode of operation, the MMCs use the same modulation employed in power converters, and the distributed cell capacitors are controlled in an appropriate way; this guarantees the complementary operation of the upper and lower arms. The control of the power flow between AC and DC can be accomplished without undesired inrush currents at the DC side. To limit inrush currents caused by an existing mismatch between the common-mode voltage and the input DC voltage, the arm inductor is used. In this mode, MMC is used like a VSC, because its upper and lower arms are used simultaneously. AC and DC currents circulate in all arms, whereas the AC current is responsible for transferring active power from the AC side to the converter; the DC current is used to transfer power from the DC/DC converter

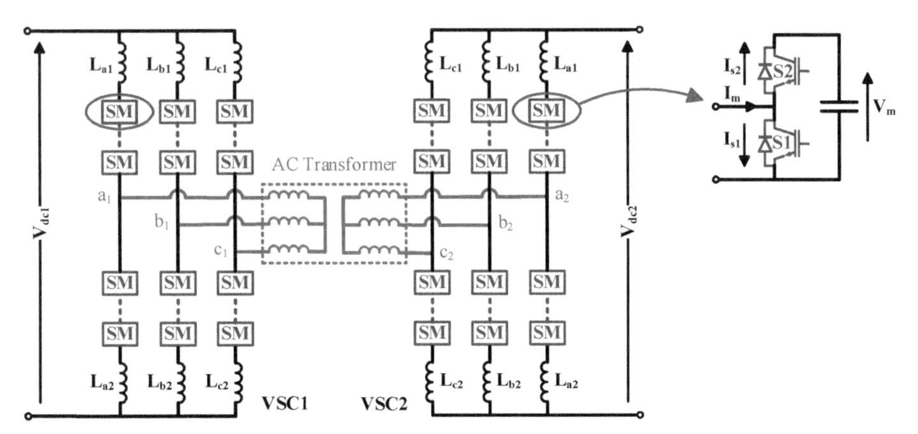


Figure 2.38 Schematic diagram of the conventional DAB-MMC as the DC/DC converter [27, 28].

to the DC side. The major advantages and disadvantages of this model are summarized as follows [27, 28]:

1. Low switching losses compared to its two-level counterpart.
2. Low voltage stress (*dv/dt*) compared to its two-level counterpart.
3. The full modulation index for AC voltage control is available during DC fault and black start.
4. Lower power density is caused by not full exploitation of the active power due to low-order harmonics.

The second mode of operation was firstly used for diode clamped multilevel inverters and later was used for MMCs. This mode uses MMCs as a two-level converter, and cell capacitors are used as a clamping package to facilitate the voltage transition at the output terminals. The fundamental current is allowed to flow through MMC cell capacitors during the voltage transition. The required energy storage capacity of cell capacitances, therefore, decreases. In this mode, upper and lower cell capacitors are bypassed when the output voltage is equal to $-V_{dc}/2$ or $+V_{dc}/2$. This feature leads to negligible voltage mismatches among the DC link voltage and the voltage of cell capacitors, and thus, a small inductor is enough to suppress and limit circulating currents. These features make this mode of operation of conventional DAB-MMC a suitable and promising DC/DC converter for MT-HVDC grids [27, 28].

#### 2.4.1.2.3.2 DAB-MMC based on controlled transition bridge

DAB-MMC using Controlled Transition Bridge (CTB) is a three-phase controlled MMC consisting of full-bridge SMs. The schematic diagram of this converter is shown in Figure 2.39 [28].

The converter in Figure 2.39 consists of $N_1$ full-bridge SMs in each arm that can generate $2N+1$ voltage levels. The operation of this topology (its main two-level bridge) at the fundamental frequency generates negligible switching losses. The related chain-link withstands just half the voltage ($V_{dc}/2$) of the DC link and the voltage of each cell capacitor is $(V_{dc}/2)/N$. $2N$ switches cooperate in each arm for current conduction, in a similar fashion

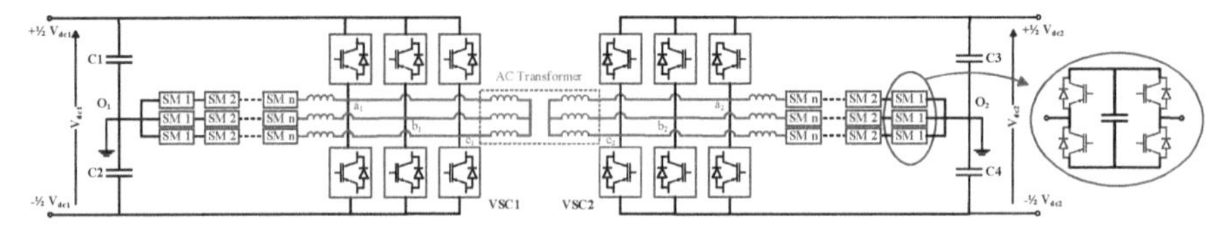


**Figure 2.39** Schematic diagram of the DAB-MMC based on CTB [28].

as the conventional two-level bridge. Since full-bridge SMs are utilized in this topology, the on-state loss is lower than that with half-bridge SMs. The performance of the chain-links and the two-level bridge is complementary, and the chain-links provide a suitable transition between positive and negative DC terminals [28].

This topology can be used as a power converter and DC/DC converter. In power converter applications, various modulation schemes can be implemented; while in DC/DC converter applications, the quasi-two-level operation is preferred, because the size and weight of the converter are highly important. This quasi-two-level operation mode provides several advantages, such as low semiconductor losses, scalability, modularity, and low *dv/dt*. The main disadvantage of this topology is the discharging of the DC capacitor, which can lead to high current stress in the event of DC faults. This is not a severe challenge because the intermediary AC link is weak [28].

#### 2.4.1.2.3.3 DAB-MMC based on transition arm converter

DAB-MMC based on Transition Arm Converter (TAC) is basically an MMC in which the typical half-bridge SMs are replaced with high voltage (HV) series-connected SMs made of IGBTs. The schematic diagram of this topology is shown in Figure 2.40 [28].

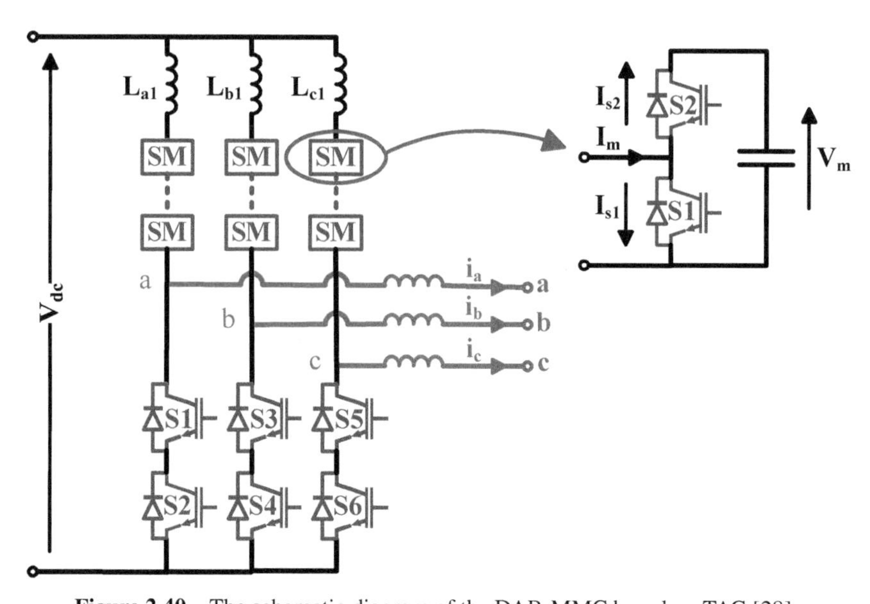


**Figure 2.40** The schematic diagram of the DAB-MMC based on TAC [28].

This topology features a lower number of semiconductor devices for the DC/DC conversion in MT-HVDC grids. The upper arm is employed as the control and provides a suitable step-level transition between the positive and negative DC terminals during AC conversion; the lower arm switches are turned on only when the negative DC voltage ($-V_{dc}/2$) must be converted at the AC link. Based on its operation, a common-mode current flows in both arms. Also, the voltage stress (*dv/dt*) across the switches occurs gradually. This characteristic leads to the elimination of snubber circuits. In addition, the quasi-two-level operation mode can be applied, which makes this topology a promising DC/DC converter [28].

### 2.4.1.2.3.4 DAB-Alternative arm MMC

DAB-Alternative Arm MMC (DAB-AAMMC) is designed by using full-bridge SM-based MMC, but with two major differences: 1) the number of the full-bridge sub-modules is reduced compared to typical MMCs; and 2) the basic operation of MMCs is modified such that each DAB-AAMMC arm is utilized for 180°, while the director switch ensures full DC link voltage blocking. The schematic diagram of the DAB-AAMMC is shown in Figure 2.41.

In this topology, a deliberate short time overlap is considered for both upper and lower arms to provide a smooth current commutation for a fault blocking capability. The switch losses are reduced compared to those of a typical full-bridge SM-based MMCs and DAB-MMC based on hybrid cascaded two-level converters.

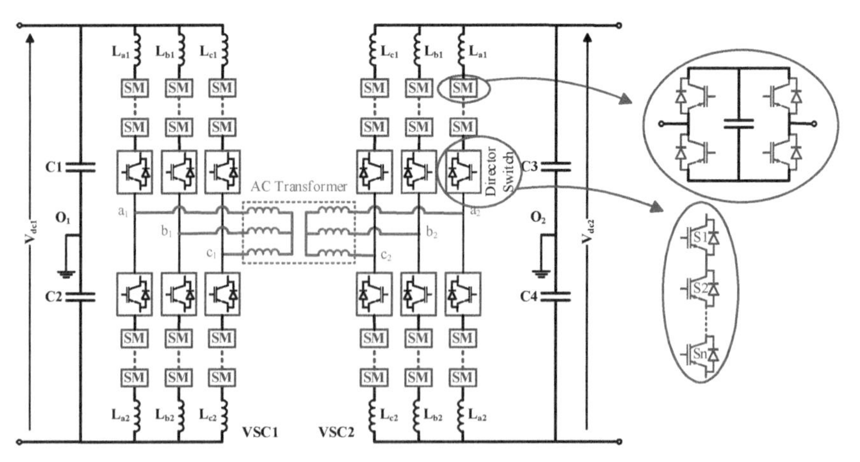


**Figure 2.41** Schematic diagram of the DAB-AAMMC [27, 28].

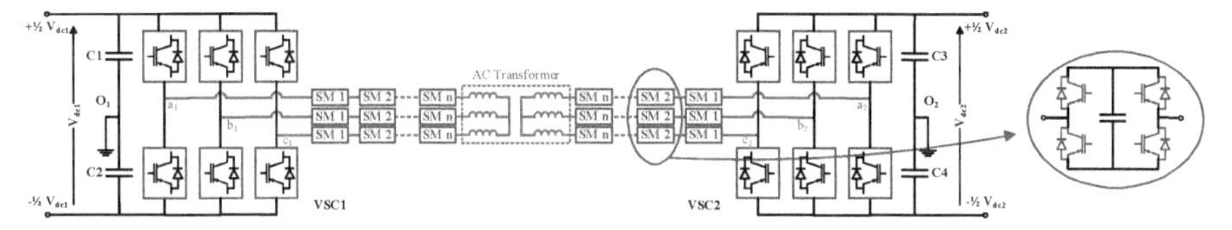


**Figure 2.42** Schematic diagram of the DAB-MMC based on HCTC [27, 28].

Similar to conventional MMCs, the quasi-two-level operation mode is applicable for this topology. This operation mode can reduce the cell capacitor's size and arm inductors. This operation mode of DAB-AAMMC makes it a suitable choice for HV DC/DC conversion applications. However, this topology has a low efficiency if compared to that of the half-bridge SM-based MMC operated in quasi-two-level mode. In addition, the concentrated DC link capacitor can increase the transient peak of DC fault currents [27, 28].

#### 2.4.1.2.3.5 DAB-MMC based on hybrid cascaded two-level converter

DAB-MMC based on Hybrid Cascaded Tow-level Converters (HCTC) is a DC/DC converter with a similar structure to DAB-MMC based on CTB. The schematic diagram of the DAB-MMC based on HCTC is shown in Figure 2.42.

This topology has a significant advantage over its CTB counterpart, which is the reverse fault blocking capability, with the same number of SMs and capacitor cells. However, there are some disadvantages that make this topology inadequate for HV DC/DC conversion applications. In its structure, more switches ($4N$) are used compared to the CTB counterpart ($2N$), because full-bridge SMs are used. Therefore, this topology has more power losses, even in quasi-two-level operation mode. Similar to CTB and AAMMC counterparts, its DC link capacitor can increase the current stress during DC side faults [27, 28].

### 2.4.2 Non Isolated DC/DC Converters

Non-Isolated converters are another family of DC/DC converters, which is larger than the isolated counterpart. This family can be classified into two major subgroups based on the presence/absence of the transformer: 1) DC Autotransformer; and 2) Transformerless DC/DC converters [27].

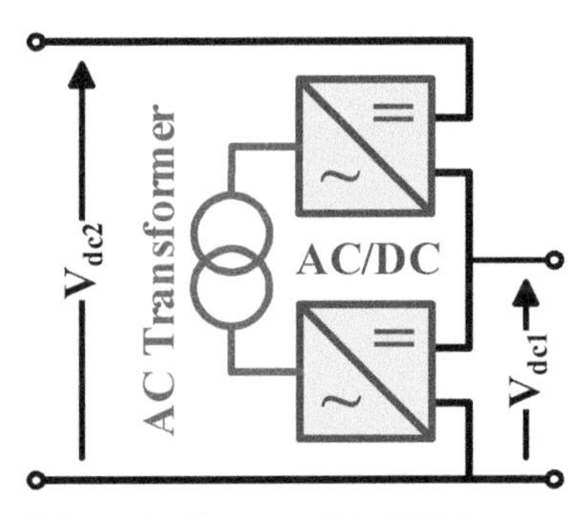


**Figure 2.43** Schematic diagram of the DC Autotransformer [27].

### 2.4.2.1 DC autotransformer

DC (or HVDC) Autotransformer consists of two HV DC/AC converters based on DC/AC/DC conversion, and their performance is similar to that of its isolated counterpart. The schematic diagram of the DC Autotransformer is shown in Figure 2.43, where the DC and AC sides of two DC/AC converters are connected in series. The DC side is connected directly, but the AC side is connected via an AC transformer. This topology is different from the DAB family previously discussed because a portion of the power is conducted through the AC link, which leads to lower power losses and transformer ratings. Series connections lead to a reduction in voltage ratings of each DC/AC converter. They are compatible with VSCs (such as two-level, three-level, and MMCs), HCVSCs, or VSC-based diode rectifiers. However, this family is more suitable for low or medium transformation ratios [27].

### 2.4.2.2 Transformerless

Transformerless DC/DC converters do not have transformers in their structure, and they will be discussed below.

#### 2.4.2.2.1 Resonant DC/DC converters

Resonant DC/DC converters are made of LC tanks. Their performance is based on the resonance of inductances and capacitances as a practical means for stepping up the DC voltage while preserving the soft-switching scheme. The schematic diagram of the resonant DC/DC converters is shown in Figure 2.44.

This family has been initially proposed as a multifunctional unit to be used in MT-HVDC grids, capable of bidirectional power flow and limiting DC faults to a predefined section of the HVDC system. It can be implemented

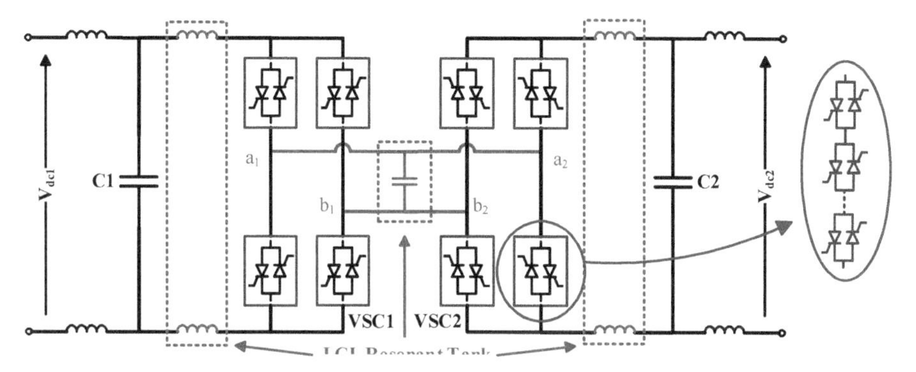


**Figure 2.44** Schematic diagram of the resonant DC/DC converters [27].

as two different subgroups based on the number of LC tanks used in their structure: 1) Single-stage resonant DC/DC converters; and 2) Multiple stage resonant DC/DC converters [27].

#### 2.4.2.2.1.1 Single-Stage resonant DC/DC converters

Single-Stage Resonant DC/DC converters operate under the DC/AC/DC approach but without isolation, i.e., there are no AC transformers in their structure. The schematic diagram of this topology is the same as that shown in Figure 2.44, where power electronics bridges have been connected to the AC side through the resonant tank. Various resonant tanks, such as LCL and LC parallel tanks with voltage doublers, can be used. However, this topology is suitable for low and medium power ranges with medium voltage ratios because of the high electrical stress on passive elements and the need for high voltage/current rating resonant elements [27].

#### 2.4.2.2.1.2 Multi stage resonant DC/DC converters

Multi Stage Resonant DC/DC converter is the most advanced topology in the resonant family. In this topology, the full-rated central resonance tank is divided into several smaller and low-rated resonant circuits. The schematic diagram of the multi stage resonant DC/DC converter is shown in Figure 2.45. This structure leads to a low-complexity design and increases its modularity compared to that of its single-stage counterpart. However, it is not a pure modular topology because its structure has an inherent issue, which is the uneven distribution of currents/voltages on switches. In this topology, each resonant tank is used sequentially, and the power is transferred tank by tank to deliver the full power to the HV terminals. The reverse process is

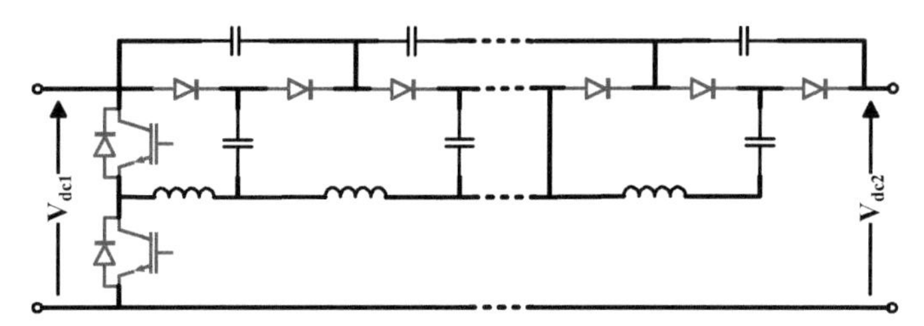


**Figure 2.45** Schematic diagram of the multi stage resonant DC/DC converter [27].

done similarly for step-down operations. This topology is suitable for high transformation ratios without considerable electrical stress on the resonant elements, unlike its single-stage counterpart [27].

#### 2.4.2.2.2 DC modular DC/DC converters

DC Modular DC/DC converters employ the modular approach of MMCs to make the DC/DC conversion. To do so, two basic approaches are adopted: 1. DC-MMC and 2. Classical Choppers. Each of these approaches is herein examined [27].

##### 2.4.2.2.2.1 DC modular multilevel converters

DC Modular Multilevel Converters (DC-MMCs) are very similar to MMC power converters because they employ chain-links of MMC's SMs to generate voltages and currents at the desired frequencies. The schematic diagram of the DC-MMCs is shown in Figure 2.46.

This topology relies on the simultaneous generation of DC and AC waveforms to realize the power transmission. AC waveforms are used to ensure the energy balance in SMs, whereas DC waveforms are used to transfer power between DC terminals. However, issues arise due to the combination of AC and DC waveforms at the output DC terminal. To prevent such issues, a passive filter or a proper control scheme, can be used (Figure 2.46(a)-(b)). This family adopts the hard-switching scheme, and power losses of switches are high; this issue has been partially compensated by low-frequency switching. Increasing the operating frequency can decrease the required filter size, but switching losses will increase; thus, a trade-off is present between the filter size and power losses. This family of DC/DC converters holds inherent advantages of MMCs (i.e., high reliability, modularity, and scalability) and is suitable for high power and voltage applications. However, there are some challenges, such as the required filter and AC circulating currents, which limit its application in low/medium power/voltage applications [27].

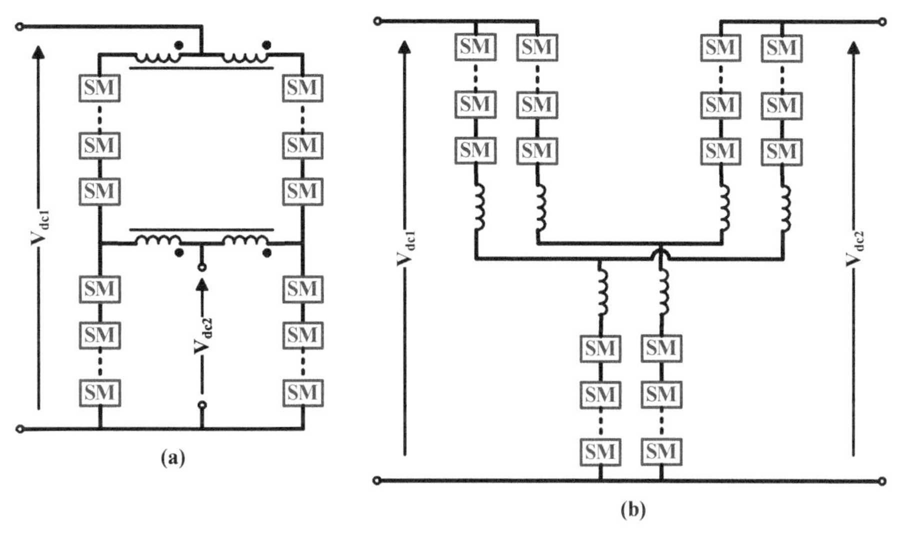


**Figure 2.46** Schematic diagram of the DC-MMCs: a) equipped with filter; and b) equipped with control scheme [27].

#### 2.4.2.2.2.2 Classical choppers

Classical Choppers belong to another family of DC modular DC/DC converters, and their operating principle relies on replacing some of the basic converter's switches with MMC's SMs, which extends the control capability. There are two topologies in this family based on the employed energy storage system: 1. Capacitive Accumulation Choppers; and 2. Inductive Accumulation Choppers. The schematic diagrams of both types of Choppers are shown in Figure 2.47.

The principle of Capacitive Accumulation Choppers is the adoption of MMC SMs as variable capacitors for energy storage purposes; existing capacitors are charged and discharged by switching the related chain-link of SMs, which are connected between DC terminals. Also, a dead time in the switching pattern is required to make this principle feasible considering DC voltage levels. The main switches operate under the soft-switching approach, but SMs are switched under the hard-switching approach. The schematic diagram of Capacitive Accumulated Choppers is shown in Figure 4.14(a).

Inductive Accumulation Choppers adopt a central inductor for energy storage, which is charged and discharged by switching the related chain-link of SMs. In this topology, a stair-case transition is performed between charging and discharging modes by using interleaved SMs, operating in

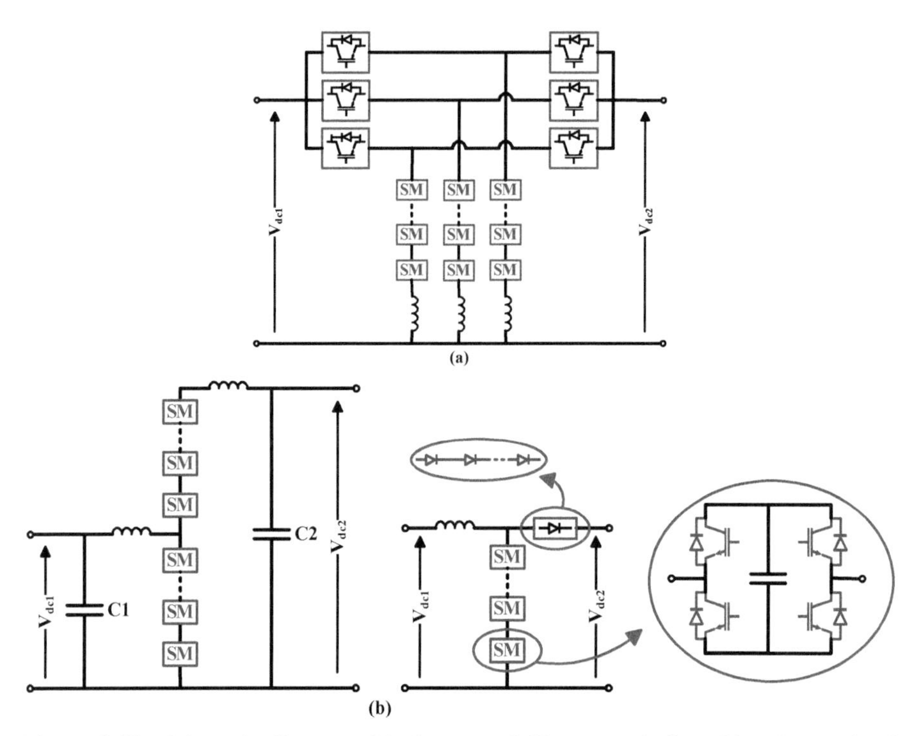


**Figure 2.47** Schematic diagram of both types of Choppers: a) Capacitive Accumulated Choppers; and b) Inductive Accumulation Choppers [27].

either the resonance mode or resonance discontinuous-connection mode. As opposed to their capacitive counterpart, SMs can be operated using the soft-switching approach but with the cost of high conduction losses due to AC circulating currents. Therefore, a compromise should be achieved to reach an optimal operation point. Inductive Accumulation Choppers are suitable for applications that require high transformation ratios at low power, but a large-size central inductor should be employed. The SMs are switched following the hard-switching approach. The schematic diagram of the Inductive Accumulated Choppers is shown in Figure 2.47(b) [27].

## 2.5 DC Power Flow Controllers

DC Power Flow Controllers (DC-PFCs), also known as DC Current Flow Controllers (DC-CFCs), are power electronic devices primarily used for power flow control in HVDC transmission systems. Besides this purpose,

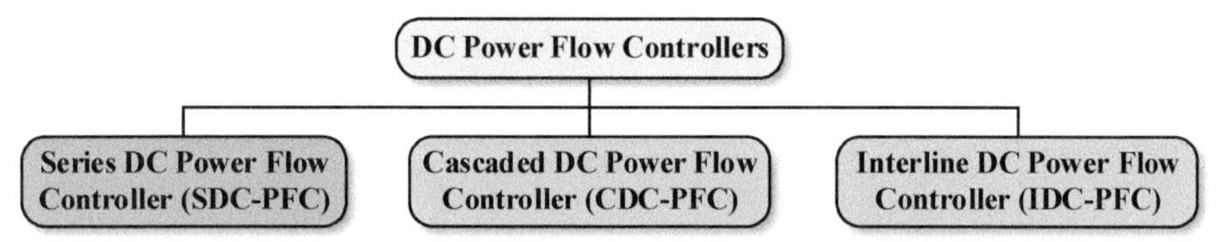


**Figure 2.48** General classification of the DC-PFCs.

they also perform fault blocking/limiting and protection functions. Therefore, DC-PFCs play a critical role in HVDC transmission systems. The importance of DC-PFCs becomes significant in the case of meshed MT-HVDC grids because power can flow in different paths. Employing suitable devices (similar to Flexible AC Transmission Systems (FACTS) in AC systems) is essential for the power flow control and for compensating purposes. Various studies have been conducted to propose and develop novel and effective topologies for DC-PFCs [30–35].

First, we briefly introduce the principle of power flow in DC transmission systems, for which, as opposed to AC systems, the capacitance and the inductance of transmission lines are not present in steady-state studies. The only parameter of DC transmission lines is their series and parallel resistances. According to the inherent feature of DC systems, there are no phasor angle and frequency in the voltage or current at DC terminals. Therefore, the power flow of a DC line depends only on the voltage difference between DC sending terminals and DC receiving ends, and the line's series resistance. It is worth mentioning that line's parallel resistance has a negligible impact on the power flow of a DC line. Therefore, the power flow of a DC line can only be controlled by changing the series resistance of a DC line or the voltage difference at the two DC ends. Various DC-PFCs have been proposed, which can be categorized into three major groups: 1) Series DC-PFC (SDC-PFC); 2) Cascaded DC-PFC (CDC-PFC); and 3) Interline DC-PFC (IDC-PFC) [36–39].

DC-PFCs can also be divided into two major groups based on the power electronics components used in their main structure: 1) Resistance-type (R-type); and 2) Voltage-type (V-type). DC-PFCs are discussed in detail below. The general classification of DC-PFCs is shown in Figure 2.48 [35–46].

### 2.5.1 SDC-PFC

SDC-PFCs are power electronics devices installed in series in DC lines to control the current flow. These devices can be R-type or V-type. The R-type SDC-PFCs "inject" resistance in a continuous or discrete mode. In

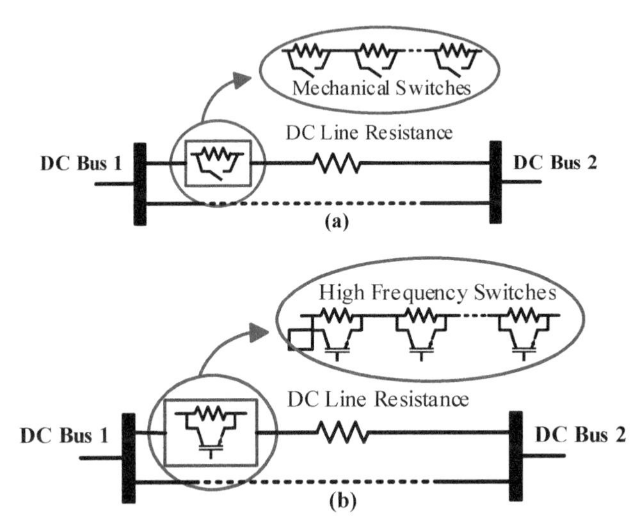


**Figure 2.49** Schematic diagram of the R-type SDC-PFCs: a) Discrete R-type SDC-PFC; b) Variable R-type SDC-PFC [42].

the discrete mode, mechanical switches are used to inject or bypass one or several resistances. By using semiconductor switches, a specific amount of resistance can be injected in the DC lines ranging between zero (bypassed resistance condition) and $R_{max}$ (the entire amount of resistance). However, this approach can only increase the total resistance of a DC line, and consequently, its current flow can only be decreased. Although the principle, implementation, and control of this type of SDC-PFCs are straightforward, the R-type SDC-PFCs is not an attractive solution for HVDC transmission systems, due to their significant inherent power losses. The schematic diagram of the R-type SDC-PFCs is shown in Figure 2.49 [42].

The V-type SDC-PFCs use power electronic converters to inject a series of DC voltages into DC lines, and the voltage difference between the two DC ends will be accordingly changed. Using this method, the current flow can be increased and decreased. The principle is the use of an internal connection with the main system or an external connection with another energy source to inject a series of voltages into DC lines. The schematic diagram of V-type SDC-PFC is shown in Figure 2.50 [43].

### 2.5.2 CDC-PFCs

CDC-PFCs are power electronics and V-type DC-PFC devices, whose purpose is the power flow control in a DC line (similar to SDC-PFCs).

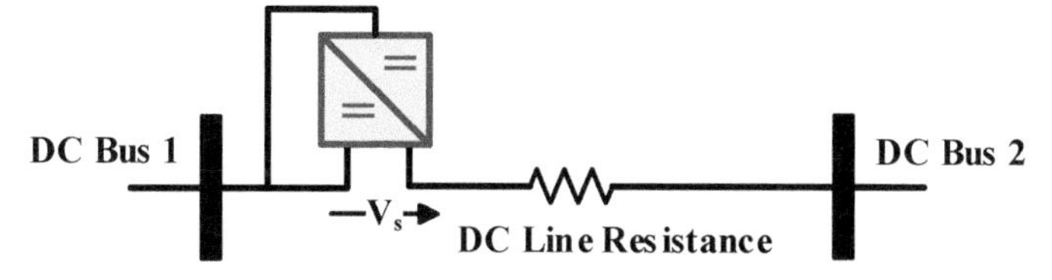


**Figure 2.50** Schematic diagram of the V-type SDC-PFC [43].

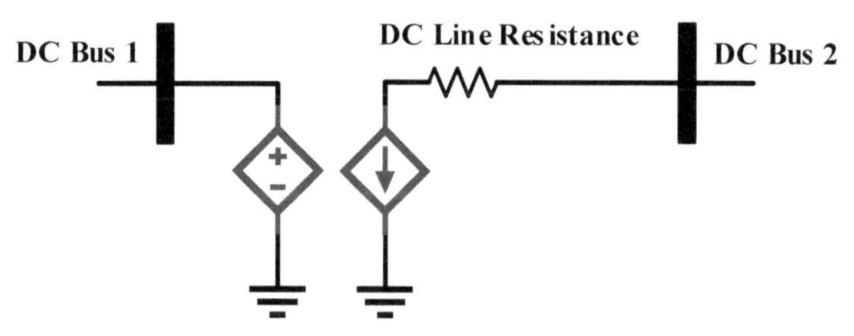


**Figure 2.51** Conceptual structure of the CDC-PFC [45].

CDC-PFCs are connected in series, their operation principle relies on changing the DC voltage with a suitable ratio, and their performance is similar to a DC transformer. DC/DC converters can be considered as CDC-PFC because of DC/DC converters' capability to control the power flow of DC lines. However, the applications of DC/DC converters are slightly different from CDC-PFC. DC/DC converters are used for an entire HVDC station, and their size and power capacity are very large; while the CDC-PFC concept has been proposed for just one HVDC line, and their size and power capacity are smaller than that of DC/DC converters. DC/DC converters can be redesigned to be used in HVDC lines. In real life, a CDC-PFC has not yet been realized, but its conceptual structure has been proposed in [45], as shown in Figure 2.51.

### 2.5.3 IDC-PFCs

IDC-PFCs are the most sophisticated family among the DC-PFCs. The primary purpose of this family is to simultaneously control power flow in two or more DC lines. The principle of operation consists of power exchanges between two or more adjacent HVDC lines through an energy hub; passive elements, such as capacitances or inductances, are used to store extra power/energy from one line, which is then injected into one (or more) adjacent line(s). In this way, the current flow of the desired DC line will be controlled. Different IDC-PFCs topologies and structures have been

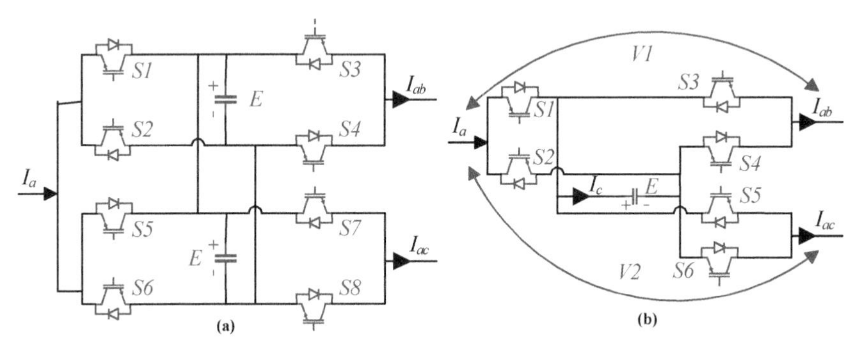


**Figure 2.52** Schematic diagram of an IDC-PFC: a) Original Structure b) Simplified Structure [32].

proposed and are under development. A leading and practical configuration has been proposed in [32]. This topology is based on merging and simplifying two simple H-bridges. The schematic diagram of an IDC-PFC is shown in Figure 2.52.

The operation of this topology is similar to its AC counterpart, Interline Power Flow Controller (IPFC). In IPFC, power compensation and power flow control are achieved using the master-slave concept, and the parameters of the main line (i.e., the master line) are independently/actively controlled, while one parameter of another line (i.e., the slave line) is indirectly/passively controlled to meet the power balance criteria of the device. According to this operating philosophy, in IDC-PFC, the power of the master line is actively controlled, and the power of the slave line is passively controlled to provide equilibrium in the power transmission of the device. Other features, such as fault blocking/limiting, composite operation, and protection capabilities, can also be added. Various topologies are under development in this active and promising research area [32, 36–39].

## 2.6 Conclusion

HVDC transmission systems are the promising technology serving as the backbone of tomorrow's power grids. B2B, P2P, and Multi-Terminal HVDC (MT-HVDC) grids have been herein discussed. Various configurations such as Monopolar, Bipolar, Homopolar, and Hybrid have been analyzed based on the principle of AC to DC power conversion and vice versa. These systems are generally based on power electronics converters and inverters. In this chapter, existing power electronics converters and inverters in HVDC transmission systems have been introduced.

Generally, power electronics converters can be classified into three major groups: 1. Power Converters; 2. DC/DC Converters; and 3. DC Power Flow Controllers.

Power converters are primarily used to convert AC to DC power and vice versa. AC/DC power converters are implemented as electrical energy sources, according to the network theory definition of Voltage Source and Current Source, and all power converters can be categorized into three major groups: 1. VSCs; 2. CSCs; and 3. HCVSCs. VSCs and its MMC family will play an essential role in tomorrow's HVDC transmission systems because of their superior performance.

DC/DC converters, or DC Transformers, are power electronics devices with the duty of providing DC voltage matching. They can divide large MT-HVDC grids into several smaller protection zones, and realize DC voltage regulation, fault isolation, and connection of Bipolar/Monopolar configurations and CSC/VSC technologies. DC/DC power converters can be categorized into two major concepts: 1. Isolated; and 2. Non Isolated. In the Isolated configuration, an AC link exists in the heart of the DC/DC converters, whereas the AC link is not present in the Non Isolated configuration. Isolated DC/DC converters, mainly the DAB family for HVDC transmission systems, are superior topologies compared to others.

DC-PFCs or DC-CFCs are power electronics devices for power flow control in HVDC transmission systems, especially in meshed MT-HVDC grids. Their operation principles have been inspired by FACTS in AC systems. According to the inherent feature of DC systems, the power flow of a DC line depends only on the voltage difference between the DC terminals and the line's series resistance. Therefore, the principle of DC-PFCs is to control or change the series resistance of a DC line or the voltage difference between the two DC ends. DC-PFCs can be categorized into three major groups: 1. SDC-PFC; 2. CDC-PFC; and 3. IDC-PFC. They can also be categorized into R-type or V-type. IDC-PFCs possess superior performance compared to other DC-PFCs, which is crucial for tomorrow's HVDC transmission systems.

## References

[1] M. Josephson, . *Edison: A Biography*. New York: J. Wiley, 1992.

[2] N.G. Hingorani, . "High-voltage DC Transmission: A Power Electronics Workhorse," ." *IEEE Spectrum*, vol. 33, no. 4, pp. 63–72, Apr. 1996, doi: 10.1109/6.486634.

[3] M. Benasla, T. Allaoui, M. Brahami, M. Dena, and V.K. Sood, . "HVDC links between North Africa and Europe: Impacts and benefits on the dynamic performance of the European system.," *Renewable Sustainable Energy Reviews*, vol. 82, pp. 3981–3991, Feb. 2018, doi: 10.1016/j.rser.2017.10.075.

[4] S. Fischer and O. Geden, . *Moving targets: negotiations on the EU's energy and climate policy objectives for the post-2020 period and implications for the German energy transition*, vol. 3/2014. Berlin: Stiftung Wissenschaft und Politik -SWP- Deutsches Institut fÃijr Internationale Politik und Sicherheit, 2014.

[5] M.A. Avila, . "United States and Mexico New Cross-border Connections.," in *2008 IEEE/PES Transmission and Distribution Conference and Exposition: Latin America*, Aug. 2008, pp. 1–5. doi: 10.1109/TDC-LA.2008.4641857.

[6] D.A.N. Jacobson, P. Wang, M. Mohaddes, M. Rashwan, and R. Ostash, . "A Preliminary Look at the Feasibility of VSC HVdc in Manitoba.," in *2011 IEEE Electrical Power and Energy Conference*, Oct. 2011, pp. 80–85. doi: 10.1109/EPEC.2011.6070258.

[7] N. Herath, X. Xu, N. Jayasekara, D. Jacobson, and S. Filizadeh, . "Interconnection of Northern Canadian Communities to the Manitoba Hydro Power System: Evaluation of AC and DC Alternatives.," in *2019 IEEE Electrical Power and Energy Conference (EPEC)*, Oct. 2019, pp. 1–6. doi: 10.1109/EPEC47565.2019.9074826.

[8] C. Zhou, C. Fang, M. Kandic, P. Wang, K. Kent, and D. Menzies, . "Large-Scale Hybrid Real Time Simulation Modeling and Benchmark for Nelson River Multi-Infeed HVdc System.," *Electric Power System Research*, vol. 197, p. 107294, Aug. 2021, doi: 10.1016/j.epsr.2021.107294.

[9] W. Lin, J. Wen, J. Liang, S. Cheng, M. Yao, and N. Li, . "A Three-Terminal HVDC System to Bundle Wind Farms With Conventional Power Plants.," *IEEE Transactions Power Systems*, vol. 28, no. 3, pp. 2292–2300, Aug. 2013, doi: 10.1109/TPWRS.2012.2228506.

[10] M.K. Bucher, R. Wiget, G. Andersson, and C.M. Franck, . "Multiterminal HVDC Networks—What is the Preferred Topology?," *IEEE Transactions Power Delivery*, vol. 29, no. 1, pp. 406–413, Feb. 2014, doi: 10.1109/TPWRD.2013.2277552.

[11] T. Vrana, S. Dennetiere, Y. Yang, J.A. Jardini, D. Jovcic, and H. Saad, . "The Cigré B4 DC Grid Test System," *CIGRE Electra*, vol. 270, Oct. 2013.

[12] N.R. Watson and J.D. Watson, . "An Overview of HVDC Technology.," *Energies*, vol. 13, no. 17, Art. no. 17, Jan. 2020, doi: 10.3390/en13174342.

[13] W. Leterme, P. Tielens, S.D. Boeck, and D. Van Hertem, "Overview of Grounding and Configuration Options for Meshed HVDC Grids," *IEEE Transactions Power Delivery*, vol. 29, no. 6, pp. 2467–2475, Dec. 2014, doi: 10.1109/TPWRD.2014.2331106.

[14] A. Korompili, Q. Wu, and H. Zhao, . "Review of VSC HVDC Connection for Offshore Wind Power Integration.," *Renewable Sustainable Energy Revolution*, vol. 59, pp. 1405–1414, Jun. 2016, doi: 10.1016/j.rser.2016.01.064.

[15] N. Joshi and J. Sharma, . "An Overview on High Voltage Direct Current Transmission Projects in India.," in *2021 6th International Conference on Inventive Computation Technologies (ICICT)*, Jan. 2021, pp. 459–463. doi: 10.1109/ICICT50816.2021.9358704.

[16] S. Norrga, L. ngquist, K. Sharifabadi, and X. Li, . *Power electronics for HVDC grids: An overview*. 2015.

[17] J. Kharade and N. Savagave, . "A Review of HVDC Converter Topologies.," *International Journal of Innovative Research in Computer and Communication Engineering*, vol. 6, pp. 1822–1830, Mar. 2017, doi: 10.15680/IJIRSET.2017.0602054.

[18] R. Marquardt, . "Modular Multilevel Converter: An Universal Concept for HVDC-Networks and Extended DC-Bus-Applications.," in *The 2010 International Power Electronics Conference - ECCE ASIA -*, Jun. 2010, pp. 502–507. doi: 10.1109/IPEC.2010.5544594.

[19] A. Nami, J. Liang, F. Dijkhuizen, and G.D. Demetriades, . "Modular Multilevel Converters for HVDC Applications: Review on Converter Cells and Functionalities.," *IEEE Transactions Power Electron*, vol. 30, no. 1, pp. 18–36, Jan. 2015, doi: 10.1109/TPEL.2014.2327641.

[20] M.N. Raju, J. Sreedevi, R.P. Mandi, and K.S. Meera, . "Modular Multilevel Converters Technology: a A Comprehensive Study on its Topologies, Modelling, Control and Applications.," *IET Power Electron*, vol. 12, no. 2, pp. 149–169, 2019, doi: 10.1049/iet-pel.2018.5734.

[21] C. Xu, J. He, and L. Lin, . "Research on Capacitor-Switching Semi-Full-Bridge Submodule of Modular Multilevel Converter Using Si-IGBT and SiC-MOSFET.," *IEEE Journal of Emerging and Selected Topics in Power Electronics*, vol. 9, no. 4, pp. 4814–4825, Aug. 2021, doi: 10.1109/JESTPE.2020.3034451.

[22] G.P. Adam, T.K. Vrana, R. Li, P. Li, G. Burt, and S. Finney, . "Review of technologies for DC grids – power conversion, flow control and protection.," *IET Power Electron*, vol. 12, no. 8, pp. 1851–1867, Mar. 2019, doi: 10.1049/iet-pel.2018.5719.

[23] J. Qin, M. Saeedifard, A. Rockhill, and R. Zhou, . "Hybrid Design of Modular Multilevel Converters for HVDC Systems Based on Various Submodule Circuits.," *IEEE Transactions. Power Delivery*, vol. 30, no. 1, pp. 385–394, Feb. 2015, doi: 10.1109/TPWRD.2014.2351794.

[24] D. Das, J. Pan, and S. Bala, . "HVDC Light for Large Offshore Wind Farm Integration.," in *2012 IEEE Power Electronics and Machines in Wind Applications*, Jul. 2012, pp. 1–7. doi: 10.1109/PEMWA.2012.631 6363.

[25] M.J. Carrizosa, A. Benchaib, P. Alou, and G. Damm, . "DC Transformer for DC/DC Connection in HVDC Network," in *2013 15th European Conference on Power Electronics and Applications (EPE)*, Sep. 2013, pp. 1–10. doi: 10.1109/EPE.2013.6631774.

[26] S. Kedia and H.J. Bahirat, . "DC-DC converter Converter for HVDC Grid Application.," in *2017 National Power Electronics Conference (NPEC)*, Dec. 2017, pp. 346–351. doi: 10.1109/NPEC.2017.8310483.

[27] J.D. Páez, D. Frey, J. Maneiro, S. Bacha, and P. Dworakowski, . "Overview of DC–DC Converters Dedicated to HVDC Grids.," *IEEE Transactions Delivery*, vol. 34, no. 1, pp. 119–128, Feb. 2019, doi: 10.1109/TPWRD.2018.2846408.

[28] G.P. Adam, I.A. Gowaid, S.J. Finney, D. Holliday, and B.W. Williams, . "Review of dc–dc Converters for Multi-Terminal HVDC Transmission Networks.," *IET Power Electron*, vol. 9, no. 2, pp. 281–296, 2016, doi: 10.1049/iet-pel.2015.0530.

[29] I. Alhurayyis, A. Elkhateb, and D. John J. Morrow, . "Isolated and Non-Isolated DC-to-DC Converters for Medium Voltage DC Networks: A Review.," *IEEE Journal of Emerging and Selected Topics in Power Electronics*, pp. 1–1, 2020, doi: 10.1109/JESTPE.2020.3028057.

[30] K. Rouzbehi, A. Miranian, J.I. Candela, A. Luna, and P. Rodriguez, . "Proposals for Flexible Operation of Multi-Terminal DC Grids: Introducing flexible DC Transmission System (FDCTS).," in *2014 International Conference on Renewable Energy Research and Application (ICRERA)*, Oct. 2014, pp. 180–184. doi: 10.1109/ICRERA.2014.7016553.

[31] S.S. Heidary H. Yazdi, J. Milimonfared, S.H. Fathi, and K. Rouzbehi, . "Optimal Placement and Control Variable Setting of Power Flow Controllers in Multi-Terminal HVDC Grids for Enhancing Static Security.," *InternationalInt. Journal J. Electrical. Power Energy System*, vol. 102, pp. 272–286, Nov. 2018, doi: 10.1016/j.ijepes.2018.05.001.

[32] J. Sau-Bassols, E. Prieto-Araujo, and O. Gomis-Bellmunt, . "Modelling and Control of an Interline Current Flow Controller for Meshed HVDC Grids.," *IEEE Transactions Power Delivery*, vol. 32, no. 1, pp. 11–22, Feb. 2017, doi: 10.1109/TPWRD.2015.2513160.

[33] S.S.H. Yazdi, K. Rouzbehi, J.I. Candela, J. Milimonfared, and P. Rodriguez, . "Analysis on Impacts of the Shunt Conductances in Multi-Terminal HVDC Grids Optimal Power-Flow.," in *IECON 2017 - 43rd Annual Conference of the IEEE Industrial Electronics Society*, Oct. 2017, pp. 121–125. doi: 10.1109/IECON.2017.8216025.

[34] S.S.H. Yazdi, S.H. Fathi, J.M. Monfared, and E.M. Amiri, . "Optimal Operation of Multi Terminal HVDC Links Connected to Offshore Wind Farms.," in *2014 11th International Conference on Electrical Engineering/Electronics, Computer, Telecommunications and Information Technology (ECTI-CON)*, May 2014, pp. 1–6. doi: 10.1109/ECTI-Con.2014.6839885.

[35] S.S.H. Yazdi, K. Rouzbehi, J.I. Candela, J. Milimonfared, and P. Rodriguez, . "Flexible HVDC transmission systems small signal modelling: A case study on CIGRE Test MT-HVDC grid.," in *IECON 2017 - 43rd Annual Conference of the IEEE Industrial Electronics Society*, Oct. 2017, pp. 256–262. doi: 10.1109/IECON.2017.8216047.

[36] M. Abbasipour, S.S. Haidary H. Yazdi, J. Milimonfared, and K. Rouzbehi, . "Technical Constrained Power Flow Studies for IDC-PFC Integrated into the MT-HVDC Grids.," *IEEE Transactions Power Delivery*, pp. 1–1, 2020, doi: 10.1109/TPWRD.2020.3032220.

[37] M. Abbasipour, J. Milimonfared, S.S. Heidary H. Yazdi, and K. Rouzbehi, . "Static Modeling of the IDC-PFC to Solve DC Power Flow Equations of MT-HVDC Grids Employing the Newton-Raphson Method.," in *2019 10th International Power Electronics, Drive Systems and Technologies Conference (PEDSTC)*, Feb. 2019, pp. 450–457. doi: 10.1109/PEDSTC.2019.8697665.

[38] M. Abbasipour, J. Milimonfared, S.S. Heidary H. Yazdi, and K. Rouzbehi, . "Power injection model of IDC-PFC for NR-based and technical constrained MT-HVDC grids power flow studies.,"

*Electrical Power Systems Research*, vol. 182, p. 106236, May 2020, doi: 10.1016/j.epsr.2020.106236.

[39] M. Abbasipour, J. Milimonfared, S.S.H. Yazdi, G.B. Gharehpetian, and K. Rouzbehi, . "New Representation of Power Injection Model of IDC-PFC within NR-based MT-HVDC Grids Power Flow Studies.," in *2020 28th Iranian Conference on Electrical Engineering (ICEE)*, Aug. 2020, pp. 1–7. doi: 10.1109/ICEE50131.2020.9260803.

[40] C.D. Barker and R.S. Whitehouse, . "A Current Flow Controller for Use in HVDC Grids.," pp. 44–44, Jan. 2012, doi: 10.1049/cp.2012.1973.

[41] W. Chen *et al.*, "A Novel Interline DC Power-Flow Controller (IDCPFC) for Meshed HVDC Grids.," *IEEE Transactions Power Delivery*, vol. 31, no. 4, pp. 1719–1727, Aug. 2016, doi: 10.1109/TPWRD.2016.2547960.

[42] T. Zhang, C. Li, and J. Liang, . "A Thyristor Based Series Power Flow Control Device for Multi-Terminal HVDC Transmission.," in *2014 49th International Universities Power Engineering Conference (UPEC)*, Sep. 2014, pp. 1–5. doi: 10.1109/UPEC.2014.6934802.

[43] S. Balasubramaniam, J. Liang, and C.E. Ugalde-Loo, . "An IGBT Based Series Power Flow Controller for Multi-Terminal HVDC Transmission.," in *2014 49th International Universities Power Engineering Conference (UPEC)*, Sep. 2014, pp. 1–6. doi: 10.1109/UPEC.2014.69 34626.

[44] W. Chen, X. Zhu, L. Yao, X. Ruan, Z. Wang, and Y. Cao, . "An Interline DC Power-Flow Controller (IDCPFC) for Multiterminal HVDC System.," *IEEE Transactions Power Delivery*, vol. 30, no. 4, pp. 2027–2036, Aug. 2015, doi: 10.1109/TPWRD.2015.2425412.

[45] K. Rouzbehi, J.I. Candela, A. Luna, G.B. Gharehpetian, and P. Rodriguez, . "Flexible Control of Power Flow in Multiterminal DC Grids Using DC–DC Converter.," *IEEE Journal Emergency Selected Top Power Electron*, vol. 4, no. 3, pp. 1135–1144, Sep. 2016, doi: 10.1109/JESTPE.2016.2574458.

[46] K. Rouzbehi, S.S. Heidary H. Yazdi, and N. Shariati S. Moghadam, "Power Flow Control in Multi-Terminal HVDC Grids Using a Serial-Parallel DC Power Flow Controller," *IEEE Access*, vol. 6, pp. 56934–56944, 2018, doi: 10.1109/ACCESS.2018.2870943.

# 3

# Optimal Multi-Objective Energy Management of Renewable Distributed Integration in Smart Distribution Grids Considering Uncertainties

**M. Zellagui[1,2], N. Belbachir[3], S. Settoul[4], and C. Z. El-Bayeh[5]**

[1]Département de Génie Électrique, École de Technologie Supérieure, Montréal, Canada
[2]Department of Electrical Engineering, University of Batna 2, Batna, Algeria
[3] Department of Electrical Engineering, University of Mostaganem, Mostaganem, Algeria
[4]Department of Electrotechnic, University of Constantine 1, Constantine, Algeria
[5]Canada Excellence Research Chairs Team, Concordia University, Montréal, Algeria
E-mail: m.zellagui@ieee.org; m.zellagui@univ-batan2.dz

## Abstract

The energy supplies problem keeps floating on the surface. For this, daily improvements are implemented to do the optimization of the power system configuration and the generators' power. Renewable Distributed Generators (RDGs) represent one of the best solutions, also a reference for those improvements. The optimal placement and sizing of RDG sources in the Smart Distribution Grid (SDG) are considered a trendy problem that usually can be solved based on the utilization of various approaches and algorithms due to their high complexity.

The presence of RDGs in the SDG can provide many benefits and advantages. These benefits may be summarized generally as, power losses minimization, voltage profiles improvement, system load-ability and reliability growth, system security, and protection enhancement. To achieve the mentioned benefits, RDGs should be optimized in location and size based on various objective functions.

A recent nature-inspired metaheuristic approach called the Marine Predators Algorithm (MPA) is used, which is based on various foraging strategies among optimal encounter rates policy in biological interaction and ocean predators. This algorithm was utilized to optimize many types of RDG units to obtain an optimal location and sizing of Photovoltaic Distributed Generator (PVDG) and Wind Turbine Distributed Generator (WTDG) units into the SDG considering uncertainties. This was performed when taking into consideration the uncertainties of the power generated from the RDG as well as the load demand variation during each of the day's hours.

This chapter proposed a new Multi Objective Indices (MOI) which is considered to minimize simultaneously five technical indices based on the power losses, the voltage deviation, and the overcurrent protection system of the SDG.

The chosen algorithm is validated on different standards IEEE 33-bus, and 69-bus distribution grids for the purpose of testing its efficiency, where also three cases of RDGs' allocation were studied. The convergence characteristics reveal that the MPA was effectively a quick technique that may arrive at the best solutions in a small iterations' number compared to the other algorithms: Particle Swarm Optimization (PSO), Ant Lion Optimization (ALO), Grey Wolf Optimizer (GWO), Grasshopper Optimization Algorithm (GOA), and Moth Flame Optimizer (MFO) algorithms.

The optimal allocation of RDG identifies the suitable results in the satisfaction of the permissible voltage limits and power loss minimization. After the installation of both RDGs, the power losses are minimized, the profile of the voltage has more augmented, and the overcurrent protection system had a considerable improvement. The simulation results confirm the feasibility of optimal power planning. In Addition, the obtained results reveal that the optimal integration of the WTDG units based on the chosen algorithm was the best choice over the PVDG units, which led to the minimization of the expected APL, RPL, VD, OT, and CTI of different test systems.

**Keywords:** Renewable distributed generation, photovoltaic distributed generator, wind turbine distributed generator, smart distribution grids, optimal integration, daily uncertainties, multi objective indices, new optimization algorithms.

## 3.1 Introduction

The actual era in sustainable development is concentrated on the quick presence of Renewable Energy Sources (RESs) driven by a large range of socio-economic objectives [1]. The higher level of automation is a transition from traditional distribution grids to the SDGs of the future with the chance to optimize the system's control and operation [2]. In the near past years, integration of the RES has been raised at a tremendous rate. There have been various factors as government motivation in terms of technological development and environmental aspects [3].

The RDGs which refers to the small-scale power generations that are normally linked to the SDG, play the main role in reducing the power losses, enhancing the voltage stability and voltage profiles, as well as the maximization of load-ability of the system [4].

The deployment of RDG components at the nonoptimal locations, which is not optimally selected, can disadvantage rather than improve the system performance. Selecting the best allocation for the RDG and the desirable size of its units in the SDG is not a simple optimization issue [5].

The optimal RDG allocation problem refers to determining the optimal location and size of RDG units to be integrated into the existing SDG depending on the various constraints [6].

In the last decade, meta-heuristic search and optimization algorithms have been frequently used for solving this problem, i.e., to acquire the optimal allocation of RDG. The advantages of these techniques are their simple application and implementation to find the near-optimal solution for such complicated optimization problems. In the literature, the variety of the existing optimization techniques applied to do the organization and installation of RDGs into SDGs [7]: applied and proposed Ant Lion Optimization Algorithm (ALOA) to solve the optimal sizing and siting problem of DG for reduction of power losses and buses voltage deviation [8], Virus Colony Search (VCS) algorithm applied to enhance the reliability indices in SDG [9], Breeder Genetic Algorithm (BGA) for determination of the optimal allocation of DG by considering the minimization of power losses as the main objective [10], Strength Pareto Evolutionary Algorithm 2 (SPEA 2) to minimize APL, annual costs of operation and emissions of pollutant gas [11], applied Teaching Learning Based Optimization (TLBO) for the aim of minimizing the APL and the VDI [12], and applied Biogeography-Based Optimization (BBO) algorithm to reduce APL with an effective power factor mode [13]. In 2019, Applied the Mixed Integer Linear Programming (MILP) to minimize energy cost and balance unsymmetrical loading [14], Gravitational Search Algorithm (GSA) to reduce power losses and maximize annual cost savings with uncertainties of DG and load demand [15], applied the Moth-Flame Optimizer (MFO) algorithm with the aim of losses reduction and enhancement of distribution network voltage stability [16], used Chaotic Differential Evolution (CDE) algorithm by minimizing different objectives comprises annual economic loss, cost of maintenance, power loss and voltage deviation at the buses [17], an Opposition based Tuned-Chaotic Differential Evolution (OTCDE) algorithm for RDG integration problem including voltage deviation index, cost index and line flow capacity index [18], Spider Monkey Optimization (SMO) algorithm to identify the optimal allocation of RDG for voltage security enhancement [19].

In 2020, applied Adaptive Modified Whale Optimization Algorithm (AMWOA) to maximize VSI and minimize the power losses [20], Phasor PSO (PPSO) algorithm to do the reduction of yearly economic loss in the practical distribution system in Portuguese [21], Water Cycle Algorithm (MOWCA) for multi objective function based technical and economic parameters [22], and Modified Gravitational Search Algorithm (MGSA) to minimize the active and reactive losses with maximizing VSI [23]. Recently in 2021, used Salp Swarm Algorithm (SSA) for active power loss reduction in the Algerian distribution system [24], proposed a hybrid FA and adaptive PSO algorithm for multi objective function based on technical, economic, and environmental issues with integration of DSTATCOM [25], Chimp Optimizer Algorithm (COA) for objective function considered to minimize the total active power loss [26], applied Whale Optimization Algorithm (WOA) for the optimal allocation of RDG for multi-objective that comprises power loss reduction, voltage profile enhancement and total cost of operation minimization [27], Slime Mould Algorithm (SMA) to minimize the active power loss, the voltage stability index, the short circuit level and the cost of annual losses considering the uncertainty of load demand in 24 hours [28], and Equilibrium Optimizer Algorithm (EOA) for the best allocation of RES units with the uncertainties [29]. This chapter presents the application of a recent nature-inspired metaheuristic approach called the Marine Predators Algorithm (MPA), which is based on various foraging strategies among optimal encounter rates policy in biological interaction and ocean predators.

This algorithm was applied to optimally locate and size many types of RES-based PVDG and WTDG units into the SDG considering uncertainties. This was performed when taking into consideration the uncertainties of the power generated from the RDG as well as the load demand variation during each of the day's hours.

This chapter proposed a new MOI which are devoted to simultaneously minimizing the total of Active Power Loss Index (APLI), the Reactive Power Loss Index (RPLI), the Voltage Deviation Index (VDI), the Operation Time Index (OTI), and improve the Coordination Time Interval Index (CTII) between the primaries and backups overcurrent relays in SDG.

The chosen approach is validated on different standard IEEE 33-bus, and 69-bus distribution grids for a reason to test its efficiency, where three cases were studied. The convergence characteristics reveal that the MPA was effectively a quick technique that may arrive at the best solutions in a small iterations' number compared to the other optimization algorithms.

## 3.2 Uncertainty Modeling of RDG Source

### 3.2.1 Modeling of Load Demand Uncertainty

The following equations represent the model of load demand uncertainties [30, 31]:

$$P_k(t) = \lambda(t) \times P_{ok} \tag{3.1}$$

$$Q_k(t) = \lambda(t) \times Q_{ok} \tag{3.2}$$

The curve in Figure 3.1 represents the daily load demand variation for 24 hours.

Investigating the previous curve shows that the load demand kept varying along the day's hours from a minimum rate of 55% registered around 5h00 am, until a maximum rate of 100% that registered around mid-day hours.

### 3.2.2 Modeling of Solar DG Uncertainty

The solar irradiance in each of the day's hours would be modeled by the Beta Probability Density Function (PDF) based on historical data [32, 33]. Along every period (1 h in this study), the solar irradiance's PDF is expressed as [34, 35]:

$$f_b(s) = \begin{cases} \frac{\Gamma(A+B)}{\Gamma(A)\Gamma(B)} s^{(\alpha-1)} & 0 \leqslant s \leqslant 1, \quad A, B \geqslant 0 \\ 0 & Otherwise \end{cases} \tag{3.3}$$

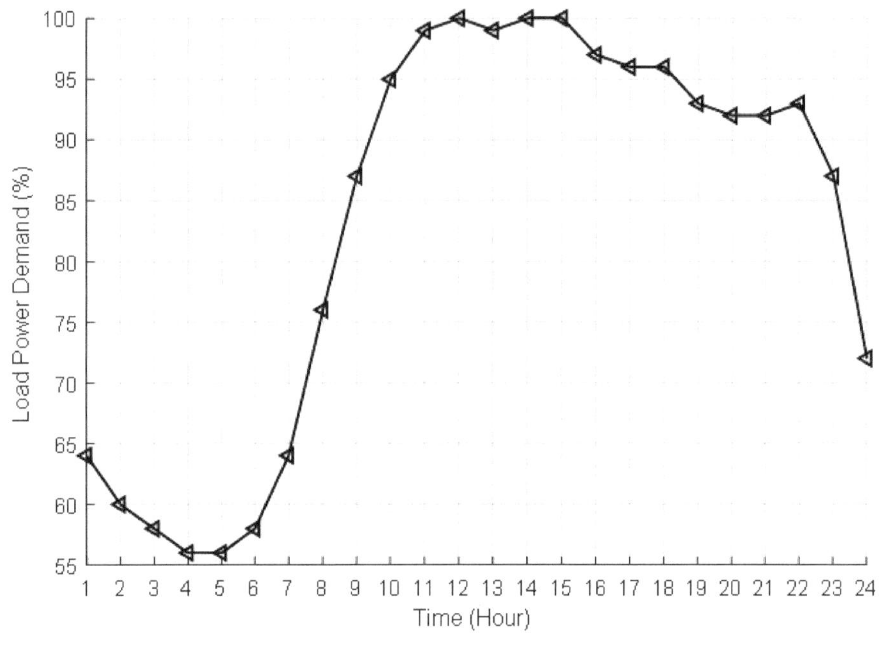


**Figure 3.1** Daily load demand varriation

where, the parameters *A* and *B* can be calculated as [36, 37]:

$$B = (1 - \mu)\left(\frac{\mu(1 - \mu)}{\sigma^2} - 1\right) \tag{3.4}$$

$$A = \frac{\mu \times B}{1 - \mu} \tag{3.5}$$

The solar irradiance state (*s*) probability along any specific hour may be calculated as:

$$P_s\{G\} = \int_{s_1}^{s_2} f_b(s)ds \tag{3.6}$$

The PV module's output power is expressed as in [34–38]:

$$P_{pvo}(s) = N \times FF \times V_y \times I_y \tag{3.7}$$

$$FF = \frac{V_{MPP} \times I_{MPP}}{V_{oc} \times I_{sc}} \tag{3.8}$$

$$V_y = V_{oc} \times K_v \times T_{cy} \tag{3.9}$$

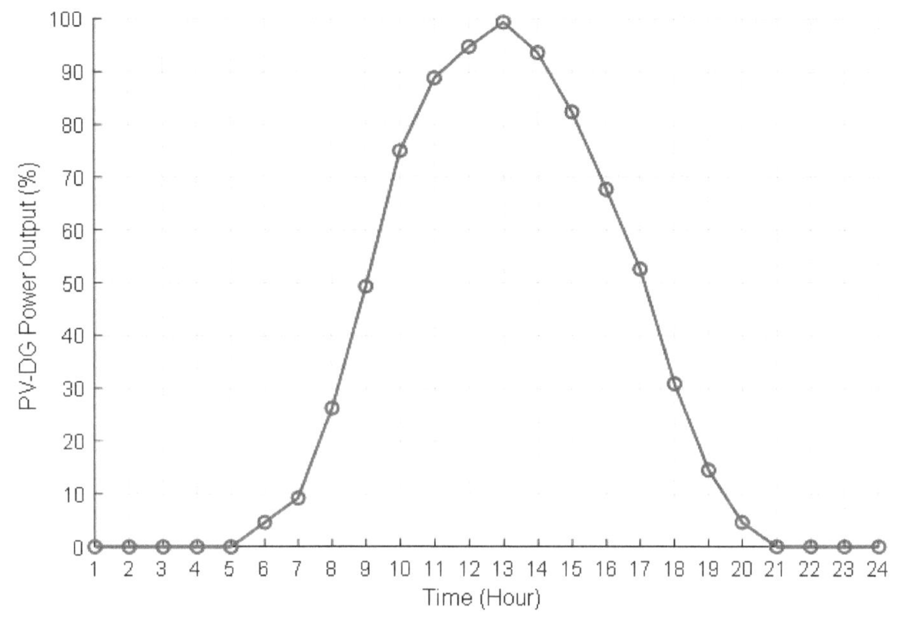


**Figure 3.2** Daily variation of PVDG output power.

$$I_y = s[I_{sc} \times K_i \times (IT_{cy} - 25)] \tag{3.10}$$

$$T_{cy} = T_A + s\left(\frac{N_{OT} - 20}{0.8}\right) \tag{3.11}$$

The RDG unit's total output power depends on the PV panel's irradiance characteristics and specification.

$$P_{PV}(t) = \int_{s_1}^{s_2} P_{PV\circ}(s) P_s \{G\}\, ds \tag{3.12}$$

Figure 3.2 illustrated the daily output power generated from the PVDG for 24 hours.

From Figure 3.2, it is clear that the output power from the PVDG was present and kept varying only in the period from 6h00 to 18h00 of the day's hours, reaching the maximum of that generation of 100% around mid-day. In addition, it is obvious that the more the solar irradiance is present and strong, the PV will generate its maximum power, until remains to zero when there is no irradiance, generally from 18h00 to 6h00 of the day's hours.

### 3.2.3 Wind Turbine DG Uncertainty Modeling

The wind turbine's output power is dependent on the wind speed, also on the power performance curve parameters. So that, once the Rayleigh PDF is generated in a specified period segment, the output power along many states can be calculated using the next equation [37, 39]:

$$P_{WT}(v) = \begin{cases} 0 & 0 \leqslant v \leqslant v_{ci} \\ P_{rated} \times \frac{(v - v_{ci})}{(v_r - v_{ci})} & v_{ci} \leqslant v \leqslant v_r \\ P_{rated} & v_r \leqslant v \leqslant v_{co} \\ 0 & v_{co} \leqslant v \end{cases} \tag{3.13}$$

Figure 3.3 shows the daily output power generated from the WTDG for 24 hours.

By analyzing Figure 3.3, the first interesting remark is that the WTDG generates power throughout every hour of the day from a minimum rate of about 20 % around 8h00 until a maximum rate of 100% around 17h00. It is

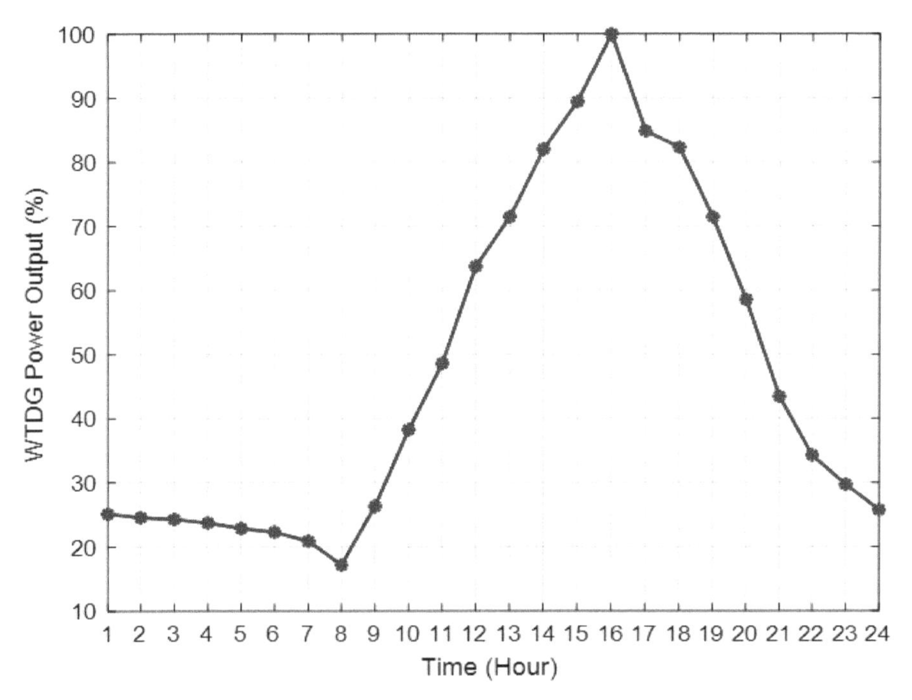


**Figure 3.3** Daily variation of WTDG output power.

obvious that the more the conditions are windy, the more the WT generates its maximum output power.

## 3.3 Multi Objective Indices Evaluation

### 3.3.1 Multi Objective Indices

The proposed MOI in this paper is consist of optimally allocating both multiple PVDG and WTDG units into the two standards test systems SDGs, by minimizing simultaneous the various technical indices of APLI, RPLI, VDI, OTI, and CTII, which would be formulated as next:

$$MOI = Minimize \sum_{i=1}^{N_{bus}} \sum_{j=2}^{N_{bus}} \sum_{i=1}^{N_{PR}} \sum_{j=1}^{N_{BR}} [APLI_{i,j} + RPLI_{i,j} + VDI_j + OTI_i + CTII_{i,j}] \tag{3.14}$$

Starting by the APLI which may be defined by [40, 41]:

$$APLI = \frac{APL_{AfterRDG}}{APL_{BeforeRDG}} \times 100\% \tag{3.15}$$

$$APL_{i,j} = \alpha_{ij}(P_iP_j + Q_iQ_j) + \beta_{ij}(Q_iP_j + P_iQ_j) \tag{3.16}$$

$$\alpha_{ij} = \frac{R_{ij}}{V_iV_j}\cos(\delta_i - \delta_j) \tag{3.17}$$

$$\beta_{ij} = \frac{R_{ij}}{V_iV_j}\sin(\delta_i - \delta_j) \tag{3.18}$$

Then, the RPLI which may be defined by [42]:

$$RPLI = \frac{RPL_{After}RDG}{RPL_{Before}RDG} \times 100\% \tag{3.19}$$

$$RPL_{i,j} = \alpha_{ij}(P_iP_j + Q_iQ_j) + \beta_{ij}(Q_iP_j + P_iQ_j) \tag{3.20}$$

$$\alpha_{ij} = \frac{X_{ij}}{V_iV_j}\cos(\delta_i - \delta_j) \tag{3.21}$$

$$\beta_{ij} = \frac{X_{ij}}{V_iV_j}\sin(\delta_i - \delta_j) \tag{3.22}$$

Thirdly, the VDI [43, 44]:

$$VDI = \frac{VD_{AfterRDG}}{VD_{BeforeRDG}} \times 100\% \tag{3.23}$$

$$VD_j = \sum_{j=2}^{N_{bus}} |1 - V_j| \tag{3.24}$$

Also, the OTI of the primaries overcurrent relays (OCR) [45, 46]:

$$OTI = \frac{OT_{AfterRDG}}{OT_{BeforeRDG}} \times 100\% \tag{3.25}$$

$$T_i = TDS_i \left( \frac{A}{M_i^B - 1} \right) \tag{3.26}$$

$$M_i = \frac{I_F}{I_p} \tag{3.27}$$

At least, the CTII of the primaries and backups OCRs [46]:

$$CTII = \frac{CTI_{AfterRDG}}{CTI_{BeforeRDG}} \times 100\% \tag{3.28}$$

$$CTI = OT_{Bachup} - OT_{primary} \tag{3.29}$$

### 3.3.2 Equality Constraints

$$P_G + P_{RDG} = P_D + APL \tag{3.30}$$

$$Q_G + Q_{RDG} = Q_D + RPL \tag{3.31}$$

### 3.3.3 Distribution Line Constraints

$$V_{min} \leq |V_i| \leq V_{max} \tag{3.32}$$

$$|1 - V_i| \leq \Delta V_{max} \tag{3.33}$$

$$|S_{ij}| \leq S_{max} \tag{3.34}$$

### 3.3.4 RDG Constraints

$$P_{RDG}^{min} \leq P_{RDG} \leq P_{RDG}^{max} \tag{3.35}$$

$$Q_{RDG}^{min} \leq Q_{RDG} \leq Q_{RDG}^{max} \tag{3.36}$$

$$\sum_{i=1}^{N_{RDG}} P_{RDG}(i) \leq \sum_{i=1}^{N_{bus}} P_D(i) \tag{3.37}$$

$$\sum_{i=1}^{N_{RDG}} Q_{RDG}(i) \leq \sum_{i=1}^{N_{bus}} P_D(i) \tag{3.38}$$

$$2 \leq RDG_{position} \leq N_{bus} \tag{3.39}$$

$$N_{RDG} \leq N_{RDG.max} \tag{3.40}$$

$$n_{RDG,i}/Location \leq 1 \tag{3.41}$$

## 3.4 Distribution Test System

The selected algorithms were validated and tested on the standards IEEE 33-bus, and 69-bus, where they are represented by the single diagrams in Figures 3.4 (a) and 3.4(b), respectively. The base voltage is 12.66 kV in both standards. The total loads are 3715.00 kW and 2300.00 kVar for the first test system, also 3790.00 kW and 2690.00 kVar for the second test system.

All of the systems' buses are protected and covered by a primary OCR and its backup, where the coordination time interval (CTI) is between them and set above 0.2 seconds. It would calculate for the first system, 32 OCRs and 31 CTIs, while for the second test system 68 OCRs and 67 CTIs, where more details about both test systems' main characteristics are mentioned in the descriptive summary in Table 3.1.

## 3.5 Analysis Results and Comparison

Table 3.1 represents the main characteristics and parameters of both smart distribution systems before RDG units' installation, while Tables 3.2 and 3.3 summarize all the results after optimization using the different algorithms for the comparison in presence of PVDG units and WTDG units into IEEE 33 and 69-bus SDGs respectively.

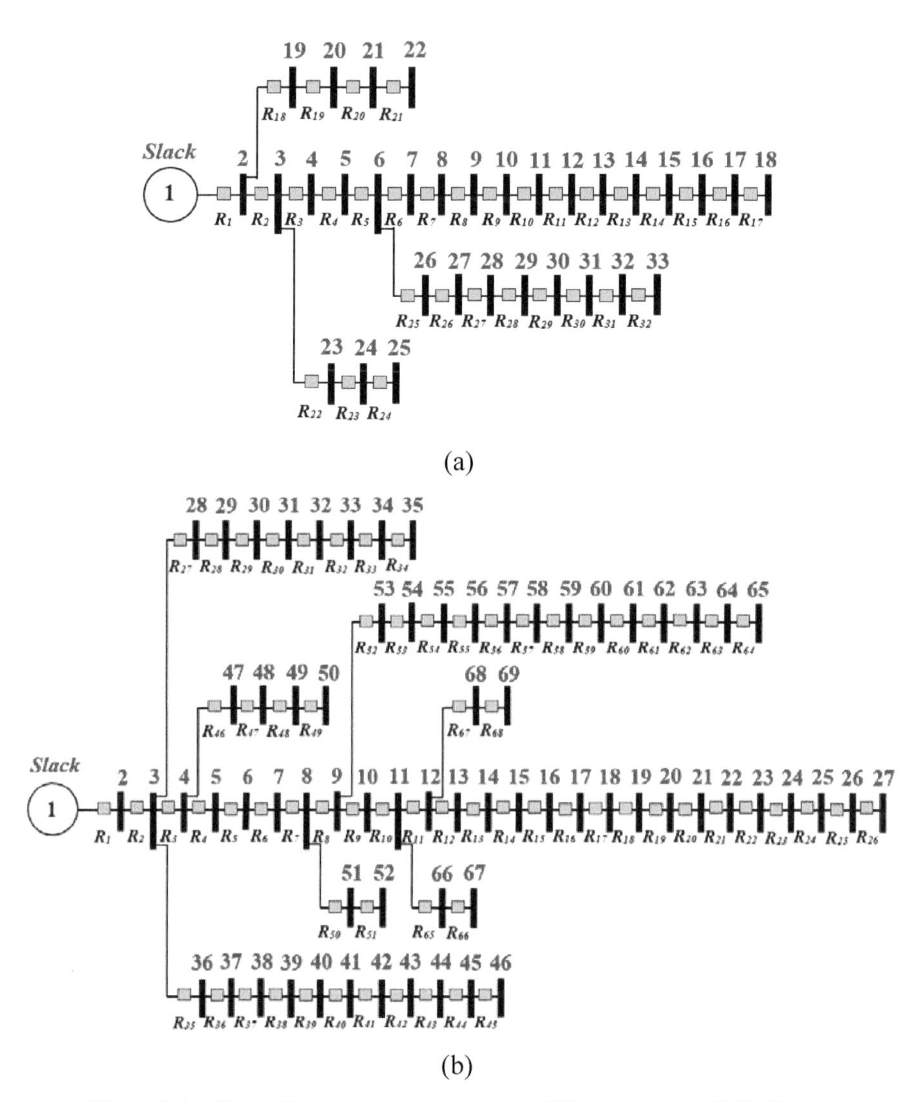

Figure 3.4 Single diagram of test systems: a). IEEE 33-bus, b). IEEE 69-bus.

**Table 3.1** The daily main characteristics of the investigated SDG test systems.

| Characteristics | Buses | Branches | Relays | $\sum APL$ (kWh) | $\sum RPL$ (kVarh) | $\sum VD$ (p.u.) | $\sum OT$ (sec) | $\sum CTI$ (sec) |
|---|---|---|---|---|---|---|---|---|
| IEEE 33-bus | 33 | 32 | 32 | 3557.02 | 2412.00 | 35.65 | 492.08 | 186.98 |
| IEEE 69-bus | 69 | 68 | 68 | 3785.31 | 1720.70 | 36.83 | 929.28 | 404.69 |

**Table 3.2** The obtained results after optimization for IEEE 33-bus.

| Algorithm Applied | *RDG Type* | *RDG Bus* | $P_{RDG}$ (MW) | $Q_{RDG}$ (MVar) | *APLI* (%) | *RPLI* (%) | *VDI* (%) | *OTI* (%) | *CTII* (%) | *MOF* (%) |
|---|---|---|---|---|---|---|---|---|---|---|
| **PSO** | **PVDG** | 6<br>14<br>24 | 1.4892<br>0.5223<br>0.7975 | — | 68.00 | 68.94 | 80.33 | 99.64 | 99.76 | 337.15 |
| | **WTDG** | 13<br>19<br>30 | 0.7845<br>0.8104<br>0.9755 | 0.3747<br>0.4057<br>0.9260 | 40.87 | 40.79 | 57.43 | 99.32 | 99.54 | 250.56 |
| **GOA** | **PVDG** | 5<br>13<br>30 | 1.0884<br>0.6179<br>0.7488 | — | 67.75 | 68.76 | 79.88 | 99.63 | 99.75 | 334.78 |
| | **WTDG** | 2<br>13<br>30 | 1.7452<br>0.7751<br>0.9596 | 0.8283<br>0.3600<br>0.9132 | 34.15 | 34.24 | 52.28 | 99.26 | 99.50 | 249.64 |
| **MFO** | **PVDG** | 3<br>14<br>30 | 1.6031<br>0.6192<br>0.8396 | — | 67.50 | 68.24 | 80.17 | 99.64 | 99.76 | 334.01 |
| | **WTDG** | 15<br>26<br>31 | 0.5906<br>0.9920<br>0.5328 | 0.2168<br>0.8863<br>0.4937 | 33.33 | 34.03 | 51.41 | 99.25 | 99.49 | 247.46 |
| **GWO** | **PVDG** | 3<br>13<br>30 | 1.5726<br>0.6556<br>0.8253 | — | 67.47 | 68.31 | 80.07 | 99.63 | 99.75 | 333.77 |
| | **WTDG** | 4<br>13<br>30 | 1.3077<br>0.6955<br>0.8358 | 0.8485<br>0.3654<br>0.7403 | 30.89 | 32.77 | 49.90 | 99.22 | 99.48 | 244.81 |
| **ALO** | **PVDG** | 12<br>25<br>29 | 0.6890<br>0.6720<br>1.0595 | — | 66.62 | 66.89 | 80.40 | 99.64 | 99.76 | 332.01 |
| | **WTDG** | 11<br>24<br>31 | 0.9233<br>0.8344<br>0.7030 | 0.5419<br>0.6358<br>0.5442 | 32.34 | 32.42 | 52.37 | 99.26 | 99.50 | 242.49 |
| **MPA** | **PVDG** | 13<br>24<br>30 | 0.6821<br>0.9188<br>0.8813 | — | 66.35 | 66.69 | 80.25 | 99.64 | 99.76 | 329.05 |
| | **WTDG** | 11<br>25<br>30 | 0.8629<br>0.6829<br>0.9048 | 0.4340<br>0.4079<br>0.8417 | 29.47 | 29.66 | 51.01 | 99.24 | 99.49 | 236.86 |

**Table 3.3** The obtained results after optimization for IEEE 69-bus.

| Algorithm Applied | *RDG Type* | *RDG Bus* | $P_{RDG}$ (MW) | $Q_{RDG}$ (MVar) | *APLI* (%) | *RPLI* (%) | *VDI* (%) | *OTI* (%) | *CTII* (%) | *MOF* (%) |
|---|---|---|---|---|---|---|---|---|---|---|
| **PSO** | **PVDG** | 9<br>50<br>61 | 1.1412<br>0.3585<br>1.3092 | — | 66.63 | 66.66 | 83.74 | 99.86 | 99.85 | 335.15 |
| | **WTDG** | 14<br>36<br>61 | 0.6253<br>0.9161<br>1.4993 | 0.4294<br>0.5917<br>1.0608 | 27.59 | 31.44 | 50.69 | 99.69 | 99.68 | 235.62 |
| **GOA** | **PVDG** | 10<br>49<br>63 | 0.9435<br>0.6523<br>1.3603 | — | 65.81 | 65.78 | 82.23 | 99.85 | 99.84 | 332.03 |
| | **WTDG** | 17<br>36<br>61 | 0.3239<br>1.6510<br>1.6077 | 0.2506<br>1.2468<br>1.2117 | 26.74 | 30.97 | 53.55 | 99.71 | 99.69 | 239.32 |
| **MFO** | **PVDG** | 18<br>53<br>63 | 0.4153<br>0.4387<br>1.3456 | — | 65.48 | 67.21 | 80.97 | 99.84 | 99.83 | 331.26 |
| | **WTDG** | 4<br>17<br>61 | 1.3595<br>0.5078<br>1.5210 | 1.0214<br>0.3317<br>1.0812 | 27.41 | 31.43 | 50.84 | 99.70 | 99.68 | 234.01 |
| **GWO** | **PVDG** | 50<br>61<br>69 | 0.5639<br>1.4870<br>0.3786 | — | 65.86 | 66.17 | 83.48 | 99.86 | 99.85 | 331.50 |
| | **WTDG** | 50<br>61<br>69 | 0.5719<br>1.4772<br>0.7195 | 0.2711<br>1.0020<br>0.4812 | 28.66 | 29.19 | 54.17 | 99.71 | 99.70 | 234.64 |
| **ALO** | **PVDG** | 22<br>28<br>61 | 0.4057<br>0.3069<br>1.5465 | — | 64.62 | 66.65 | 80.97 | 99.84 | 99.83 | 330.15 |
| | **WTDG** | 11<br>18<br>61 | 0.3645<br>0.4158<br>1.4693 | 0.2186<br>0.2849<br>1.0536 | 26.60 | 30.50 | 49.16 | 99.69 | 99.67 | 231.81 |
| **MPA** | **PVDG** | 17<br>49<br>61 | 0.4691<br>0.6761<br>1.4871 | — | 64.57 | 65.03 | 80.79 | 99.84 | 99.83 | 325.09 |
| | **WTDG** | 23<br>49<br>61 | 0.4573<br>0.7479<br>1.5370 | 0.2967<br>0.5090<br>1.0939 | 26.78 | 27.53 | 50.84 | 99.70 | 99.68 | 228.85 |

From Table 3.2 and Table 3.3, the best MOI minimization was provided by the MPA for both cases studied of both RDG units. For the IEEE 33-bus and by analyzing the results depicted in Table 3.2, the buses 13, 24, and 30 were selected as the best locations for the three independent PVDG units with an injected active power of 0.6821, 0.9188, and 0.8813 MW, respectively, besides, gives the minimum value of MOI which equal to 329.05 %, which is represented as the sum of the minimized indices until 66.62 % of APLI, 66.89 % of RPLI, 80.40 % for VDI, 99.64 % and 99.76 % for both OTI and CTII. The next best MOI is recorded by the ALO algorithm as it is equal to 332.01%, which is bigger than the MPA by 2.96 %. Also, by analyzing the result of each index of MOI in the case of the ALO algorithm separately, the APLI, RPLI, and VDI are reduced by 66.62 %, 66.89 %, and 80.40 %, which are bigger values compared to what the MPA provided, while the percentage of OTI and CTII is 99.64 % and 99.76% which is the same as MPA. The rest of the algorithms reduced MOI as follows with the order from the best until the worst: GWO to 333.77 %, MFO to 334.01 %, GOA to 334.78 %, and finally PSO to 337.15 %. On the other hand, the best MOI for WTDG case is 236.86 %, also obtained by MPA, with a reduced rate of 29.47 % for APLI, 29.66 % for RPLI, 51.01 % for VDI, 99.24 %, and 99.49 % for OTI and CTII, respectively, knowing that the best locations were 11, 25 and 30 with respect to the active and reactive powers of 0.8629, 0.6829 and 0.9048 MW, 0.4340, 0.4079 and 0.8417 MVar. The ALO recorded a value of MOI bigger than MPA by 5.63 %, as well as in the case of PVDG, PSO has the worst MOI value which is higher than MPA by 13.7 %, the obtained MOI's values for the other algorithms are greater than MPA by 7.95 % for GWO, 10.6 % for MFO and 12.78 % for GOA. The MOI value in the case of WTDG is lower by a difference of 92.19 % than PVDG.

For the IEEE 69-bus, from Table 3.3 the best placements obtained by MPA for the PVDG case were buses 17, 49, and 61 while their powers were 0.4691, 0.6761, and 1.4871 MW with respect to each bus. The MPA reduced the APLI and RPLI to 64.57 % and 65.03 %, while the VDI to 80.79 % and the OTI and CTII to 99.84 % and 99.83 %, which is, represent the total MOI's value of 325.09 %. This value is lower than ALO by 5.06 % where it represents the minimum value after MPA, also lower than GWO, MFO, and GOA by 6.41 %, 6.17 %, and 6.94 %, knowing that those algorithms have close values to the MOI, while the worst results were obtained by PSO algorithm which was bigger than the optimal value by 10.06 %. On the other side, buses 23, 49, and 61 were determined by MPA as optimal placements for the WTDG case, while their optimal powers were 0.4573, 0.7479, and 1.5370

MW, 0.2967, 0.5090, and 1.0939 kVar, respectively. The best MOI recorded by MPA and got reduced to 228.85 %, represented as the minimization in APLI and RPLI to 26.78 % and 27.53 %, VDI to 50.84 % while OTI and CTII to 99.70%, and 99.68 %, respectively. The optimal MOI value is lower than ALO by 2.96 %, while the values of GWO and MFO are close to each other and greater by 5.79 % and 5.16 % than the MPA value. In this case, the worst value is obtained by GOA which is bigger than MPA by 10.47 %. The optimal MOI value for the WTDG case is better and lower than the case of PVDG by 96.24 %.

Figures 3.5 and 3.6 represented the boxplot of MOI minimization of all the applied algorithms for 20 runs in each for the optimal presence of PVDG and WTDG units in the two test systems.

The selected algorithms were implemented in MATLAB Software (version 2017. b) by a PC that has the processor of Intel Core i5 with 3.4 GHz and 8 GB of RAM. Meanwhile, in order to improve the comparison between the applied algorithms, a boxplot of MOI minimization was implemented and illustrated in the next Figures.

It is obvious that the algorithms showed a great performance in delivering very good results which were obtained after 20 independents run, in terms of getting the best MOI minimization, with superiority of the MPA algorithm among all of them.

The MPA was capable to find the fairly best results of MOI, including the minimum median for all RDG's cases studied into both SDGs. The best results after 20 runs recorded by the MPA algorithm were provided from the case WTDG units in both test systems, where also can observe mostly a very small difference between the best and worst of its result.

Figures 3.7 and 3.8 demonstrate the convergence characteristics for the MOI minimization after applying the selected algorithms for the optimal integration of the two RDGs into both test distribution systems. After doing the analysis of the convergence curves in the mentioned figures when applying the selected algorithms for both RDGs' optimal integration in the two test systems SDGs, also for a number of maximum iterations equal to 100 and a population size of 10, shows that the MPA was the best approach that produced the best solutions and results.

The MPA delivered the minimum of MOI for both cases studied of RDGs' integration into the two test systems with much better and superior results for the case of WTDG units, whereas it minimized the MOI until a rate of 236.86 % for the first SDG and until a rate of 228.85 % for the second SDG, comparing to the other algorithms. It is also clear that the MPA had a quick

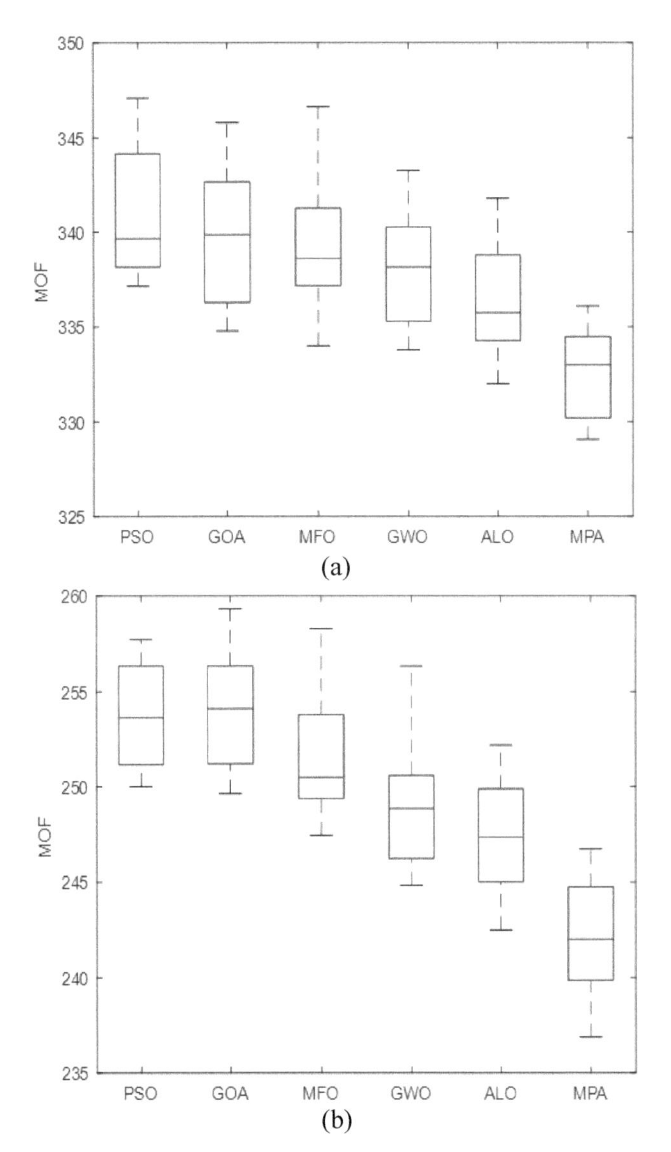


**Figure 3.5** Boxplot of MOI after applying the algorithms for the IEEE 33-bus: a). PVDG case, b). WTDG case.

and smooth convergence characteristic when reaching the optimal solutions also it early settles down almost within 60 iterations for both cases of RDGs into the two test distribution systems.

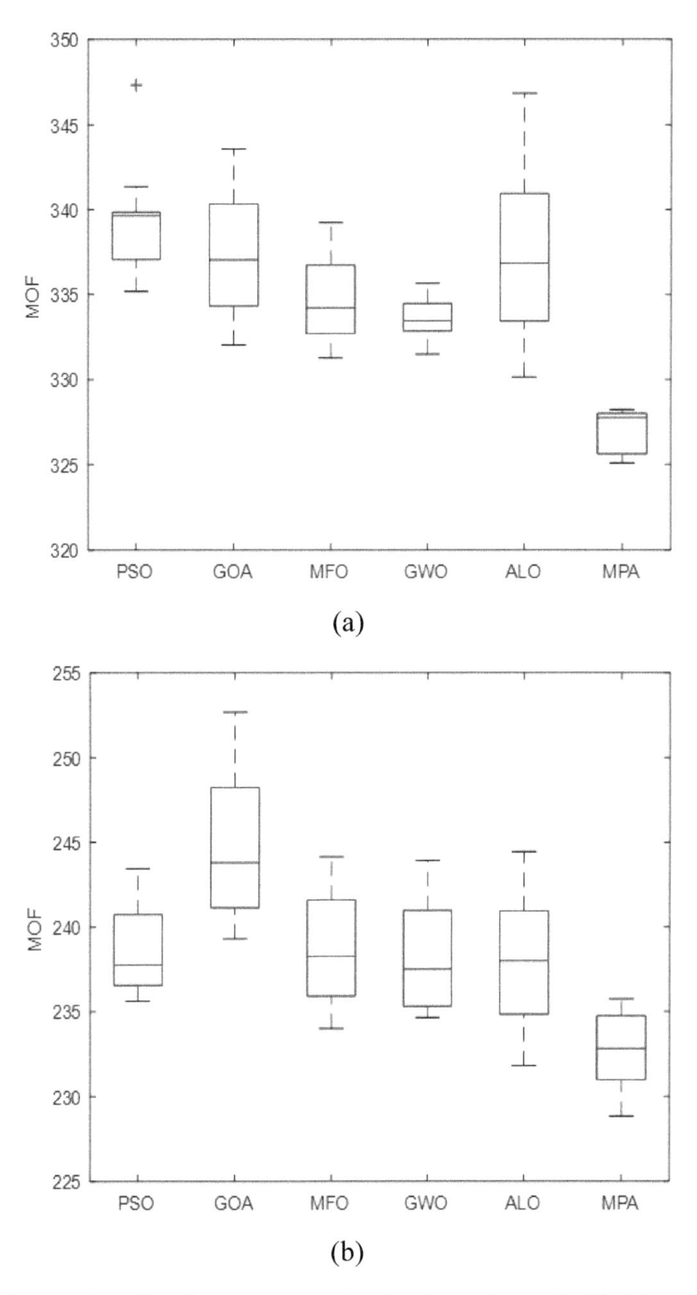


**Figure 3.6** Boxplot of MOI after applying the algorithms for the IEEE 69-bus: a). PVDG case, b). WTDG case.

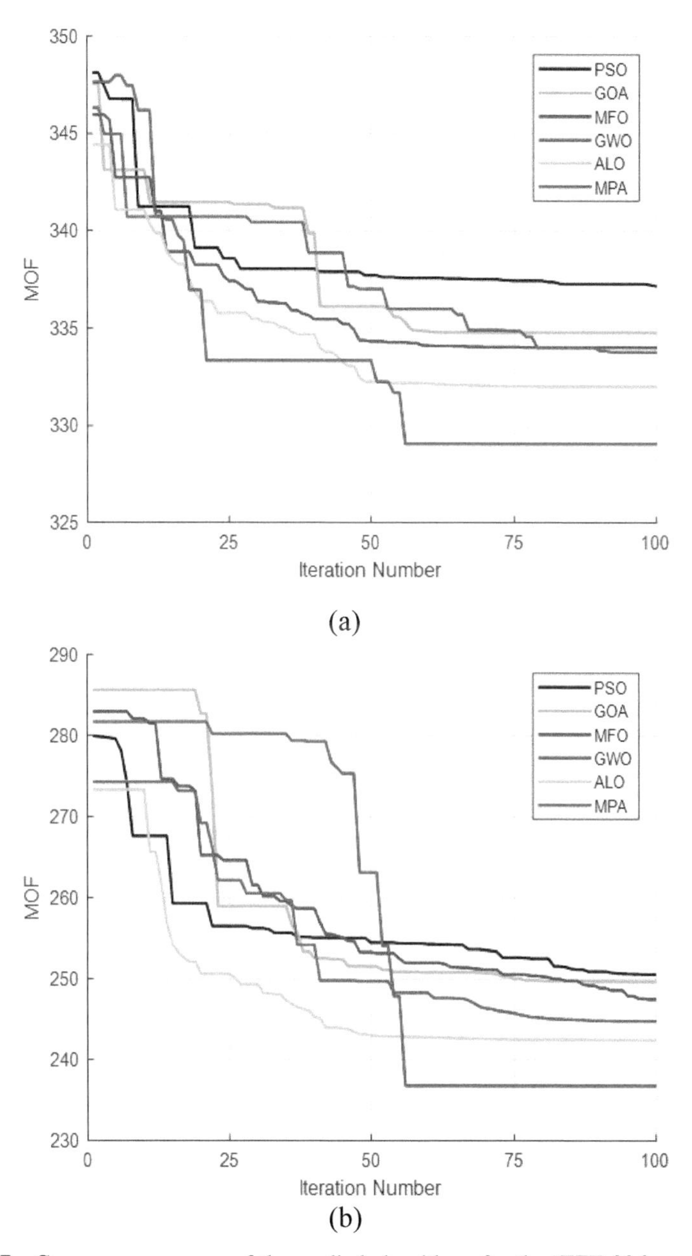


**Figure 3.7** Convergence curves of the applied algorithms for the IEEE 33-bus: a). PVDG Case, b). WTDG Case.

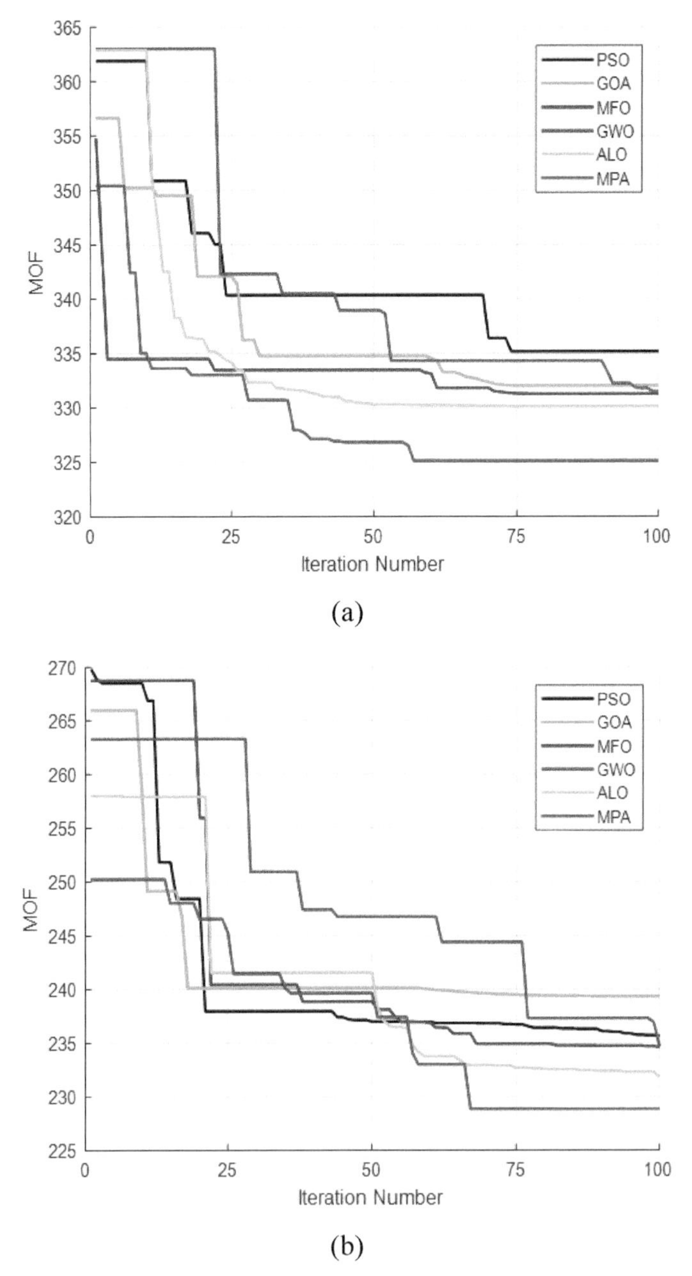


**Figure 3.8** Convergence curves of the applied algorithms for the IEEE 69-bus: a). PVDG Case, b). WTDG Case.

Figure 3.9 contains the daily total voltage deviation variation in 24 hours, for all RDG's cases studied in both test systems, represented in p.u.

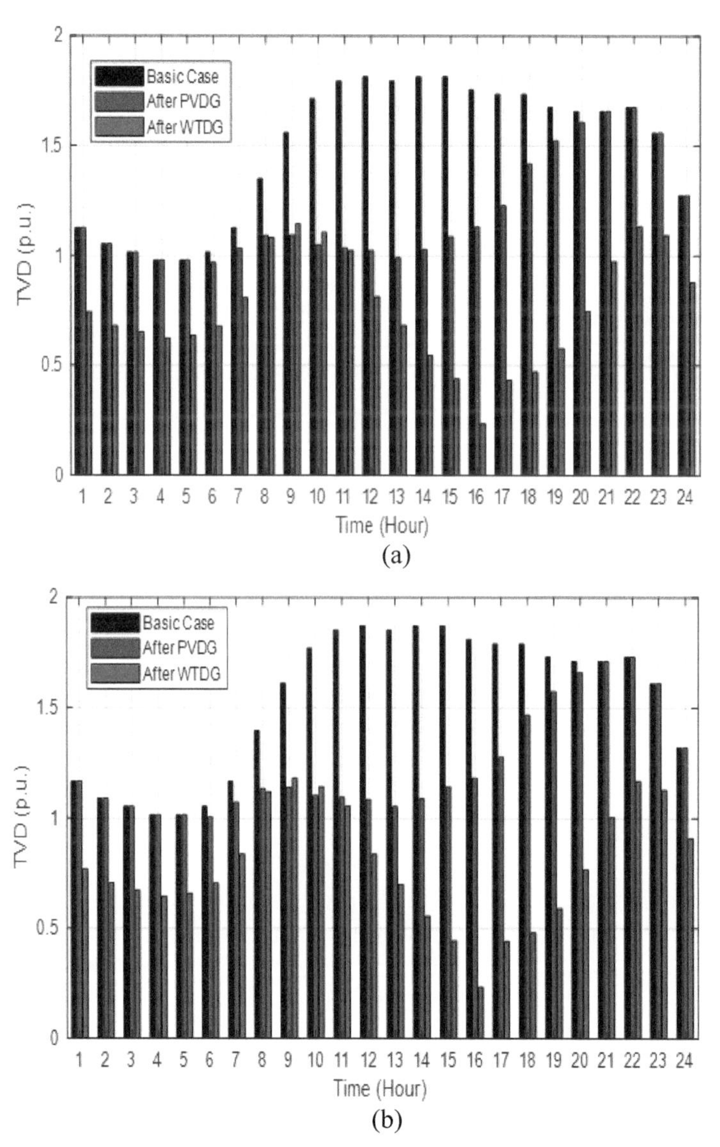


**Figure 3.9** Total voltage derivation variation for both test systems: a). IEEE 33-bus, b). IEEE 69-bus.

From Figure3.9, in the basic case, the TVD after 7h00 has an increment from 1 p.u. to almost 1.8 p.u. whereas by 11h000 was still great than 1.5 p.u. and keep until 23h00. The observation would be valid for two test systems, this increment is due to the call of load, while the consumption recorded between 8h00 to 23h00 is great than 85 %. On the other hand, due to the absence of output power from PVDG units before 6h00 and after 20h00, it is observed that no minimization in TVD for this period. Contrarily, out of the mentioned period (i.e., from 6h00 to 20h00), there was a small reduction in the TVD values, which starts showing up at 6h00, while the minimization appears clearly between 9h00 to 18h00 of the day's hours, due to the fact that the PVDG units offer more than 50% of its power in this period.

Other than that, the WTDG units provided energy along the 24h which excellently contributed to minimizing the TVD in the whole day as shown in the previous figure. After 10h00 the WTDG units provide more than 40% of its power generation and last until 21h00 which represents the period with the high reduction in TVD, knowing that in this period high energy consumption is recorded. The peak of wind turbine output was around 16h00, where it is observed that the minimum value of TVD was recorded in that hour. Another observation, the sum of TVD values along 24 hours for the first and the second test systems respectively that obtained after WTDG units of 18.19 p.u., and 18.73 p.u. is clearly lower and better than the values from PVDG units of 28.61 p.u., 29.76 p.u.

Figures 3.10 and 3.11 illustrate the daily voltage profiles variation of both test systems in 24 hours for both cases of optimal RDGs integration.

From both Figures 3.10 and 3.11, it is obvious that the daily voltage profiles have been ameliorated after the optimal presence of both RDG units into both test systems SDGs.

The case of WTDG units integrating was the best choice for both test systems, that enhanced the daily voltage profiles almost along the 24 hours as long as the WT provided its active and reactive generation without any interruptions, especially around mid-day where the voltages profiles were at their maximum values, and this was synchronous with the maximum generation from WT. The case of PVDG units integrating also provided good results but not as good as the case of WTDG units, for reason that the PV generates its only active power mostly between 6h00 to 18.h00 of the day's hours. Meanwhile, it is seen that the best results of daily voltage profiles ameliorating were obtained at 13h00 which is the time that the PVDG units provided their maximum power generation.

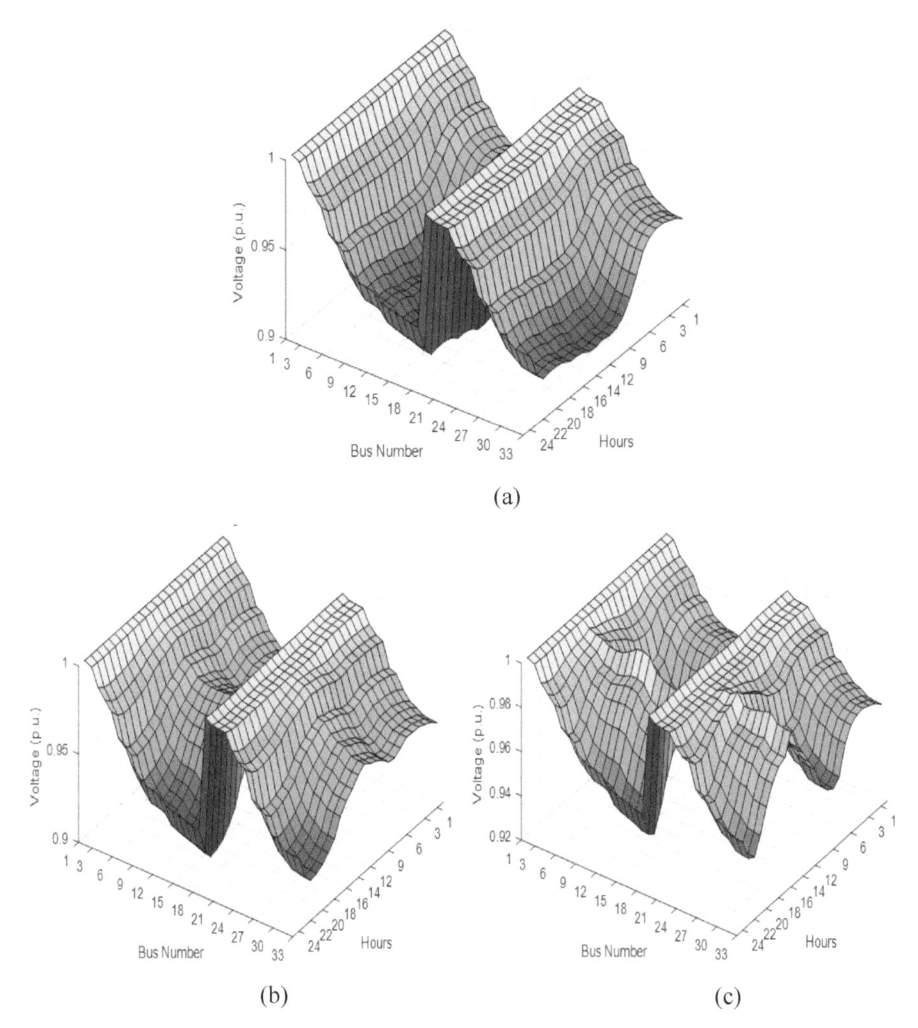


**Figure 3.10** Daily voltage profiles variation for the IEEE 33-bus: a). Basic Case, b). After PVDG, c). After WTDG.

Figures 3.12 and 3.13 demonstrate the daily branch active power loss variation for both cases of RDGs units' optimal integration in the two standards SDGs studied.

It is seen that the optimal installation of the two RDGs units had a clear impact on both test systems' technical parameters as the daily active power losses which are mentioned in the 3D graphics shown in the next Figures of 3.12 and 3.13.

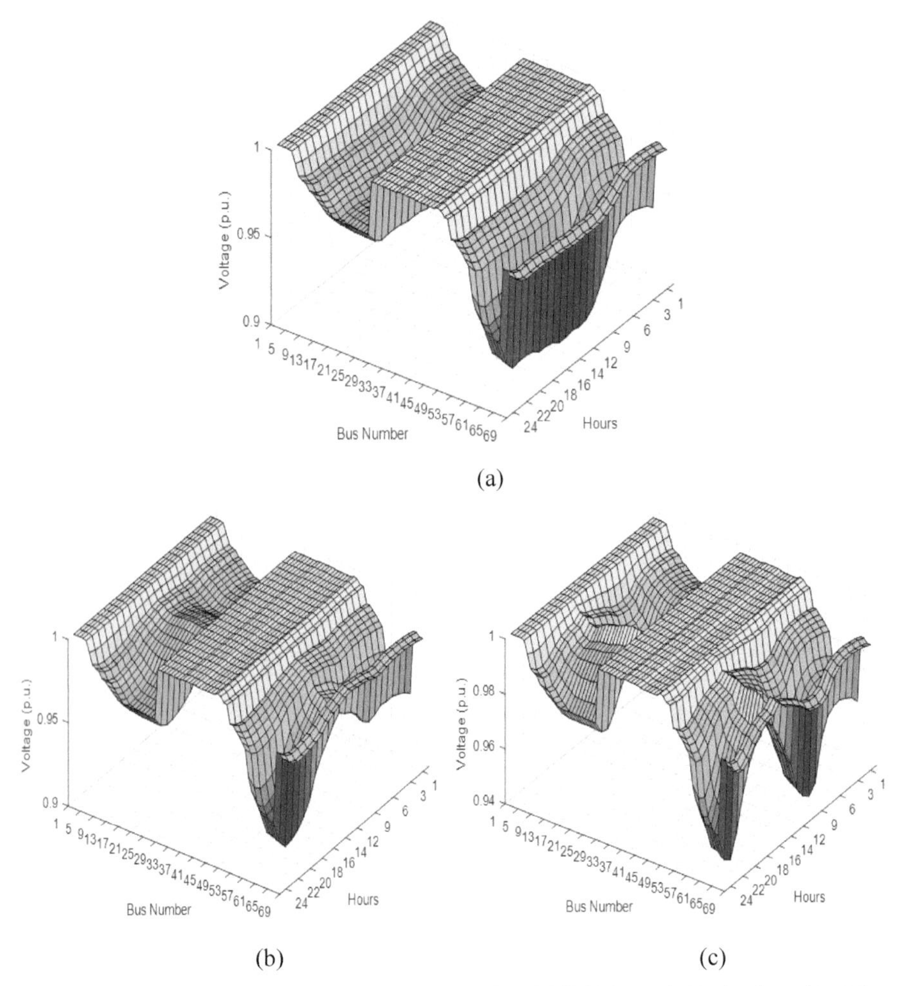


**Figure 3.11** Daily voltage profiles variation for the IEEE 69-bus: a). Basic Case, b). After PVDG, c). After WTDG.

It is clear that the daily active power loss had a significant minimization in both SDGs after that optimal integration of the two RDGs with a strong and superior impact for the case of WTDG units that registered along the day's hours in almost every branch of the two standards test systems.

Also, the case of WTDG units caused the minimization of the total daily active power losses from 3557.02 kWh to 1048.50 kWh with a reduction's rate until 29.47 % for the first SDG, and from 3785.31 kWh to 1013.90 kWh

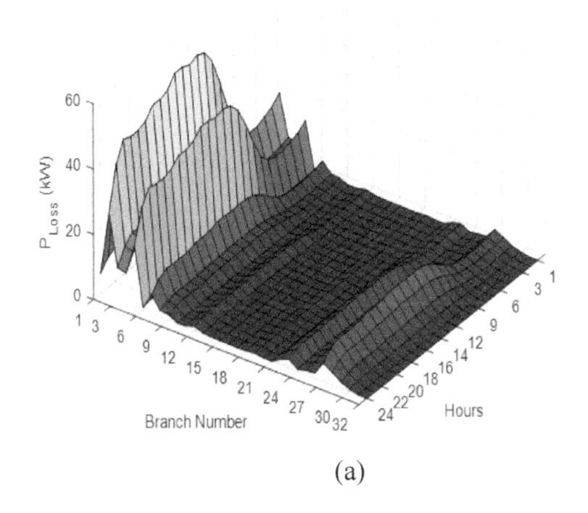


(a)

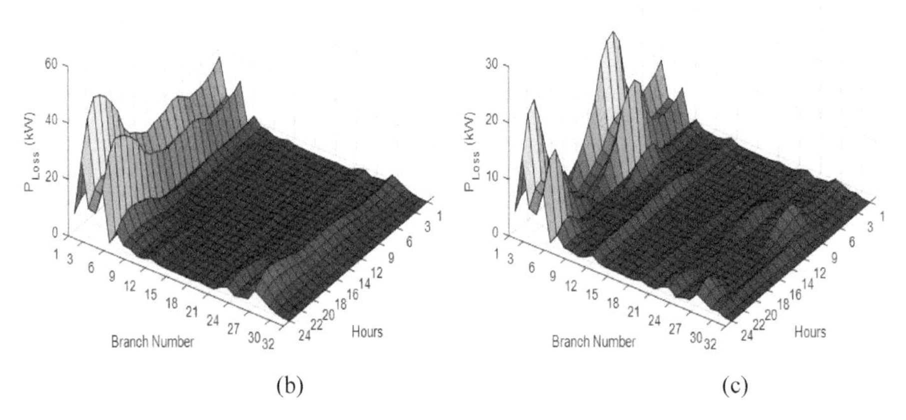


(b) (c)

**Figure 3.12** Active power loss variation for the IEEE 33-bus: a). Basic Case, b). After PVDG, c). After WTDG.

with a reduction's rate until 26.78 % for the second SDG, for reason that WT generates both active and reactive powers to the SDGs for almost all the day's hours, contrarily to the PV which delivers only active power to the SDGs and mostly in a specific period, which is generally between 6h00 to 18.h00 of the day's hours.

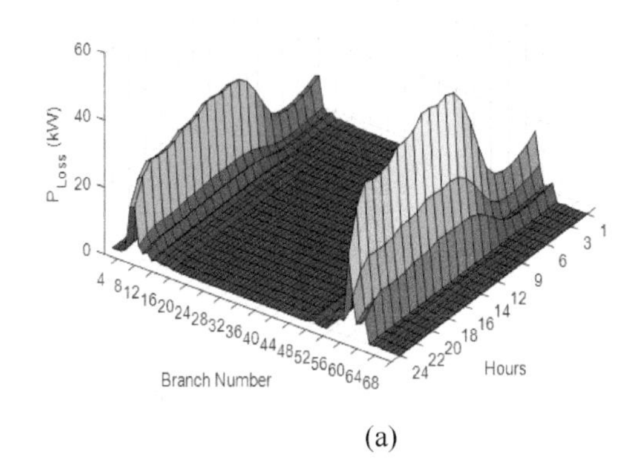


(a)

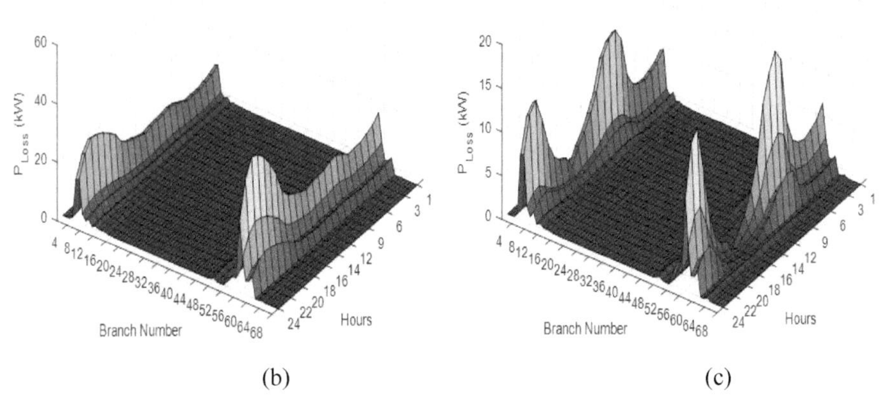


(b) (c)

**Figure 3.13** Active power loss variation for the IEEE 69-bus: a). Basic Case, b). After PVDG, c). After WTDG. c). After WTDG.

Figure 3.14 illustrates the daily variation of the overcurrent relays' operation time for both cases studied of RDGs' optimal presence in the two test systems.

The principal task and function of the overcurrent relays are to sense and identify the fault current that happens through the lines, and to do the rapid removal, separation, and protection of the targeted system's parts. Doing the minimization of the OCRs' operation time would be very favorable and

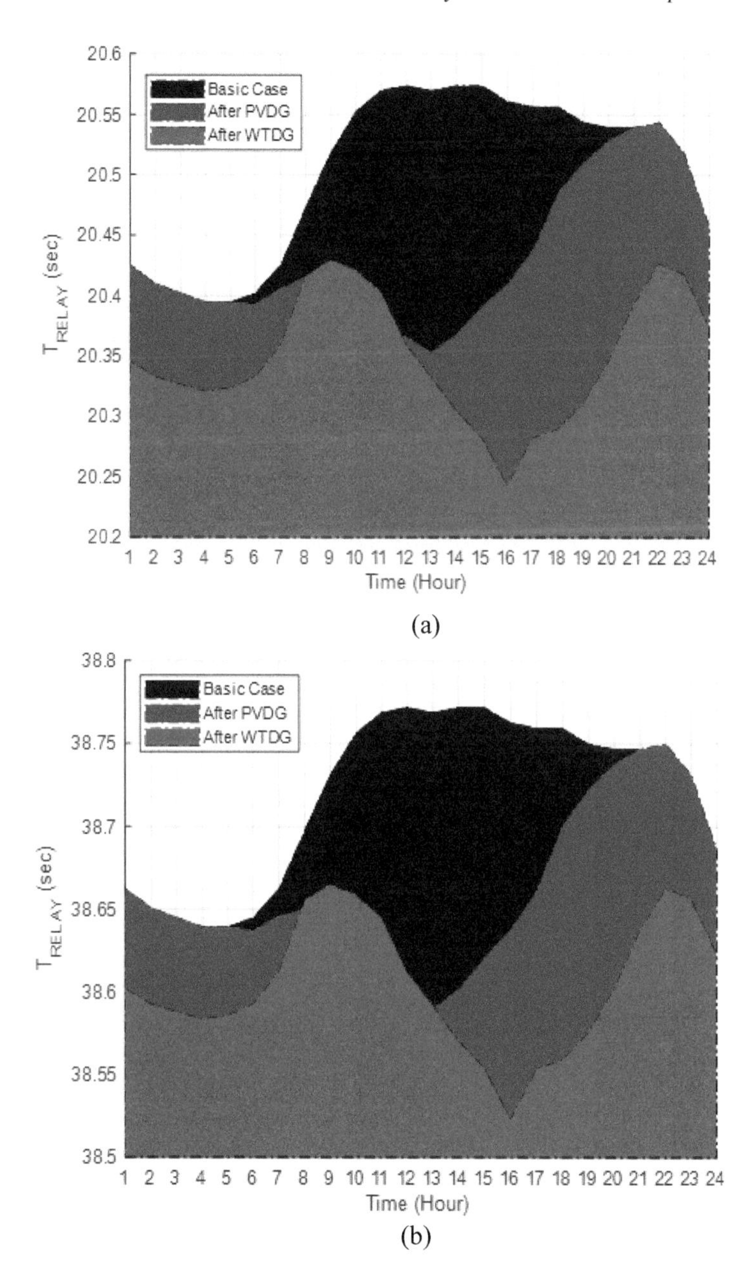


**Figure 3.14** The total operation time of overcurrent relay variation for both test systems:a). IEEE 33-bus, b). IEEE 69-bus.

beneficial in many aspects as protecting the system's parts, extending the system's equipment lifetime, and maintaining the continuity of service.

Applying the MPA when minimizing the MOI for the optimal integration of both RDGs in the two test systems obviously led to the minimization of the daily operation time of the overcurrent relays for both SDGs, with noticing superior and best results from the case of WTDG units.

As mentioned in Figure 3.14, the integration of the WTDG units causes the minimization of that operation time along the day's hours also from a daily total value of 492.08 seconds until 488.38 seconds with a reduction's rate until 99.24 % for the first test system, while from 929.28 seconds until 926.50 seconds with a reduction's rate until 99.70 %. The daily generation of both active and reactive powers from the WT was the main reason for those better and superior results over the PV which delivers only active powers exclusively when the solar irradiance is present.

## 3.6 Conclusion

This paper was consisted of the analysis of the MOI minimization represented as the sum of various technical indices of APLI, RPLI, VDI, OTI, and CTII, by optimally locating and size of both DGs based on renewable PV and WT sources in the two standards test systems IEEE 33-bus and 69-bus, when using plenty recent metaheuristics algorithms, considering the daily uncertainties of the load demand and the power output generation from RDGs units.

Among all the applied algorithms, the MPA showed the best effectiveness and reliability in providing the finest optimal results for both test systems SDGs, including strong behavior and fast convergence characteristics.

The obtained results clarified the efficiency of the optimal integration of both RDGs units in providing a clear improvement for both test systems SDGs' performances, with much better and superior achievements from the case of WTDG units that present along the day's hours, which obviously reduced the active and reactive power losses until 29.47 % and 29.66 % respectively, for the first test system, and until 26.78 % and 27.53 % respectively, for the second test system, ameliorated the voltage profiles and clearly improved the overcurrent protection system, as long as it is capable of delivering both active and reactive powers for a long period along the day compared to the only active power generated from the PV which is limited in the period when the solar irradiance is presence.

Based on the previous results and discussions, it would be recommended to choose the application of WTDG units to the smart distribution grids for their immense technical benefits. The next work would concentrate on studying the smart distribution grid's performance when integrating RDGs units considering the seasonal uncertainties of the load demand and the RDGs' output powers including various new technical and economical indices.

## References

[1] S.K. Rangu, P.R. Lolla, K.R. Dhenuvakonda, and A.R. Singh. "Recent Trends in Power Management Strategies for Optimal Operation of Distributed Energy Resources in Microgrids: A Comprehensive Review." *International Journal of Energy Research*. vol. 44, no. 13, pp. 9889–9911, 2020. https://doi.org/10.1002/er.5649

[2] Y. Jiang, C.C. Liu, and Y. Xu. "Smart Distribution Systems." *Energies*. vol. 9, no. 4, 297, 2016. https://doi.org/10.3390/en9040297

[3] U. Agarwal, and N. Jain. "Distributed Energy Resources and Supportive Methodologies for Their Optimal Planning Under Modern Distribution Network: A Review." *Technology and Economics of Smart Grids and Sustainable Energy*. vol. 4, no. 1, pp. 1–21, 2019. https://doi.org/10.1007/s40866-019-0060-6

[4] U. Sultana, A.B. Khairuddin, M.M. Aman, A.S. Mokhtar, and N. Zareen. "A review of Optimum DG Placement Based on Minimization of Power Losses and Voltage Stability Enhancement of Distribution System." *Renewable and Sustainable Energy Reviews*. vol. 63, pp. 363–378, 2016. https://doi.org/10.1016/j.rser.2016.05.056

[5] H. Doagou-Mojarrad, G.B. Gharehpetian, H. Rastegar, and J. Olamaei. "Optimal Placement and Sizing of DG (Distributed Generation) Units in Distribution Networks by Novel Hybrid Evolutionary Algorithm." *Energy*. vol. 54, pp. 129–138, 2013. https://doi.org/10.1016/j.energy.2013.01.043

[6] Y. Latreche, H.R.E.H. Bouchekara, F. Kerrour, K. Naidu, H. Mokhlis, and M. S. Javaid. "Comprehensive Review on the Optimal Integration of Distributed Generation in Distribution Systems." *Journal of Renewable and Sustainable Energy*. vol. 10, no. 5, pp. 1–33. 2018. https://doi.org/10.1063/1.5020190

[7] Z. Abdmouleh, A. Gastli, L. Ben-Brahim, M. Haouari, and N.A. Al-Emadi. "Review of Optimization Techniques Applied for the Integration of Distributed Generation From Renewable

Energy Sources." *Renewable Energy*. vol. 113, pp. 266–280, 2017. https://doi.org/10.1016/j.renene.2017.05.087

[8] M.J. Hadidian-Moghaddam, S. Arabi-Nowdeh, M. Bigdeli, and D. Azizian. "A Multi-Objective Optimal Sizing and Siting of Distributed Generation Using Ant Lion Optimization Technique", *Ain Shams Engineering Journal*. vol. 9, no. 4, pp. 2101–2109, 2018. https://doi.org/10.1016/j.asej.2017.03.001

[9] S.J.A. Hosseini, M. Moradian, H. Shahinzadeh, and S. Ahmadi. "Optimal Placement of Distributed Generators With Regard to Reliability Assessment Using Virus Colony Search Algorithm." *International Journal of Renewable Energy Research*. vol. 8, no. 2, pp. 714–723, 2018.

[10] Y.A Rahman, S. Manjang, A. Yusran, and A. Ilham. "Distributed Generation's Integration Planning Involving Growth Load Models by Means of Genetic Algorithm." *Archives of Electrical Engineering*. vol. 67, no 3, pp. 667–682, 2018. https://doi.org/10.24425/123671

[11] I.B. Hamida, S.B. Salah, F. Msahli, and M.F. Mimouni. "Optimal Network Reconfiguration and Renewable DG Integration Considering Time Sequence Variation in Load and DGs" *Renewable Energy*. vol. 121, pp. 66–80, 2018. https://doi.org/10.1016/j.renene.2017.12.106

[12] S. Barik, and D. Das. "Determining The Sizes of Renewable DGs Considering Seasonal Variation of Generation and Load and Their Impact on System Load Growth." *IET Renewable Power Generation*. vol. 12, no. 10, pp 1101–1110, 2018. https://doi.org/10.24425/123671

[13] S. Ravindran, and T.A.A. Victoire, "A Bio-Geography-Based Algorithm for Optimal Siting and Sizing of Distributed Generators With an Effective Power Factor Model." *Computers & Electrical Engineering*. vol. 72, pp 482–501, 2018. https://doi.org/10.1016/j.compeleceng.2018.10.010

[14] H. Mehrjerdi. "Simultaneous Load Leveling and Voltage Profile Improvement in Distribution Networks by Optimal Battery Storage Planning." *Energy*. vol. 181, pp 916–926, 2019. https://doi.org/10.1016/j.energy.2019.06.021

[15] V.V.S.N. Murty, and A. Kumar. "Optimal DG Integration and Network Reconfiguration in Microgrid System With Realistic Time Varying Load Model Using Hybrid Optimization." *IET Smart Grid*. vol. 2, no. 2, pp 192–202, 2019. https://doi.org/10.1049/iet-stg.2018.0146

[16] M. Sabri, A. Ghallaj, H. Sheikhbaglou, and D. Nazarpour. "Optimal Multi-Indices Application of Distributed Generations in Radial Distribution Networks Based on Moth-Flame Optimizer."

*Russian Electrical Engineering*. vol. 90, no. 3, pp. 277–284, 2019. https://doi.org/10.1016/j.asej.2020.07.009

[17] S. Kumar, K.K. Mandal, and N. Chakraborty. "Optimal DG Placement by Multi-Objective Opposition Based Chaotic Differential Evolution for Techno-Economic Analysis." *Applied Soft Computing Journal*. vol. 78, pp. 70–83 2019. https://doi.org/10.1016/j.asoc.2019.02.013

[18] S. Kumar, K.K. Mandal, and N. Chakraborty. "A Novel Opposition-Based Tuned-Chaotic Differential Evolution Technique for Techno-Economic Analysis by Optimal Placement of Distributed Generation."*Engineering Optimization*. vol. 51, pp. 1–20, 2019. https://doi.org/10.1080/0305215X.2019.1585832

[19] G. Deb, K. Chakraborty, and S. Deb. "Spider Monkey Optimization Technique-Based Allocation of Distributed Generation for Demand Side Management." *International Transactions on Electrical Energy Systems*. vol. 29, pp. 1–17, 2019. https://doi.org/10.1002/2050-7038.12009

[20] A. Uniyal, and S. Sarangi. "Optimal Network Reconfiguration and DG Allocation Using Adaptive Modified Whale Optimization Algorithm Considering Probabilistic Load Flow." *Electric Power Systems Research*. vol. 192, e106909, 2020. https://doi.org/10.1016/j.epsr.2020.106909

[21] Z. Ullah, M.R. Elkadeem, S. Wang, and S.M.A, Akber. "Optimal Planning of RDS Considering PV Uncertainty With Different Load Models Using Artificial Intelligence Techniques." *International Journal of Web and Grid Services*. vol. 16, no. 1, pp 63–80, 2020. https://doi.org/10.1504/IJWGS.2020.106126

[22] A.A. Saleh, T. Senjyu, S. Alkhalaf, M.A. Alotaibi, and A.M. Hemeida. "Water Cycle Algorithm for Probabilistic Planning of Renewable Energy Resource, Considering Different Load Models." *Energies*. vol. 13, no. 21, 5800, 2020. https://doi.org/10.3390/en13215800

[23] A. Aid. "Allocation of Distributed Generations in Radial Distribution Systems Using Adaptive PSO and Modified GSA Multi-Objective Optimizations." *Alexandria Engineering Journal*. vol. 59, no. 6, pp. 4771–4786, 2020. https://doi.org/10.1016/j.aej.2020.08.042

[24] S. Settoul, M. Zellagui, and R. Chenni. "A New Optimization Algorithm for Optimal Wind Turbine Location Problem in Constantine City Electric Distribution Network Based Active Power Loss Reduction." *Journal of Optimization in Industrial Engineering*. vol. 14, no. 2, 13–22, 2021. https://doi.org/10.22094/joie.2020.1892184.1725

[25] M. Zellagui, A. Lasmari, S. Settoul, R. A. El-Sehiemy, C.Z. El-Bayeh, and R. Chenni. "Simultaneous Allocation of Photovoltaic DG and DSTATCOM for Techno-Economic and Environmental Benefits in Electrical Distribution Systems at Different Loading Conditions Using Novel Hybrid Optimization Algorithms." *International Transactions on Electrical Energy Systems*. vol. 31, no. 8, e12992, 2021. https://doi.org/10.1002/2050-7038.12992

[26] A. Fathy, D. Yousri, A.Y. Abdelaziz, and H.S. Ramadan. "Robust Approach-Based Chimp Optimization Algorithm for Minimizing Power Loss of Electrical Distribution Networks Via Allocating Distributed Generators." *Sustainable Energy Technologies and Assessments*. vol. 47, 101359, 2021. https://doi.org/10.1016/j.seta.2021.101359

[27] S. Settoul, R. Chenni, M. Zellagui, and H. Nouri. "Optimal Integration of Renewable Distributed Generation Using the Whale Optimization Algorithm for Techno-Economic Analysis." *Lecture Notes in Electrical Engineering*. vol. 682, pp. 513–532, 2021. https://doi.org/10.1007/978-981-15-6403-1_35

[28] M. Zellagui, S. Settoul, A. Lasmari, C.Z. El-Bayeh, R. Chenni, and H. A. Hassan. "Optimal Allocation of Renewable Energy Source Integrated-Smart Distribution Systems Based on Technical-Economic Analysis Considering Load Demand and DG Uncertainties." *Lecture Notes in Networks and Systems*. vol. 174, pp. 391–404, 2021. https://doi.org/10.1007/978-3-030-63846-7_37

[29] A. Ramadan, M. Ebeed, S. Kamel, A.Y. Abdelaziz, and H.H. Alhelou. "Scenario-Based Stochastic Framework for Optimal Planning of Distribution Systems Including Renewable-Based DG Units." *Sustainability*. vol. 13, no. 6, 3566, 2021. https://doi.org/10.3390/su13063566

[30] A.A. Abou El-Ela, R.A. El-Sehiemy, E.S. Ali, and A.M. Kinawy. "Minimization of Voltage Fluctuation Resulted From Renewable Energy Sources Uncertainty in Distribution Systems." *IET Generation, Transmission & Distribution*. vol.13, no. 12, pp 2339–2351, 2019. https://doi.org/10.1049/iet-gtd.2018.5136

[31] A.A. Elsakaan, R.A. El-Sehiemy, S.S. Kaddah, and M.I. Elsaid. "Optimal Economic-Emission Power Scheduling of RERs in MGs With Uncertainty." *IET Generation, Transmission & Distribution*. vol.14, no.1, pp 37–52, 2020. https://doi.org/10.1049/iet-gtd.2019.0739

[32] Z.M. Salameh, B.S. Borowy, and A.R.A. Amin. "Photovoltaic Module-Site Matching Based on the Capacity Factors."

*IEEE Transactions on Energy Conversion*. vol. 10, no. 2, pp 326–332, 1995. https://doi.org/10.1109/60.391899

[33] J.H. Teng, S.W. Luan, D.J. Lee, and Y.Q. Huang. "Optimal Charging/Discharging Scheduling of Battery Storage Systems for Distribution Systems Interconnected with Sizeable PV Generation Systems." *IEEE Transactions on Power Systems*. vol. 28, no. 2, pp 1425–1433, 2013. https://doi.org/10.1109/TPWRS.2012.2230276

[34] D.Q. Hung, N. Mithulananthan, and K.Y. Lee. "Determining PV Penetration for Distribution Systems with Time Varying Load Models." *IEEE Transactions on Power Systems*, vol. 29, no. 6, pp 3048–3057, 2014. https://doi.org/10.1109/TPWRS.2014.2314133

[35] Y.M. Atwa, E.F. El-Saadany, M.M.A. Salama, and R. Seethapathy. "Optimal Renewable Resources Mix for Distribution System Energy Loss Minimization." *IEEE Transactions on Power Systems*. vol. 25, no. 1, pp 360–370, 2010. https://doi.org/10.1109/TPWRS.2009.2030276

[36] D.K. Khatod, V. Pant, and J. Sharma. "Evolutionary Programming Based Optimal Placement of Renewable Distributed Generators." *IEEE Transactions on Power Systems*. vol. 28, no. 2, pp 683–695, 2013. https://doi.org/10.1109/TPWRS.2012.2211044

[37] A. Soroudi, M. Aien, and M. Ehsan. "A Probabilistic Modelling of Photo Voltaic Modules and Wind Power Generation Impact on Distribution Networks." *IEEE System Journal*. vol. 6, no. 2, pp 254–259, 2012. https://doi.org/10.1109/JSYST.2011.2162994

[38] S. Eftekharnejad, V. Vittal, G.T. Heydt, B. Keel, and J. Loehr. "Impact of Increased Penetration of Photovoltaic Generation on Power Systems." *IEEE Transactions on Power Systems*. vol. 28, no. 2, pp 893–901, 2013. https://doi.org/10.1109/TPWRS.2012.2216294

[39] M.H. Haque. "Evaluation of Power Flow Solutions with Fixed Speed Wind Turbine Generating Systems." *Energy Conversion and Management*. vol. 79, pp 511–518, 2014. https://doi.org/10.1016/j.enconman.2013.12.049

[40] N. Belbachir, M. Zellagui, A. Lasmari, C.Z. El-Bayeh, and B. Bekkouche. "Optimal PV Sources Integration in Distribution System and its Impacts on Overcurrent Relay-Based Time-Current-Voltage Tripping Characteristics." ATEE, *International Symposium on Advanced Topics in Electrical* Engineering., Bucharest, Romania, 25–27 March 2021. https://doi.org/10.1109/ATEE52255.2021.9425155

[41] N. Belbachir, M. Zellagui, S. Settoul, C.Z. El-Bayeh and B. Bekkouche. "Simultaneous Optimal Integration of Photovoltaic Distributed Generation and Battery Energy Storage System in Active Distribution Network Using Chaotic Grey Wolf Optimization." *Electrical Engineering & Electromechanics*. vol. 2021, no. 3, pp. 52–61, 2021. https://doi.org/10.20998/2074-272X.2021.3.09

[42] A. Bayat, and A. Bagheri. "Optimal Active and Reactive Power Allocation in Distribution Networks Using a Novel Heuristic Approach." *Applied Energy*. vol. 233, pp. 71–85, 2019. https://doi.org/10.1016/j.apenergy.2018.10.030

[43] Y. Latreche, H.R.E.H. Bouchekara, F. Kerrour, K. Naidu, H. Mokhlis, and M.S. Javaid. "Comprehensive Review on the Optimal Integration of Distributed Generation in Distribution Systems." *Journal of Renewable and Sustainable Energy*. vol. 10, 5303, 2018. https://doi.org/10.1063/1.5020190

[44] N. Belbachir, M. Zellagui, A. Lasmari, C.Z. El-Bayeh, and B. Bekkouche. "Optimal Integration of Photovoltaic Distributed Generation in Electrical Distribution Network Using Hybrid Modified PSO Algorithms." *Indonesian Journal of Electrical Engineering and Computer Science*. vol. 24, no. 1, pp. 50–60, 2021. https://doi.org/ 10.11591/ijeecs.v24.i1.pp50-60

[45] N. Belbachir, M. Zellagui, S. Settoul, and C.Z. El-Bayeh. "Multi-Objective Optimal Renewable Distributed Generator Integration in Distribution Systems Using Grasshopper Optimization Algorithm Considering Overcurrent Relay Indices." MPS, *International Conference on Modern Power Systems.*, Cluj, Romania, 16–17 June 2021. https://doi.org/10.1109/MPS52805.2021.9492567

[46] M. Zellagui, N. Belbachir, and C.Z. El-Bayeh. "Optimal Allocation of RDG in Distribution System Considering the Seasonal Uncertainties of Load Demand and Solar-Wind Generation Systems." EUROCON, *International Journal of Smart Technology and Learning.*, Lviv, Ukraine, 6–8 July 2021. https://doi.org/10.1109/EUROCON52738.2021.9535617

# 4

# Security Challenges in Smart Grid Management

**S. Nithya[1], K. Vijayalakshmi[2], and M. Parimala Devi[3]**

[1,2]Assistant Professor, Department of EEE, SRMIST, Ramapuram, Chennai
[3]Associate Professor, Department of ECE, Velalar College of Engineering and Technology, Erode, Tamil Nadu
E-mail: nithisavidhina@gmail.com; vijayalk1@srmist.edu.in; parimaladevi.vlsi@gmail.com

## Abstract

The transformation from the power grid to the smart grid has become one of the greatest technological evolutions in the last few years. The reliability of electric supply is the most credible feature wanted by both the developing and developed countries. The smart grid enables us to reduce the emission of carbon as there is an integration of renewable energy resources in it. The smart grid is a system that monitors and manages energy use through a network of computers and power infrastructures. The smart grid is an interlink combining different domains and it has to withstand the following threats such as natural disasters, intentional attacks, financial risk hackers, transportation, storage, and all other personal or information leak problems. It also includes Smart meter problems such as privacy invasion, reliability, overcharging, hacking and other health issues. To overcome the above issues traditional security analysis approaches such as certification and internal quality assurance are essential, but they fall short when it comes to critical systems. To evaluate smart grid systems, industry and government must be innovative. Even though services such as Google Power Meter are opt-in, customers have little control over how power data is used by utility providers. To address these issues with smart grid applications, various algorithms can be proposed.

## 4.1 Introduction

A smart grid is an IOT-enabled application that permits electricity and information sharing between utilities and their customers. The bidirectional data flow makes the grid "smart". In particular, smart grids include smart meters, renewable energies, and consumer smart devices. It uses technological tools to acquire and act on information, such as relevant data on supplier and customer behavior, in an automated manner to enhance the efficiency, dependability, economics, and sustainability of energy production and distribution.

Smart grids are currently being employed in transmission circuits, from power plants to electricity users in homes and businesses. The term "grid" refers to the networks that transport electricity from power plants to customers. It consists of cables, substations, transformers, and switches among other things. The major benefits are significant increases in energy efficiency on the electrical grid as well as in the homes and businesses of energy consumers.

The central control unit manages all units linked to the traditional intelligent grid efficiently. The central control facility not only improves energy management within the building but also helps to reduce peak-time electricity use. This decrease shows substantial savings in energy.

A smart grid also helps the shift from traditional power to renewable energy easier. The grid simplifies its incorporation into the grid if the system includes a renewable energy source. The smart grid enables greater integration of widely disparate renewable energy sources such as wind and solar.

The smart grid provides a new path to a greener future. It not only provides higher energy benefits, but it also creates new job possibilities for young people. Converting conventional operating units into smart ones capable of communicating with the smart grid, for example, offers up a world of new and intriguing possibilities. With ideas and developments made by younger, dynamic brains, the worldwide market for intelligent instruments is in trend.

## 4.2 The Demand for a Smart Grid

India is surely one of the fastest-growing countries in the world with over one billion inhabitants and a current GDP growth rate of over 8%. Despite its strong economic progress, the country nevertheless faces fundamental challenges such as a power deficit, with over 40% of its rural residents without electricity.

Although India has nearly quadrupled its energy generating capacity in the last decade, adding over 85 GW, its grid systems have lost more than 30 GW of this generated electricity. This has caused great anxiety among many in India's energy business who are concerned about the efficiency of electricity distribution.

The World Resources Institute estimates that the world's largest electrical transfer and distribution loss in India is 27%. This is a massive waste of one of the most ecologically harmful commodities available. These findings lead to the conclusion that India needs new technologies to improve the monitoring and management of energy transmission and distribution.

A smart grid is a computerized electrical infrastructure that allows for the collection and dissemination of data about power use by suppliers and customers. Electricity services will become more dependable, efficient, cost-effective, and ecologically sensitive as a result of this. Smart Grid is a type of smart grid.

## 4.3 Benefits of Smart Grid

- The smart grid reduces operating costs, saves energy, and enhances dependability.
    1. Smart grid applications can more efficiently regulate the power flow. They can detect surges, blackouts, and technical energy losses.
    2. They can also detect waste and fraud, resulting in cost savings that will be passed on to the customers.
    3. Peak loads or variations may be handled instantly and automatically.
- The integration of renewables is improved via a smart grid.
    1. Renewable energy provides healthier living areas and smart grid systems can also assist in this area.
    2. An enhanced energy infrastructure would enable stakeholders, including the acquisition of asset owners, producers, service providers, and local and central government authorities, to manage regionally spread energy, such as wind farms, solar plants, and hydroelectric stations, strategically.

- Saving energy by lowering usage
  1. One of the benefits of smart grids is that they can tell us the energy usage at all times, so that consumers may be better informed about their true consumption.
  2. In addition, the contracted power may be modified with improved consumer monitoring to match the customer's genuine needs. These two variables reduce the consumption of consumers and adapt their contractual capacity to their real requirements.
- Improved customer service and more accurate invoices
  1. Another significant benefit of tele management systems is the increased accuracy of billing. They are always based on actual monthly consumption rather than estimates. saving money over the traditional technique of manual energy meters readings.
  2. Problems can be more easily diagnosed and remedies can be applied faster, increasing customer service, to get installation information remotely.
- Enhanced competition
  1. Having actual load curve data encourages marketing firms to modify their rates based on energy consumption. With more data, marketing firms may develop better offers that are more in line with their consumers' realities, boosting competitive alternatives through a larger range of offers (hourly tariffs, energy packages, etc.).
  2. More competition leads to more competitive pricing, which benefits customers.
- Carbon emission reduction
  1. All of the foregoing advantages include lowering consumption, which lowers CO2 emissions.

As a result, smart grids contribute to a more sustainable future. All of this will have a direct impact on the future integration of electric car charging systems into the electrical grid. As utilities acquire more control over their networks, the adoption of renewable energy technologies becomes easier.

## 4.4 Smart Grid Operation

Smart grids are equipped with sensors that capture and send data. This data allows for the automatic adjustment of electricity flows. Grid managers, who are situated remotely, are kept up to date on the situation in real-time and can react quickly if an issue arises. Furthermore, this grid may connect with any smart meters and, for example, automatically switch on users' home appliances when there is a lot of power in the grid and therefore decrease costs.

Unfortunately, smart grids are using the "IoT" technology to put intelligence and monitoring into each node, unlike traditional grids built merely to distribute the power from the producer to the consumer. The smart grid combines energy and data in two dimensions. The close-to-earth data exchange allows the system to know when and how power is utilized and IoT devices immediately adapt to network circumstances. The services increase the smartness of the electrical grid and make the new technology "self-healing."

The grid is more responsive with IoT and can precisely isolate failures and react utilizing communication diagnostics. Then the power flow may be reconfigured, sections of the system reenergized, and the power infrastructure protected. This implies that outages are shorter and recovery is faster.

## 4.5 Smart Grid Security Challenges

1. A cyber-attack on smart grid systems, for example, might render a metropolis completely black
2. Smart meter security flaws might lead to fraud or data leaks.
3. Potential attackers include criminals who might deactivate alarm systems through organizations of cyber attackers who deal with blackmailing hostile nation-states.
4. Consumers may even be interested in hacking smart meters to save money, attempting to make the smart meter show lower power use than is the case.
5. The placement of such metering equipment in private houses has consequences for residents' privacy.

Physical infrastructure destruction, data poisoning, denial of service, malware, and infiltration are the most common threats [1] to the smart grid's functioning and security. Data breaching and malicious control of personal gadgets and appliances are the most common threats to consumers.

## 4.6 Literature Review

Various smart grid security dangers and difficulties have been identified in [1, 2, 3] and many other research studies, including the potential of adversaries breaching a large volume of sensitive consumer information.

To execute SG, numerous communication methods, electronic power systems, charging stations, etc., which are seen as the core of SG [11–13], will have to be used. Because the SGs and their communication systems have significantly integrated topology, they are more sensitive to cyber-attacks.

Smart grids can be exposed to cyber assaults, relying on a communication network and smart meters [4]. Consequently, adequate defense methods must be implemented [5, 6, 7]. One of the most popular methods for attacking cyber-physical systems is False Data Injection (FDI). For a dependable grid, it is important to detect cyber assaults and related defense methods. For example, FDI and jamming assaults in intelligent grids have been suggested using a rapid detection algorithm [9]. Research [10] has demonstrated that smart metering consumption data shows the overtime use in households of all electrical equipment and allows thieves to deduce from them.

## 4.7 Key Points that Require Special Attention

Smart grids will improve power usage and transmission control aiding consumers, electricity suppliers, and grid controllers. Enhanced operations and services, from the other side, will cost the whole power supply network additional difficulties, particularly in the areas of telecommunications and safety systems. Some have previously been mentioned in earlier sections, while others are new but as important. The sequence in which they are presented has nothing to do with their value.

### 4.7.1 Requirements for Data and Information Security

ICT will serve as the future smart grids' core nervous system. Flows of data and information will overwhelm all spheres. Data security and accessibility will be critical in grid operating elements (such as generation, transmission, and distribution automation). Similarly, confidentiality must be guaranteed throughout transmission and storage of end-user data, such as consumption data or even personal data at billing systems. In some situations, the Smart Grid application determines the safety component of a given data item. In demand response systems, for instance, consumption information from smart meters might be employed.

### 4.7.2 Extensive use of "Smart" Devices

A huge network of electronics and data processing devices will be deployed as part of the smart grid, producing a huge mesh. The most notable example is undoubtedly smart meters and gadgets, as well as the AMI communication network, in general, Substation automation and transformer center smartening, on the other hand, would bring a significant number of IED and associated ICT technologies into the distribution industry. Installing, deploying, and Maintaining a robust and dependable solution would be a huge task for grid operators who are unfamiliar with it and lack the necessary technologies and internal processes. Furthermore, this solution/infrastructure must be safe in terms of all the interconnections in place, procedures, and even the devices themselves.

### 4.7.3 Grid Perimeter and Physical Security

In smart grids, physical security concerns must be given particular attention. The network linkage with DSO and DER Information Networks at the ICT level of households, buildings, and industries will considerably increase the grid security perimeter. Installation in customer residences, buildings, or businesses will not, for example, be directly controlled by DSO or retail suppliers. Transformer centers and distribution substations, on the other hand, are growing more interesting to cyber attackers as the distribution grid becomes more automated. In many situations, transformer centers are not properly safeguarded against physical threats and serve as a point of presence for the DSO's ICT infrastructure. In most transformer centers, the secure facility or locker/closet is housed physically. However, many contractors have a key duplication that permits access to such systems and the lockers are not resistant enough. Anyone who has physical access to these facilities, whether authorized or unauthorized, can access the communication network of DSO

### 4.7.4 Protocols of a Legacy and (in) Secure Communication

Security has never been taken into account in many of the communication protocols presently used for power production, transmission and distribution control, and automation. Many of them began as serial protocols without message authentication. Therefore, all gadgets will accept connections from any other device. that attempts to contact them. Moreover, no protocols use encryption or message integrity protections which expose communication to eavesdropping, hijacking, and handling. Although these vulnerabilities exist for many years, additional factors have improved the true risk. Many ICS providers have started opening up their protocols and issuing protocol specifications to enable interoperable suppliers. Organizations also migrate to common networking protocols, such as TCP/IP (i.e. Modbus/TCP, IEC

104, etc.) or new standardized open protocols like OPC, to save costs and enhance performance. Common historical communication protocols such as IEC 101, although without a security mechanism, are now accessible in a version encapsulated by TCP/IP. Operators will need to be able to deal with them 24 hours a day, seven days a week. Annex II: Smart Grid Security. In the future years, security flaws in smart grids will be addressed by innovative techniques such as the employment of compensating measures such as protocol tunneling. A whole new set of communication protocols has instead evolved to deal with new smart grid applications. This is true for AMI-related protocols like PRIME, Meters & More, DLMS/COSEM, and so on. Luckily, new security methods are being created, including end-to-end authentication and encryption cryptography. Nonetheless, to properly execute the greatest level of security, cryptographic material (such as keys, certificates, and so on) must be maintained efficiently and effectively. As previously indicated, Smart grids will have a high number of smart devices, so it is a sophisticated and challenging procedure to handle cryptographic information.

### 4.7.5 Many Stakeholders and Synergies with other Services

The intelligent grid infrastructure is complicated and it needs a big number of various stakeholders in its basic definition to work together to deliver a functional solution with the physical and logical structure. A limited number of actors have traditionally been in the power system (i.e., bulk generators, TSOs, and DSOs). However, first of all, due to the deregulation of electrical services and the redesign by the smart grid of the power system concept, a huge number of players are devoted to the provision of energy and related value-added services. There are several sorts of parties engaged, but more importantly many more midfielders: end customers, small-scale power companies, energy dealers, state-of-the-art energy service providers, EV companies, etc. It is difficult to quickly coordinate the activities of such a diverse group of stakeholders, each with their organizational processes, business priorities, information communication requirements, standards of reference regulations and the best practices, and so on, for a reliable, secure, and high level of power delivery. The notion of enhanced measurement, on the other hand, will not just influence the electricity industry. Other utilities like gas/heating and water, for example, are expected to adopt smart meters to read and analyze usage data remotely in the future years. AMIs that services reach a company or house. For example, for every smart meter type, a single AMI can be utilized (e.g., gas, heating, water, electricity).

Data is subsequently supplied to the AMI operator's back-office systems (e.g., DSO, energy retailer, gas distributor, etc.). Infrastructure that is adaptable, interoperable, and well-connected is therefore needed which can allow the sharing of information across different utilities. However, this will result in an even more complicated system, which needs to be addressed for safeguarding the intelligent grid not just by all-new power grid operators but also by other utilities.

### 4.7.6 A Lack of Clarity about the Smart Grid Concept and its Security Requirements

With smart grids, a plethora of new technologies and concepts are developing. They are arriving all at once, although the ultimate image of smart grids is not yet clearly defined. As a result, security standards must be developed while taking into account many areas involved and their significance for national security or personal data privacy. As a result, creating a reference design that outlines the essential features of smart grids is critical. In addition, risk assessment methodologies, safety best practices, and standards must be developed as an essential aspect in addressing system interoperability and security.

### 4.7.7 Lack of Awareness among Smart Grid Stakeholders:

Many of the preceding difficulties will be impossible to overcome unless manufacturers, grid operators, and other stakeholders make a genuine commitment. In reality, C-level people are sensitive to the cyber security concerns that they face in the short and long term as one of the most challenging problems of crucial infrastructures. This is especially true for smart grids, where ICT will play a critical role. As a result, public awareness campaigns are required. Requesting compliance with particular security standards, performing risk evaluations, conducting penetration testing, promoting professional events, and actively involving CSIRTs/CERTs are some examples of such efforts.

### 4.7.8 Supply Chain Security

The vulnerability of current supply chains is one of the major security concerns highlighted while tackling CI security. There would be in reality a very real danger that a particular step of the supply chains, including the microchips, embedded software, SCADA and control applications, and operating systems, get compromised by hostile agents. These hostile agents may change the electrical circuitry or replace fake components with altered circuits. In the firmware of various IEDs, controllers, or smart meters the backdoor, logic bombs, and other malicious software might also be inserted. This may lead to hostile nations or terrorists, or any other danger to a backdoor that could control the data systems impacted remotely or employ pre-installed logic bombs which could inflict dreadful damage. For intelligent grid protection, the security of the supply chain is crucial. This applies particularly to applications and components which may be of national security importance. Electronic components and applications must be designed, manufactured, assembled and distributed, and properly regulated. It is necessary to take into consideration the economic dimension of the problem and to set economically viable security objectives. The key to tackling the malicious firmware problem is to protect the worldwide supply chain.

### 4.7.9 Encourage the Interchange of Risk, Vulnerability, and Threat Information

The smart grid's security features could be a useful tool for quickly sharing information security management best practices and solutions, such as security incidents and response strategies, common security flaws and vulnerabilities and remedial actions taken, priority actions to secure ICS, and so on. It is critical to first build a network of connections at all levels, and then to make the appropriate procedures to foster confidence and facilitate information sharing. During these early stages, public agencies such as national critical infrastructure protection centers might play a vital role.

### 4.7.10 International Cooperation

While the priority objectives/objectives of intelligent grids throughout the world may differ, there are many similar elements. It might thus be quite useful for the European nations to share their ideas and perspectives with other areas worldwide, such as America, Japan, Australia, Canada, India, etc. (i.e., safety priorities, needs, methods, etc.). International collaboration would also be necessary to make Europe a world leader in formulating and developing security standards that influence the grid in the future. For manufacturers, the major reference is for international organizations like IEEE, IEC, ISO, or ITU, and their standards and technical documents while building their systems and applications.

### 4.7.11 Utility Security Management

To protect smart Grids, system suppliers play a crucial role. Protecting a power grid would be a very tough task if the product is integrated without security features or if safety standards are not taken into account throughout the development cycle. The safety of electricity grids, however, does not just depend on safe goods. The procedure depends heavily on power utilities like DSO and TSOs. It is a constant process. The safety of its current systems, notably ICS and future infrastructure installations must be evaluated by the network operators. They also need to evaluate and plan additional investments to strengthen the safety position, build security policies and procedures, train staff, and, not least, develop an information security management system to assure the achievement and continuous improvement of all of the

## 4.8 Smart Grid Security Policies

Policies often include information protection requirements such as confidentiality, integrity, and availability. The researchers included accountability as a separate policy, even though it may be seen as an issue of integrity because it is necessary for the smart grid.

### 4.8.1 Confidentiality

Keeping approved limitations on information access and disclosure in place, including safeguards for personal privacy and proprietary information. The attribute of not disclosing sensitive information to unauthorized persons, institutions, or procedures.

### 4.8.2 Integrity

Ensure that information is not modified or destroyed incorrectly, particularly ensure that information is not repudiated or authentic. Data integrity is the attribute, not unauthorized data modification. The data integrity 104 is covered Abdullah Umar, Yash Pal Singh, and Adla Sanober are data that is not modified or destroyed unauthorized or undiscovered in storage and processing processes and during transit.

### 4.8.3 Availability

Ensure that access and usage of information are prompt and trustworthy.

### 4.8.4 Accountability

Is the security objective requires a company to be tracked exclusively to the entity? This promotes non repudiation, dissuasion, isolation of faults, detection and prevention of intrusion, recovery of post action, and legal action.

## 4.9 Corrective Strategies to Improve Smart Grid Protection

It is necessary to identify all links to smart grid systems. Disconnect unneeded connections to the intelligent grid network. To ensure the maximum level of safety, it is advisable to disconnect data from other networks because of the significance of smart grid data, especially when the link provides a means to or from the Internet. Security assessments, vulnerability analyses, and security tests should be carried out on any remaining smart grid connections to safeguard the grid and assist risk management procedures using a robust approach to smart grid routes. Firewalls and intrusion detection systems (IDS) must be installed. Because smart grid control servers were created on an open-source platform or even on a commercial operating system, attacks may expose default network services. As a result, to increase security in smart grid systems, unneeded services and network daemons, such as billing systems or automatic meter readings, email, and other Internet services, must be removed [21]. Encryption is being used in the smart grid to safeguard data. However, if it is not managed properly, it can be time and expense-intensive, resulting in increased storage and bandwidth use. Another complicated problem that requires greater attention and effort is key management. In cryptography, there are two sorts of keys: symmetric keys and asymmetric keys. In the symmetric key, cryptographic communication is encrypted and decrypted using the same key. There are two sorts of key pairs used in asymmetric keys. One key encrypts the communication, while the other decrypts it [23].

## 4.10 Important Areas to Safeguard the Grid

### 4.10.1 Powerful Digital Identities

Each connected device should have its own distinct digital identity that can be used to identify it. If each device has its own unique identity, even if a device is hacked, only that device is affected. Among existing identity management methods, the public key infrastructure (PKI) is most likely the best option. A PKI system enables flexible policies and processes for handling digital certificates over a wide range of domains. A PKI certificate is used to associate a device/user identification with its public key. The public key may then be used to validate the identification and execute security activities. The existing PKIS, on the other hand, was designed for a shared information system and was unable to meet the tight SG restrictions. The SG is a large infrastructure made up of several subsystems for distributed power production, transmission, and distribution. Because we automate the connections among various subsystems while handling numerous resource requirements in real-time (for example, the latency of communication, computing delay, and flow bandwidth), identity management and processing requirements are quite different from traditional information systems. As a result, we must create PKIs to satisfy the specific SG requirements.

### 4.10.2 Mutual Verification

This implies that any two linked devices may only "talk" to each other after completing a digital challenge that only those two devices are aware of. To secure communication between smart grid and service providers, symmetric key-based cryptographic methods, such as the advanced encryption standard, can be employed. Each pair of communicating parties is given a unique secret key under these methods. Each secret key is only valid for secure communication between each pair of devices; it cannot be used to communicate securely with other network devices. Developing a lightweight and quick group-based authentication method that performs a full authentication procedure on the initial handshake and then uses the authentication token for subsequent handshakes; lowering energy usage through lowering communication overhead.

### 4.10.3 Encryption

When transferred between devices and not moving to protect them against being disturbed, data should always be encrypted. Intelligent grids were vulnerable to several assaults. The assaults are described in detail below [1].

1. Eavesdropping Attack:

Eavesdropping occurs when an adversary intercepts communication between a smart meter and the grid.

2. Analysis of Traffic:

In a traffic analysis attack, the attacker attempts to analyze the message or its communication pattern.

3. Replay Attack:

The attacker's assault authenticates the user in the network based on the old communication between authenticating parties.

4. The Man-in-the-Middle Attack:

An attacker attempts to change the message or remove the message content in a man-in-the-center attack before it is transmitted to the recipient.

5. Denial-of-service Attack:

In this assault, the attacker overwhelmed the target system's resources or bandwidth. Authenticated users, therefore, do not have access to network devices and resources.

6. The Malware attack:

An attacker is adding harmful programs in this attack such as worms, viruses, trojan horses, which are used for nefarious tasks such as stolen, deleted, altered, and encrypted sensitive information.

Various encryption and steganography methods in the smart grid are employed to overcome these assault [24, 25]. Only authenticated users with a private key may un-code the secret message in the cryptography algorithm in this way. The network's privacy must be communicated in a significant way to key-based attacks. The key on the steganography algorithm of the network is therefore secured. The algorithm for steganography hides the secret data existence and users just authenticate the private decryption key.

### 4.10.4 Continuous Security Updates

A secure smart grid should constantly grow and frequently update its safety by updating keys and challenges to digital mutual authentication every two to three years. With the wide range of ubiq- ubiq, remotely accessible networked devices used to monitor and control the grid, vulnerabilities including new smart devices will be easier to find for assailants accessing various parts of the grid, thus significantly increasing the attack surface of the power grid. In addition, the attackers can use the additional functionality from new devices, such as the multiple smart meter remote connection feedback option, which results in significant security concerns for the system. The implications of modern information technology and communication technologies on the power grid as a cyber-physical system are being studied.

There is also a slew of additional privacy issues associated with smart grid deployments. Fine-grained energy consumption data gathered by emerging technologies such as smart meters, smart appliances, and electric automobiles will pose significant privacy risks to customers, owing to the bigger scale, the more detailed, and more frequent gathering of usage data.

A secure SG must therefore be authentic, accessible, and confidential; have a high degree of integrity and efficiency, acceptability, dependability, robustness, adaptability, resilience (self-healing), and support developing electrical markets. Furthermore, it must have a high level of accessibility and observability, controllability, and allow the increased deployment of renewable energy sources as well as enhanced system performance while delivering cost savings.

In the suggested security frameworks different researchers have essentially used numerous information and communication technologies (ICT), including the Internet, ZigBee, wireless mesh networks, WiMax, 3G, 4G, Wi-Fi, and Bluetooth. The appropriate approach was used to either optimize or model. Additionally, the aim is to increase the distribution, flexibility in operation, and resiliency of the electricity grid to various security threats through emerging technologies such as micro grid, virtual plants (VPPs), distributed intelligence, smart metering, and demand response technologies, and distributed and renewable energy sources.

## 4.11 Conclusion

The intelligent grid defines a modern technical field of resilience, performance, and efficiency throughout the electricity sector. These systems also manage sensitive data, which is critical to the industry, therefore the networks' communications must be protected. As a result, security functions are equally essential to the smart grid for IT systems that use traditional networks, for example, cryptographic. Cryptographic approaches nonetheless require secured keys to safeguard smart grid communications. This ensures that the network meets the three key safety requirements: genuineness, integrity, and secrecy. Safety is an endless war of wits, pitting attackers against asset owners. The cyber-security of the Smart Grid is no exception. The Smart Grid is a broad and complicated system that is spread across a broad area geographically and it will be difficult to secure from intruders.

## References

[1] S. Goel, Y. Hong Y "Security Challenges in Smart Grid Implementation," In: (2015) *Smart GridSecurity. SpringerBriefs in Cybersecurity. Springer, London*. https://doi.org/10.1007/978-1-4471-6663-4_1

[2] B. Khelifa and S. Abla. "Security Concerns in Smart Grids: Threats, Vulnerabilities and Counter Measures", *Paper presented at the 3rd International Renewable and Sustainable Energy Conference (IRSEC)*, 2015 (2015), pp. 1–6

[3] T. Li, J. Ren, and X. Tang. "Secure Wireless Monitoring and Control Systems for Smart Grid and Smart Home." in *IEEE Wireless Communications*, vol. 19, no. 3, pp. 66–73, June 2012, DOI: 10.1109/MWC.2012 .6231161.

[4] Y. Mo, T.H.J.Kim, K. Brancik, D. Dickinson, H. Lee, A.Perrig, and B. Sinopoli, B. "Cyber-Physical Security of Smart Grid Infrastructure." *Proceedings IEEE* 100(1), pp. 195–209 (2012). https://doi.org/10.1109/JPROC.2011.2161428

[5] H. He, and J. Yan. "Cyber-Physical Attacks and Defenses in the Smart Grid: A Survey." *IET Cyber-Physical System Theory Application*, 1(1), pp. 13–27 (2016). https://doi.org/10.1049/iet-cps.2016.0019

[6] Y. Liu, P. Ning, and M.K. Reiter. "False Data Injection Attacks Against State Estimation in Electric Power Grids." *In: Proceedings of the 16th ACM Conference on Computer and Communications Security*, pp. 21–32 (2009). https://doi.org/10.1145/1952982.1952995

[7] D.B. Rawat, and C. Bajracharya. "Cyber Security for Smart Grid Systems: Status, Challenges, and Perspectives."*SoutheastCon* 2015, pp. 1–6 (2015). https://doi.org/10.1109/SECON.2015.7132891

[8] Y.Z. Lun, A. D'Innocenzo, I. Malavolta, and M.D. Di Benedetto. " Cyber-Physical Systems Security: A Systematic Mapping Study." pp. 1–32 (2016)

[9] M.N. Kurt, Y. Yilmaz, and X. Wang. "Real-Time Detection of Hybrid and Stealthy Cyber-Attacks in Smart Grid." *IEEE Transactions* on *Information Forensics* and *Security*. 14(2), pp. 498–513 (2018). https://doi.org/10.1109/TIFS.2018.2854745

[10] Quinn and L. Elias. " Privacy and the New Energy Infrastructure." February 15, 2009, SSRN.

[11] A. Moghadasi, M. Moghaddami, A. Anzalchi, A. Sarwat, and O. A. Mohammed,"Prioritized Coordinated Reactive Power Control of Wind Turbines Involving Statcom Using Multi-Objective Optimization." in

2016, *IEEE/IAS 52nd Industrial and Commercial Power Systems Technical Conference (I CPS)*, May 2016, pp. 1–9.

[12] M. Moghaddami, A. Anzalchi, A. Moghadasi, and A. Sarwat, "Pareto Optimization of Circular Power Pads for Contactless Electric Vehicle Battery Charger." in 2016, *IEEE Industry Applications Society Annual Meeting*, Oct 2016, pp. 1–6.

[13] M. Moghaddami, A. Anzalchi, and A.I. Sarwat, "Finite Element Based Design Optimization of Magnetic Structures for Roadway Inductive Power Transfer Systems." in 2016, *IEEE Transportation Electrification Conference and Expo (ITEC)*, June 2016, pp. 1–6.

[14] ENISA. "Smart Grid Security—Annex II. Security Aspects of the Smart Grid." 2012-04-25. www.enisa.europa.eu/activities/Resilience-and-CIIP/critical-infrastructure-and-services/smartgrids-and-smart-metering/ENISA_Annex%20II%20-%20Security%20Aspects%20of%20Smart%20Grid.pdf

[15] S. Wang, L. Cui, J. Que, D. Choi, X. Jiang, S. Cheng, and L. Xie. "A Randomized Response Model for Privacy-Preserving Smart Metering." *IEEE Transactions Smart Grid* 3(3), 2012, pp.1317–1324.

[16] NRG Expert Chapter 13—Security. Global smart grid report, 2011, pp. 172–179.

[17] J. Steven, G. Peterson, and D. Frinckle. "Smart-Grid Security Issues." *IEEE Security and Privacy*,8(1), 2010, pp.81–85.

[18] J. Shapiro. "Cyber Security and Smart Grid." *In: Presentation at the clean air through energy efficiency (CAFEE) conference, Dallas*, Nov 2011, pp. 8–11.

[19] F. Aloul, A.R. Al-Ali, R. Al-Dalky,M. Al-Mardini, and W. El-Hajj. "Smart Grid Security: Threats, Vulnerabilities, and Solutions." *International Journal of Smart Grid and Clean Energy* 1(1), 2012, pp.1–6.

[20] Echelon. "Protect Your Grid: Echelon's Answer for a Safe, Secure Grid." *White paper*,2012.

[21] A. Monticelli. "State Estimation in Electric Power Systems: A Generalized Approach Springer." Berlin, 1999.

[22] S. Mohagheghi, J. Stoupis, and Z. Wang, "Communication Protocols and Networks for Power Systems – Current Status and Future Trends." *In Proceedings of Power Systems Conference and Exposition (PES '09)*, 2009.

[23] A. Aggarwal, S. Kunta, and P.K. Verma, "A Proposed Communications Infrastructure for the Smart Grid." *In Proceedings of Innovative Smart Grid Technologies Conference Europe (ISGT)*, 2010.

[24] O.G. Abood, M.A. Elsadd, and S.K. Guirguis. "Investigation of Cryptography Algorithms Used for Security and Privacy Protection in Smart Grid." *In the 2017 Nineteenth International Middle East Power Systems Conference (MEPCON)*, Dec 2017, pp. 644–649. IEEE.

[25] M.R. Asghar, G. Dán, D. Miorandi, and I. Chlamtac. "Smart Meter Data Privacy: A Survey." *IEEE Communications Surveys & Tutorials*, 19(4), 2017, pp.2820–2835.

5

# Differential Protection Scheme along with Backup Blockchain System for DC Microgrid

**E. Fantin Irudaya Raj[1,*], K. Manimala[2], and M. Appadurai[3]**

[1,*]Assistant Professor, Department of Electrical and Electronics Engineering
[2]Professor, Department of Electrical and Electronics Engineering
[3]Assistant Professor, Department of Mechanical Engineering
[1,2,3]Dr. Sivanthi Aditanar College of Engineering, Tamilnadu, India
E-mail: fantinraj@gmail.com; smonimala@gmail.com;
appadurai86@gmail.com
*Corresponding Author

## Abstract

The advances in Intelligent Smart Meters, the Internet of Things (IoT), and communication technologies converge the conventional power grids into Smart Electric Grids. In the conventional one, the communication was unilateral. However, in a Smart Electric Grid platform, all system members, from different generating units to different consumers, can communicate bilaterally in real-time via modern technologies. At the same time, the new Smart Electric Grid has some challenges. There might be a chance for personal information leakage, cyber-attack, and managing small customers or prosumers are the few crucial challenges in Smart Grid. Introducing Blockchain technology to the Smart Electric Grid is an attractive solution to the problems mentioned earlier. The Blockchain platform comprises cryptographic security measures, a decentralized consensus mechanism, and a distributed ledger. In the present work, protection schemes of the DC microgrids based upon Blockchain technology are discussed in detail. Compared with AC microgrids, DC microgrids provide more efficiency and are more suited for Renewable energy sources. It is due to the reduction in conversion stages and the lack of skin effect. The differential protection scheme along with a backup Blockchain system for DC microgrid is proposed. Detection and isolation of fault with and without cyber-attack are explained with simulation results. Differential protection schemes detect and isolate the high impedance faults, and the Blockchain backup system enhances the trustworthiness of the proposed scheme by mitigating the communication failure impact.

**Keywords:** DC Microgrid; Blockchain, Differential Protection Scheme, Relay, Cyber-attack, Smart grid.

## 5.1 Introduction

As electronic loads, energy storage devices, and electric vehicles become more prevalent in the power system, DC microgrids are becoming more popular. The adoption of renewable energy sources also enhances the use of these microgrids [1]. In comparison to AC microgrids, these will be more efficient and have better regulation. The conversion steps are minimized between the producing point and the DC loads, and there is no skin effect. The modern electronic power converters maintain a high level of stability while fulfilling the stringent and modern power quality requirements [2]. Furthermore, converter performance efficiency has notably increased to a similar level to AC power transformers, making DC microgrids suitable for use in power systems.

To enable a secure, safe, and reliable DC microgrid, a significant effort in developing a standardized protection plan for DC microgrids is necessary [3]. Additionally, converters with capacitive filters at the input and output in DC microgrids result in huge transient currents during a fault. As a result of the high rising current during a fault, converters are more likely to be damaged. Therefore, faults in DC microgrids should always be isolated using the quickest possible protection technique. The ring architecture is the best framework for DC microgrids because of their increased reliability and resiliency. The conventional protection approach would not operate in these systems due to the bidirectional current flow. As a result, a communication-based strategy is among the important protection approaches necessary for the localization and fault identification in DC microgrids with ring architecture [4].

Different DC applications, including aviation, marine, and HVDC transmission, are urged to use communication-based and high-speed differential protection schemes [5]. These protection measures, on the other hand, are not immediately applicable in low voltage DC microgrids. Several types of protection equipment, like circuit breakers, processors, relays, and synchronized communication lines, must be installed in the DC microgrid to establish a high-speed differential protection scheme [6]. The differential relay method serves as the primary safety mechanism in the event of a communication breakdown, with a voltage-based protection approach serving as a backup [7].

Other intelligent and modern protection systems necessitate the contact and data sharing between protection devices to upgrade the value of the system's actual voltage and current at that time to isolate and detect fault currents. Furthermore, renewable energy sources are examined to determine if they should be ignored or included in the fault contribution [8]. As a result, DC microgrids must use communication linkages to provide an adaptive and real-time protection strategy. In DC microgrids, the physical and cyber components' interdependencies outnumber the control mechanisms [9]. In a close cyber-physical system, this problem is defined as a malfunction or dilemma in the cyber-physical system that causes adverse repercussions [10].

Data transmission protocols for DC microgrids have been developed after years of research to address this issue. The IEC 61850 standard establishes rules for communication protection techniques in DC microgrids, emphasizing their importance [11]. However, in fault situations, this standard places strict limitations on communications signals. As a result, developing a communication-based protection system is a complex process. Suppose a communication-based protection approach is utilized, which leverages intelligent electronic devices and multiple agents to isolate and detect the fault in a much shorter time. In that case, it may become more sophisticated [12].

Unquestionably, the robustness and availability of communication lines are critical considerations while creating an adaptive protection method. As a result, one of the difficulties with DC microgrid protection methods is resolving the communication failure. As previously said, penetration of communication lines in protection systems enhances these systems' vulnerability to cyber-attacks, which is regarded as one of the most recent challenges in DC microgrids. As a result, DC microgrid protection methods should take this into account, depending upon the difficulty of their design [13]. False data injection is one of these issues, leading to inaccurate data detection in SCADA-equipped systems and a significant inaccuracy in power system state estimation. It also causes destructive directives and even a blackout [14]. Several studies have been conducted in recent years to secure power grids from false data injection threats. Some researchers considered securing some measurements and state variables to prevent false data injection attacks in power systems [15].

Several components, like intelligent electronic devices, communication connections, switches, and intelligent measurement devices, are installed in the DC microgrid protection system. These will transmit trip and data signals to circuit breakers during a fault, allowing them to locate and isolate the fault by disconnecting that faulty part from the entire power system. DC microgrids are more vulnerable to cyber-attacks because of the extensive use of processors and communication links [16]. Hackers can disable the system's protection in this situation by disrupting communication channels or switches. In this instance, a backup protection strategy that takes into account cyber-attacks is essential. Several ways are provided for defending against and detecting cyber threats based on communication capabilities [17].

Even with phasor measuring units placed in the system, the present communication links used to monitor and input intelligent electronic equipment in DC microgrids are ineffectual against cyber-attacks [18]. These gadgets are vulnerable to cyber-attacks because they rely on the global positioning system. Given the scattered nature of measuring units and intelligent electronic devices in these systems, it is necessary to establish a backup plan for all functional components [19]. The DC microgrid protection arrangement can be viewed as a decentralized modern communication network infrastructure with intelligent electronic devices, distributed data acquisition, information monitoring, and knowledge storage on demand and system sides.

The blockchain concept was established in 2008 to enable peer-to-peer direct wire transfers without any third party [20]. Every sort of data is compressed into a block, which will then be injected into existing blockchains. The network is then synchronously updated with all of the blocks. As a result, each peer keeps a copy of the same ledger [21]. Until recently, blockchain technology was mainly used to address financial difficulties, such as establishing sustainable local energy markets. Blockchain technology has primarily been utilized to address financial difficulties in recent years, including demand response programs, trade privacy, and sustainable local energy markets. However, because of blockchains' ability to secure communication connections and intelligent electronic devices, this new technology can be used to construct a backup protection strategy to defend DC microgrids from cyber-attacks [22].

In contrast to earlier work, the present chapter proposes a blockchain-based DC microgrid protection scheme. The following are the significant contributions: 1. The suggested protection technique uses a differential protection mechanism to protect the DC microgrid against fault current in the first stage. Communication links are used between each protection device and intelligent electronic equipment in the given protection technique. As a result, the proposed approach isolates the problematic part during the fault with the shortest operating time, 2. Using blockchains, the suggested approach increases the self-protection of contemporary DC microgrids against cyber threats. Cyber-attacks can only be carried out locally in traditional power grids by gaining access to measurement data and protective mechanisms. Still, hackers can remotely control critical components by extending the infiltration of communication channels. By utilizing a backup blockchain-based communication infrastructure, the suggested method protects the protective mechanism from cyber threats.

## 5.2 DC Microgrids

DC microgrids work in a similar way to their AC counterparts in terms of operation. The critical distinction is that the DC bus network interconnects the dispersed generators and loads rather than the AC bus, which interconnects the distributed generators and loads in the network. The operating voltage of these DC buses is typically between 350 and 400 volts. These are the greatest solutions for various power systems, including those in office buildings, aircraft, ships, and rural areas. Additionally, the majority of these systems' resources and loads are powered by direct current. Therefore, the primary DC bus can be branched into multiple low voltage buses to meet the low voltage needs for electronics-based loads. High voltage gain DC-DC converters in DC-type microgrids, on the other hand, make it easier to connect low voltage power sources like solar modules to the high voltage DC bus. The scope of their voltage gain/power handling operation can be used to classify these converters. A small DC microgrid is depicted in Figure 5.1.

The following are the crucial advantages of DC microgrid: 1. Controllability system that works without complications like frequency control, reactive power control, harmonics, and synchronization; 2. Due to the high voltage at low amperes, the cabling is relatively small; 3. Even in remote regions, the power supply is more reliable; 4. Because there is no reactive current, the transmission efficiency is higher; 5. Aside from the cost savings from renewable energy, lower-cost converter systems can give further cost savings; 6. They are an excellent choice for powering high-performance electrical machinery due to their higher conversion efficiency. The main disadvantages of DC microgrid: 1. Converting an existing AC system to DC, there is more complexity and cost; 2. The increased initial investment can be a deterrent to their adoption.

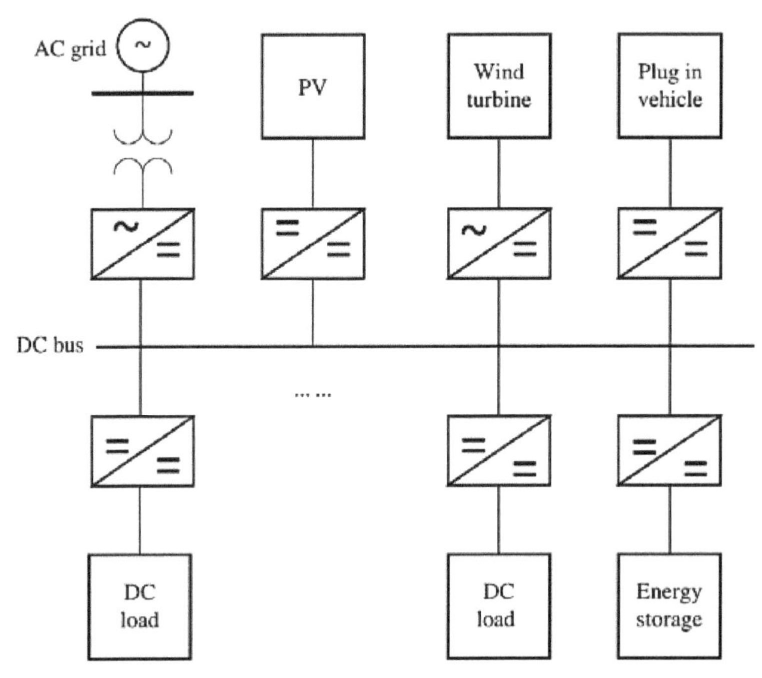


**Figure 5.1** Schematic of small DC microgrid

### 5.2.1 Various Power Sources in DC Microgrids

Different kinds of power resources exist in DC microgrids. Due to the DC voltage generated by photovoltaic (PV) systems and fuel cells, these systems are relevant sources for DC microgrids when a DC/DC converter is used. Various energy sources, such as microturbines and wind turbines (WTs), produce AC voltage and require an AC/DC converter to connect to the DC bus. Table 5.1 summarizes the DC microgrid's energy resources. Appropriate power resources should be chosen based on the DC microgrid's intended use.

**Table 5.1** The important DC microgrid's energy resources

| Sl.No | Energy Resources | Efficiency | Voltage Type |
|---|---|---|---|
| 1 | Fuel Cell | 80% to 90% | DC |
| 2 | Small Hydro Power | 90% to 98% | AC |
| 3 | Biomass | 60% to 75% | AC |
| 4 | Photovoltaic | 40% to 45% | DC |
| 5 | Wind Turbine | 50% to 80% | AC |

**Table 5.2** Different energy storage system technologies

| Sl.No | Energy Storage System Technologies | Life-Span | Capacity | Efficiency |
|---|---|---|---|---|
| 1 | Supercapacitors | Around 20 years | 0 to 0.3 MW | 60% to 65% |
| 2 | Lithium-ion battery | Up to 15 years | Up to 1 MW | 85% to 90% |
| 3 | Lead-acid battery | Up to 15 years | Up to 40 MW | 70% to 90% |
| 4 | Nickel Cadmium battery | Up to 20 years | Up to 50 MW | 60% to 65% |
| 5 | Thermal Energy Storage | 20 to 40 years | Up to 600 MW | 30% to 60% |

### 5.2.2 Energy Storage Systems in DC Microgrids

Wind turbines and solar panels, for example, have a transient response and must be used in conjunction with an energy storage system. Furthermore, an energy storage system provides backup power, load balancing, and improved power quality [23]. Table 5.2 shows a comparison of different energy storage system technologies. It's worth noting that some technologies, like thermal energy storage, have a longer lifespan and larger capacity. However, the best choice for these systems is the low capacity of low voltage DC microgrids and the necessity of efficient lithium-ion and lead-acid batteries.

### 5.2.3 Power Converters used in DC Microgrids

Both AC/DC and DC/DC converters connect DC and AC components to the main DC bus in a DC microgrid. Managing power in both directions while dealing with issues like voltage drops and faults is a monumental task for these converters. DC/DC converters are often easier to design than AC/DC converters, which results in lower cost and higher efficiency. These converters also can restrict the amount of current that flows through the device during a fault state [24]. Despite this, the fault isolation period in a DC microgrid should be shorter than in an AC microgrid because of the rapid rise in fault current and the poor fault current tolerance of converters.

## 5.3 Challenges in Protection of Smart Grid

In addition to the benefits of DC microgrids, converter-based DC microgrids require a protection system. Because there are few practical guidelines for protecting DC systems, concentrating on these systems' issues is vital for developing effective protection techniques. The primary problems of DC microgrid protections are investigated in this part.

### 5.3.1 Protection from Cyber-Attacks

As the influence of communication and processor-based protective mechanisms has increased in recent years, DC microgrid systems have become more vulnerable to cyber threats. As a result, it is crucial to establish a backup protection scheme for the major protection devices in order to safeguard the system against cyber-attacks.

### 5.3.2 Converters with a Low Tolerance

The fault current in DC microgrids reaches its peak value in a short time during a fault. And from the other hand, the converters' tolerance for high fault spikes is limited. As a result, the converters that connect the DC microgrid and the grid are subjected to a considerable fault current. As a result, clearing the fault current requires some operating time.

### 5.3.3 Inefficacy of AC Circuit Breakers

The fault current is interrupted by AC circuit breakers when they trip at the cross zero point. Because a cross zero point does not exist in DC microgrids, however, AC circuit breakers cannot be used in DC systems. Furthermore, faults in DC microgrids should be detected and isolated faster than faults in AC systems to prevent damage to voltage source converters [25–26]. As a result, power electronic devices such insulated gate bipolar transistors (IGBT) and integrated gate-commutated thyristors (IGCT) are the ideal solutions for fixing the issue.

### 5.3.4 Fault Current in both Directions

Despite the presence of conventional power networks, DC microgrids often have a ring-like structure. These systems have a bidirectional fault current, which means the radial system protection technique can't be used in DC microgrids. Moreover, the short-circuit level in DC microgrids changes due to topological variances and the uncertainty of renewable energy supply.

## 5.4 Cyber Attacks

Various types of cyber-attacks are explored in this section and their impact on security systems and communication links. Cyber-attacks on communication

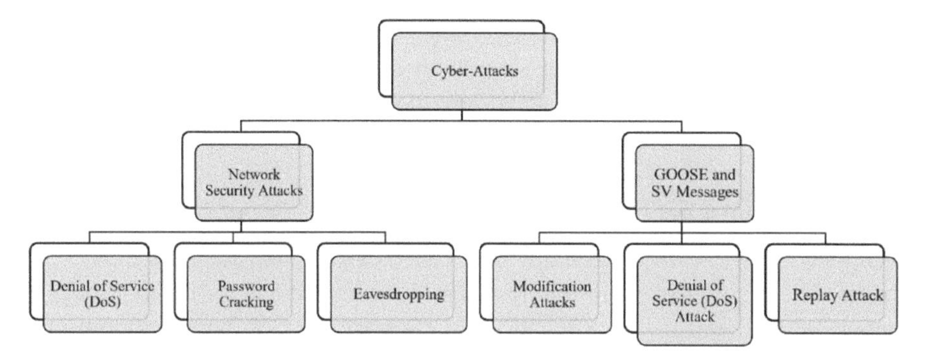


**Figure 5.2** Classification of different types of cyber attacks

networks disrupt measurement units and relay signals, interfering with the circuit breakers' ability to clear the fault [27–28]. As indicated in Figure 5.2, cyber-attacks are separated into Network security attacks and GOOSE and SV messages.

### 5.4.1 Network security cyber attacks

This type of attack is directed at communication lines used to access and modify data. Several network security cyber-attacks include Denial of Service (DoS) attacks, password cracking efforts, and eavesdropping attacks [29].

1. DoS Attacks
   Attackers repeatedly send a bogus synchronisation signal to the relay in order to undermine the operator's connection to the relays, which destroy communication linkages between protection devices. The DoS attack, for example, is carried out simultaneously on relays via FTP, HTTP, and Telnet.
2. Password Cracking Efforts
   It is another important cyber-attack in recent times. To get system access, hackers break the passwords on all devices using password cracking techniques. By gaining access to the security system, an attacker sends erroneous signals to the circuit breakers and isolating the functional components.
3. Eavesdropping Attacks
   It is a local attack involving the local network's usage to steal packets transmitted within the protection system. For example, address resolution protocol (ARP) cache poisoning is one sort of cyber-attack. The attacker uses the ARP communication technique to change the

IP address to the bogus MAC address, enabling the attacker to catch all packets. Another technique is switch port stealing, which involves sending bogus signals to the MAC address of the circuit breaker. Thus, it will allow the attacker to access the system.

### 5.4.2 GOOSE and SV Messages

The GOOSE and SV messages are the IEC 61850's two communication protocols. The trip signal is communicated to the circuit breakers via GOOSE messages, which clear the fault location. In contrast, the SV messages communicate the sensor or measuring unit's measured values to the protective devices. These signals should be transmitted in fewer than four milliseconds, according to the standards. Both messages are transmitted via the Phasor Measuring Unit (PMU) for long distances and the Ethernet network for short distances in the protective systems. There are several sorts of these attacks, including Modification attacks, Semantic attacks, Replay attacks, High-status number attacks, and High-rate flooding attacks [30].

1. Modification Attack
   Without the relay's authorization, the message between the relay and the circuit breakers is modified. The attacker captures the GOOSE messages and modifies them to create a new message, which can subsequently be used to control circuit breakers. Attackers also inject a fake analog signal into the DC microgrid controller to make SV packets, resulting in false triggers and outages. Malware scripts are another method that attackers might use to modify a system. Because this is a localized attack, it must be deployed within the network. This method intercepts messages transmitted between relays and uses the IEC 61850 network to send GOOSE messages.
2. Semantic Attacks
   The GOOSE message's status number is fixed, and the attack determines the rate at which the status changes.
3. High-Status Number Attacks
   The circuit breaker receives a single bogus GOOSE message with a high-status number. It induces too many false triggers and isolates the healthy elements from the power system.
4. High Flood Rating Attack
   Several fraudulent GOOSE messages with a larger number of statuses have been distributed. The bogus GOOSE message then uses a higher status number than the circuit breaker's expected status number.

5. Replay Attacks
   In this cyber-attack, the attacker captures the SV packet containing the measured values and transfers it to another Protection Device. The attacker captures this signal during normal operating mode and directs a trip signal to the circuit breaker for GOOSE messages.

## 5.5 Blockchain

Blockchain is a decentralized peer-to-peer computer network that maintains data as an immutable chain of records. The principles of cryptography are used to link each block of data to the next. The chain cannot be amended once the block has been inserted, but additional blocks can be attached to the chain along with the timestamp [31]. Figure 5.3 depicts the fundamental structure of a blockchain.

The Blockchain is a novel technique of transmitting data from one end to the other in a completely automated and safe manner. Assume that a transaction between two parties is required. One party starts the transaction by creating a new, tamper-proof record of data known as a block. Every new block will be joined to the chain after it has been successfully verified by all of the computers in the peer-to-peer network. Each block contains transaction data, the preceding block's cryptographic hash value, and the hash value and time stamp of the current block. In blockchain terminology, a node refers to a computer that stores Blockchain and is part of a distributed and decentralized network. Every node in the peer-to-peer network has a copy of the Blockchain that is identical to the original. Data is always uploaded to the database once validated by the Blockchain's consensus process, which turns the blockchain

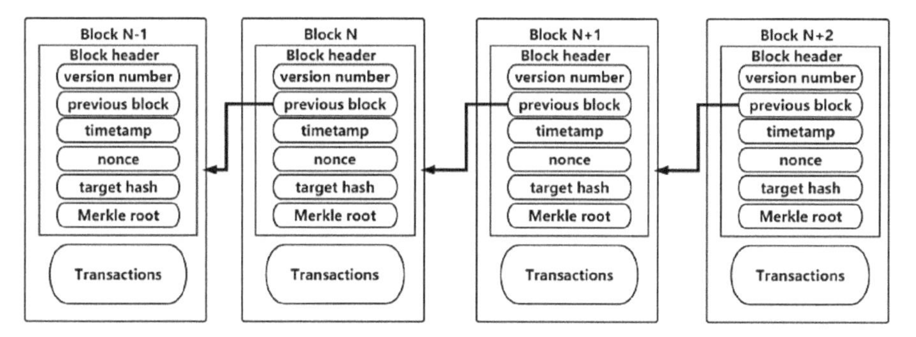


**Figure 5.3** Fundamental structure of a Blockchain

node into a decision-maker. All blockchain nodes usually connect directly without using any third-party system using peer-to-peer network technology.

The following are some of the characteristics of blockchains:

1. Decentralization
   A central unit must validate transactions in classic centralized transaction systems. On the other hand, a peer-to-peer transaction can be completed on the blockchain system without the requirement for a central unit to validate it. As a result, Blockchain lowers server expenses dramatically.
2. Persistency
   In a distributed network, tampering is practically difficult because each transaction is disseminated over the network and requires validation and recording in distributed blocks. Nodes will also check and validate each transaction. As a result, inaccurate data will be detected.
3. Anonymity
   Users can communicate with the blockchain network by using newly created addresses that have been created. Additionally, users have the option of creating multiple addresses in order to avoid being tracked. This algorithm ensures that blockchain transactions are kept private.
4. Auditability
   Users can connect to all nodes in the distributed network to check and trace previous records once they have approved and published the transaction on the Blockchain with a timestamp.

## 5.6 Blockchain-Based DC Microgrid Protection Approach

For DC microgrids, a unique blockchain-based differential protection approach is presented. When a cyber-attack occurs, the proposed method does not shut down the security system but instead employs a blockchain-based backup mechanism to detect the problem. The recommended differential protection system, which employs a relay and a high-speed communication connection, detects internal faults while disregarding external issues in the first stage. A blockchain system serves as a backup communication system in the event of a cyber-attack or a communication link failure, protecting the system.

### 5.6.1 Differential Fault Identification

Internal and exterior faults should be distinguished by a protective mechanism, with the internal faults being isolated alone. The fault current will flow

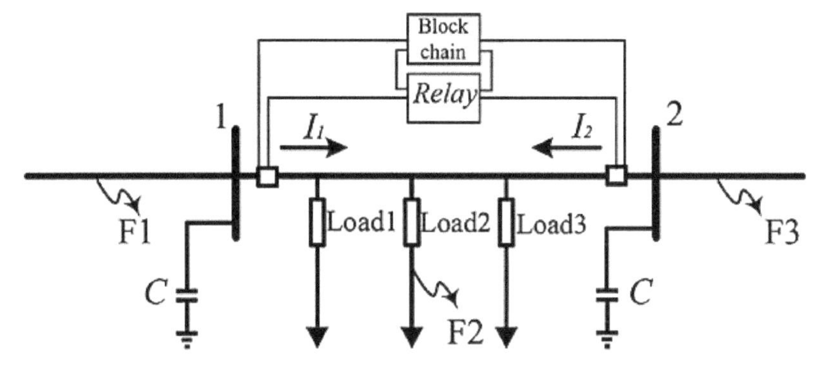


**Figure 5.4** Enactment of the differential protection scheme proposed

in the direction of the fault location during the fault; as a result, the fault type can be determined by comparing the fault current flow directions on the differential protection system's two sides [32–33]. Figure 5.4 depicts the protection method's implementation as well as several fault locations.

The differential relay derives the value of line currents from two measurement devices located at the line's two ends in this configuration. During the fault, a variable is defined on each side of the relay, $S1$ for bus 1 and $S2$ for bus 2; the value of $S1$ and $S2$ for clockwise direction will be 1; while the value of $S1$ and $S2$ for counterclockwise direction will be $r1$. The internal fault can be calculated by,

$$S1 = -S2\ (Internal\ Fault) \tag{5.1}$$

$$S1 = S2\ (External\ Fault) \tag{5.2}$$

$$ST = \frac{|S1S2 - 1|}{2} \tag{5.3}$$

Where $ST$ is equal to one for Internal Faults and becomes Zero for External Faults.

The fault current transient is divided into two components during a fault. Capacitors begin to discharge through the defective channel during the initial stage of capacitor discharge. As depicted in Figure 5.5, this phase begins with the early seconds of the fault and ends with a high peak. After the voltage has been decreased to zero, the freewheeling diode action begins as the second step in the process [34]. The voltage of the converter reverts at this point, and the converter's diode begins to conduct.

At both ends of the lines, the capacitors are equivalent to converter capacitors. During the early stages of a problem, these capacitors inject a

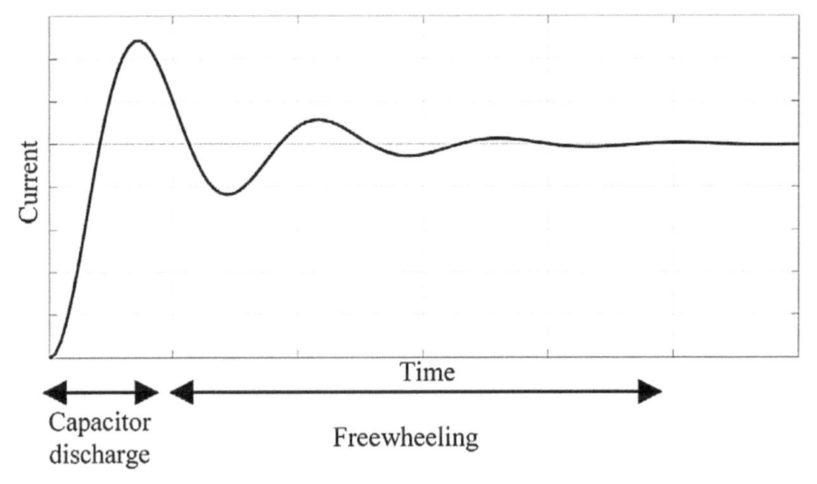


**Figure 5.5** Capacitor and freewheeling stages

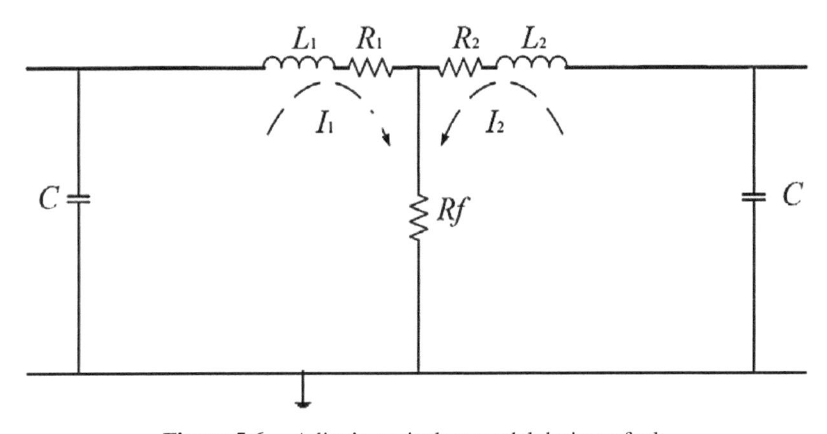


**Figure 5.6** A line's equivalent model during a fault

large amount of fault current into the fault area. During a fault, Figure 5.6 can be utilized to model a line in a bidirectional DC microgrid.

In this system, the differential value of the fault current derivative may be computed for both sides of the line by,

$$\sigma = \frac{di_1}{dt} - \frac{di_2}{dt} = \frac{L_1}{R_f}\frac{d^2 i_1}{dt} + \frac{R_1 + 2R_f}{R_f}\frac{di_1}{dt} + \frac{i_1}{CR_f} \tag{5.4}$$

Where, $L_1$ and $R_1$ is the value of inductance and resistance from bus 1 to the fault location. $i_2$ and $i_1$ are the measured current of bus 2 and 1, respectively. $C$ is the capacitance of the converters. $R_f$ is the fault resistance. The differential relay transmits the trip signal to the circuit breakers if the value of $\sigma$ changes to a value greater than a threshold. Calculate the threshold value for the least possible value of fault current. Under overload conditions, according to AC protection regulations, a load can be fed with 25 per cent more power than usual [35]. As a result, the maximum overload current is $1.25I_{Load}$, where $I_{Load}$ is the load's nominal power, and any current that exceeds this limit is deemed as a fault current. Equation (5.5) expressed the formula which is used to calculate the maximum value of fault resistance.

$$R_f = \frac{4V_n}{I_{Load}} \tag{5.5}$$

Where $V_n$ is the system's nominal voltage. The minimum value of threshold $\sigma_{min}$ is attained by,

$$\sigma_{min} = 2\frac{di_1}{dt} + \frac{I_{Load}}{4CV_n}i_1 \tag{5.6}$$

As a result, the relay's trip signal is generated as follows:

$$\sigma_{min} - ST\sigma \leq 0 \ (Trip) \tag{5.7}$$

### 5.6.2 Blockchain-based Backup and Protection System

A new blockchain-based application for administering the protection system during cyber-attacks and communication failure is proposed in this chapter. The signal from the measuring unit to the relay and the signal from the Blockchain to the relay are identical when there is no communication failure or cyber-attack [36]. During a cyber-attack on the main communication channel, however, the encrypted value from the Blockchain will be different. The relay then identifies the cyber-attack and only evaluates the blockchain data. As a result, the values of $i_2$ and $i_1$ in Equations (5.1) to (5.6) will modify the Blockchain's current decoded value.

When adopting blockchain technology, however, a small delay would be introduced to the protection system during a cyber-attack. Furthermore, increasing the reliability of the protection system against cyber-attacks by adding another measuring unit to each bus will increase the cost.

## 5.7 Results and Discussion

As illustrated in Figure 5.4, a MATLAB simulation was conducted for a line of a DC microgrid. The fault current is bidirectional since both sides of the transmission line are connected to a DC microgrid with power supply. The suggested technique is tested in various circumstances, including high impedance failures and communication link based cyber-attacks.

### 5.7.1 Detecting and Isolating the Faults without Considering Cyber-Attack

An interval line to ground fault with a fault resistance of 1 Ohm occurs at 33 % of the line from bus 1. Figure 5.7 depicts the fault current as viewed from bus 1. As illustrated in Figure 5.8, the indicated differential relay discovers and isolates the fault in 1.4 milliseconds. During the early stage of the fault, the capacitor discharge results in a significant increase in line current, as illustrated in Figure 5.8. Because the peak fault current in this situation is

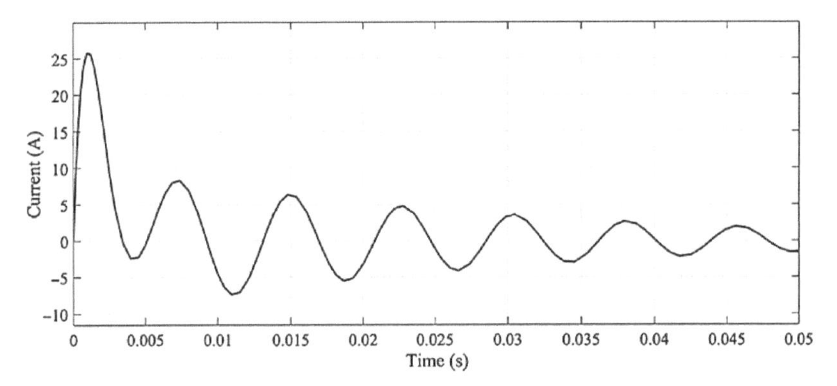


**Figure 5.7** Fault current without the use of a method of isolation.

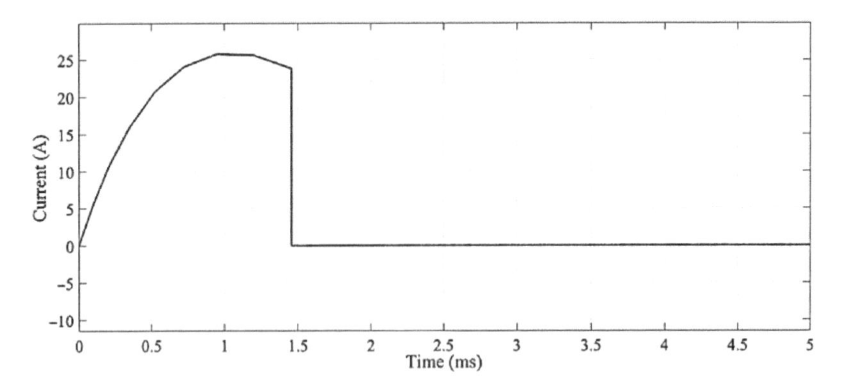


**Figure 5.8** Identifying and isolating the fault with 1Ohm fault resistance.

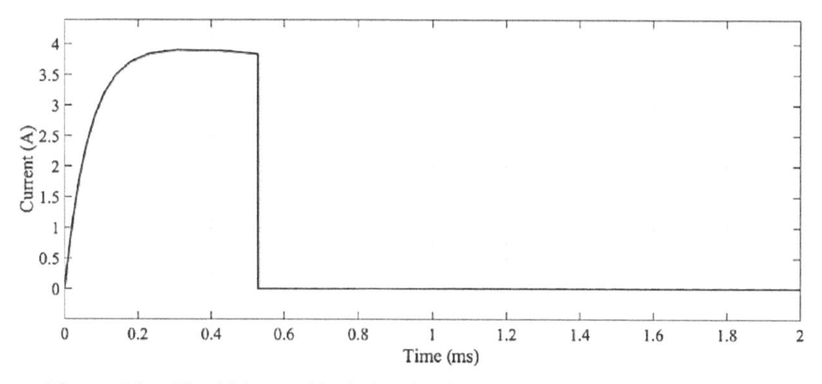


**Figure 5.9** Identifying and isolating the fault with 10 Ohms fault resistance.

nearly 25A, In 1.4 milliseconds, the differential relay detects the fault and transmits the trip signal to both circuit breakers on the line. As indicated in Figure 5.9, a line-center high impedance fault with a fault resistance of 10 Ohms occurred in another instance. The fault current is increased to around 4A, and the fault is identified and cleared in less than 0.5 milliseconds.

### 5.7.2 Detecting and Isolating Faults when considering a Cyber-Attack

As previously said, a cyber-attack is another hazard to a protection scheme. In this work, a differential relay is linked with Blockchain to protect the system from cyber-attack. A cyber-attack on the differential relay's communication link has occurred, as indicated in Figure 5.10. The blockchain system recognizes the cyber-attack and sends the encrypted data to the differential relay as a result.

The differential relay identifies and isolates a fault in the F2 with a fault resistance of 1 Ohm in less than 29 milliseconds if it happens during the cyber-attack. Due to the encryption by Blockchain and the discovery of a cyber-attack, a delay in the operation of the protection scheme is seen in this case, as shown in Figure 5.11. Additionally, because of the 10 Ohm fault resistance, a high impedance failure occurs in the middle of the protected line, as illustrated in Figure 5.12. The results indicate that the proposed protection technique rapidly finds and isolates the issue. Due to the fact that it is a blockchain-based system, a cyber-attack will have no effect on the protection system.

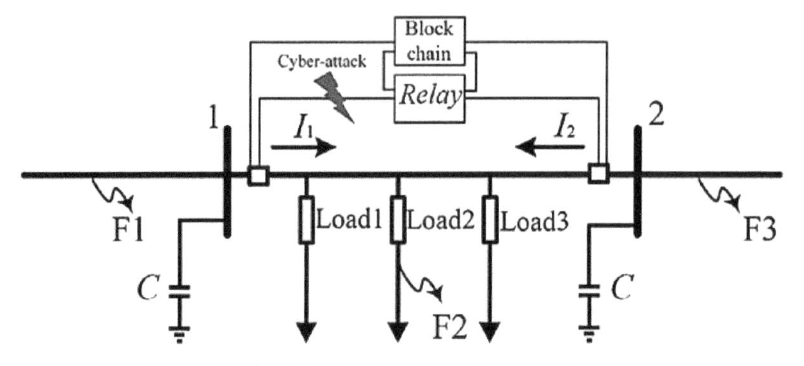


**Figure 5.10** A Cyber-Attack on the protection system

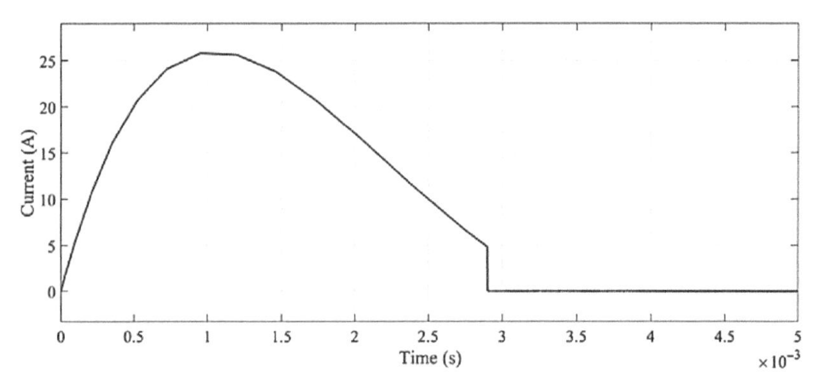


**Figure 5.11** Identifying and isolating the fault during cyber-attack with fault resistance 1 ohm

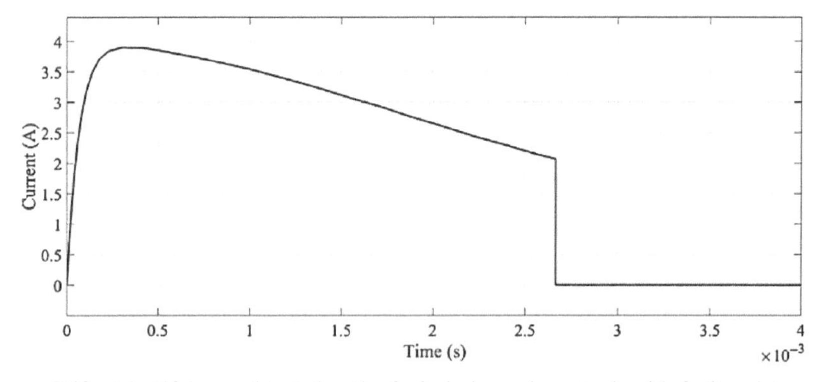


**Figure 5.12** Identifying and isolating the fault during cyber-attack with fault resistance 10 ohms

## 5.8 Conclusion

The current chapter suggests a differential protection strategy for DC microgrids based on blockchain technology. In the proposed protective method, a differential relay coupled to a blockchain system protects the system from cyber-attacks. The differential relay detects internal faults and isolates the faulty line before the capacitor discharge. Additionally, a novel threshold selection strategy is proposed that enables the differential relay to detect faults with a high impedance. Simulation is used to evaluate and validate the proposed method. The simulation results demonstrate that the proposed protection strategy successfully finds and isolates high impedance faults when a variety of cyber-attack scenarios are considered. By mitigating the impact of communication failure, the backup blockchain system boosts the protection system's reliability. The high impedance fault is identified in less than 0.5 milliseconds without considering the cyber-attack during the fault, and in less than 2.5 milliseconds when considering a cyber-attack. The findings suggest that the proposed blockchain-based differential protection approach is proficient against cyber-attacks.

## References

[1] Bayati, Navid, A. Hajizadeh, and M. Soltani. "Protection in DC Microgrids: A Comparative Review." *IET Smart Grid* 1, no. 3 (2018), pp. 66–75.

[2] Bayati, Navid, A. Hajizadeh, and M. Soltani. "Blockchain-Based Protection Schemes of DC Microgrids." In *Blockchain-Based Smart Grids*, pp. 195–214. Academic Press, 2020.

[3] Wang, Li, C. Lin, H. Wu, and A.V. Prokhorov. "Stability Analysis of a Microgrid System with a Hybrid Offshore Wind and Ocean Energy Farm Fed to a Power Grid Through an HVDC Link." *IEEE Transactions on Industry Applications* 54, no. 3 (2017), 2012–2022.

[4] Khan, Rabia, and N.N. Schulz. "Network Loss Analysis of Low-Voltage Low-Power DC Microgrids for Rural Electrification." In *2020 IEEE/PES Transmission and Distribution Conference and Exposition (T&D)*, pp. 1–5. IEEE, 2020.

[5] Li, Meng, Y. Luo, K. Jia, T. Bi, and Q. Yang. "Frequency-Based Current Differential Protection for VSC-MVDC Distribution Lines." *International Journal of Electrical Power & Energy Systems* 117 (2020), p. 105626.

[6] M. Appadurai, and E. Raj. Epoxy/Silicon Carbide (sic) Nanocomposites Based Small Scale Wind Turbines for Urban Applications. *International Journal of Energy and Environmental Engineering*, 2021, pp. 1–16.

[7] E. Raj, and F. Irudaya. "Available Transfer Capability (ATC) Under Deregulated Environment." *Journal of Power Electronics & Power Systems* 6, no. 2 (2016), pp.85–88.

[8] Saidi, Kais, and A. Omri. "Reducing CO2 Emissions in OECD Countries: Do Renewable and Nuclear Energy Matter?." *Progress in Nuclear Energy* 126 (2020), p. 103425.

[9] Sortomme, Eric, S.S. Venkata, and J. Mitra. "Microgrid Protection Using Communication-Assisted Digital Relays." *IEEE Transactions on Power Delivery* 25, no. 4 (2009), pp. 2789–2796.

[10] Konstantopoulos, C. George, A.T. Alexandridis, and C.P. Panos. "Towards the Integration of Modern Power Systems into a Cyber–Physical Framework." *Energies* 13, no. 9 (2020), p. 2169.

[11] Saleh, and Khaled. "Protection of Multi-Terminal Remote DC Microgrids With Overhead Lines Against Temporary and Permanent Faults." *IET Generation, Transmission & Distribution* 14, no. 15 (2020), pp. 2879–2889.

[12] Ali, Ikbal, S.M.S. Hussain, T.A. Tak, and T.S. Ustun. "Communication Modeling for Differential Protection in IEC-61850-Based Substations." *IEEE Transactions on Industry Applications* 54, no. 1 (2017), pp. 135–142.

[13] Beheshtaein, Siavash, R.M. Cuzner, M. Forouzesh, M. Savaghebi, and J.M. Guerrero. "DC Microgrid Protection: A Comprehensive Review." *IEEE Journal of Emerging and Selected Topics in Power Electronics* (2019).

[14] Liu, Yao, P. Ning, and M.K. Reiter. "False Data Injection Attacks Against State Estimation in Electric Power Grids." *ACM Transactions on Information and System Security (TISSEC)* 14, no. 1 (2011), pp. 1–33.

[15] Kosut, Oliver, L. Jia, R.J. Thomas, and L. Tong. "Malicious Data Attacks on the Smart Grid." *IEEE Transactions on Smart Grid* 2, no. 4 (2011), pp.645–658.

[16] Wei, Dong, Y. Lu, M. Jafari, P. Skare, and K. Rohde. "An Integrated Security System of Protecting Smart Grid Against Cyber Attacks." In *2010 Innovative Smart Grid Technologies (ISGT)*, IEEE, 2010, pp. 1–7.

[17] Acosta, M.R. Camana, S. Ahmed, C.E. Garcia, and I. Koo. "Extremely Randomized Trees-Based Scheme for Stealthy Cyber-Attack Detection in Smart Grid Networks." *IEEE access* 8 (2020), pp.19921–19933.
[18] Sengan, Sudhakar, V. Subramaniyaswamy, V. Indragandhi, P. Velayutham, and L. Ravi. "Detection of False Data Cyber-Attacks for the Assessment of Security in Smart Grid Using Deep Learning." *Computers & Electrical Engineering* 93 (2021): p.107211.
[19] Wang, Pengyuan, and M. Govindarasu. "Multi-Agent Based Attack-Resilient System Integrity Protection for Smart Grid." *IEEE Transactions on Smart Grid* 11, no. 4 (2020), pp. 3447–3456.
[20] Berdik, David, S. Otoum, N. Schmidt, D. Porter, and Y. Jararweh. "A Survey on Blockchain for Information Systems Management and Security." *Information Processing & Management* 58, no. 1 (2021), p. 102397.
[21] Upadhyay, and Nitin. "Demystifying Blockchain: A Critical Analysis of Challenges, Applications and Opportunities." *International Journal of Information Management* 54 (2020), p. 102120.
[22] Dehghani, Moslem, M. Ghiasi, T. Niknam, A. Kavousi-Fard, M. Shasadeghi, N. Ghadimi, and F. Taghizadeh-Hesary. "Blockchain-Based Securing of Data Exchange in a Power Transmission System Considering Congestion Management and Social Welfare." *Sustainability* 13, no. 1 (2021), p. 90.
[23] Sinha, Smita, and P. Bajpai. "Power Management of Hybrid Energy Storage System in a Standalone DC Microgrid." *Journal of Energy Storage* 30 (2020), p. 101523.
[24] Gui, Yonghao, R. Han, J.M. Guerrero, J.C. Vasquez, B. Wei, and W. Kim. "Large-Signal Stability Improvement of DC-DC Converters in DC Microgrid." *IEEE Transactions on Energy Conversion* (2021).
[25] Zhang, Guangxiao, and X. Tong. "The Last Circuit Breakers Identification in Hybrid AC/DC Power Grids Based on Improved Tarjan Algorithm." *IEEE Transactions on Power Delivery* 35, no. 6 (2020), pp. 2992–3002.
[26] Jain, Dr, Dr M.D. Sangale, and E. Raj. "A Pilot Survey of Machine Learning Techniques in Smart Grid Operations Of Power Systems." *European Journal of Molecular & Clinical Medicine* 7, no. 7 (2020), pp. 203–210.
[27] Raj, E.F. Irudaya, and M. Balaji. "Analysis and Classification of Faults in Switched Reluctance Motors Using Deep Learning Neural

Networks." *Arabian Journal for Science and Engineering* 46, no. 2 (2021), pp. 1313–1332.
[28] Ustun, Taha Selim, Shaik Mullapathi Farooq, and SM Suhail Hussain. "Implementing Secure Routable GOOSE and SV Messages Based on IEC 61850-90-5." *IEEE Access* 8 (2020), pp. 26162–26171.
[29] Dong, Changyin, H. Wang, D. Ni, Y. Liu, and Q. Chen. "Impact Evaluation of Cyber-Attacks on Traffic Flow of Connected and Automated Vehicles." *IEEE Access* 8 (2020), pp. 86824–86835.
[30] Hussain, S.M. Suhail, S.M. Farooq, and T.S. Ustun. "A Method for Achieving Confidentiality and Integrity in IEC 61850 GOOSE Messages." *IEEE transactions on Power Delivery* 35, no. 5 (2020), pp. 2565–2567.
[31] Pal, Amitangshu, A. Jolfaei, and K. Kant. "A Fast Prekeying-Based Integrity Protection for Smart Grid Communications." *IEEE Transactions on Industrial Informatics* 17, no. 8 (2020), pp. 5751–5758.
[32] Prasad, C. Durga, M. Biswal, and A.Y. Abdelaziz. "Adaptive Differential Protection Scheme for Wind Farm Integrated Power Network." *Electric Power Systems Research* 187 (2020), p. 106452.
[33] M. Appadurai, and E.F.I. Raj. "Finite Element Analysis of Composite Wind Turbine Blades." In *2021 7th International Conference on Electrical Energy Systems (ICEES)*, IEEE, 2021, pp. 585–589.
[34] E.F.I. Raj, and M. Appadurai. "Minimization of Torque Ripple and Incremental of Power Factor in Switched Reluctance Motor Drive." In *Recent Trends in Communication and Intelligent Systems: Proceedings of ICRTCIS 2020*, Springer Singapore, 2021, pp. 125–133.
[35] M. Deivakani, S.V.S. Kumar, N.U. Kumar, E.F.I. Raj, and V. Ramakrishna. "VLSI Implementation of Discrete Cosine Transform Approximation Recursive Algorithm." In *Journal of Physics: Conference Series*, vol. 1817, no. 1, p. 012017. IOP Publishing, 2021.
[36] Song, Kang, and C.Li. "Blockchain-Enabled Relay-Aided Wireless Networks for Sustainable e-Agriculture." *Journal of Cleaner Production* 281 (2021), p. 124496.

# 6

# Planning Active Distribution Systems Using Microgrid Formation

**Shah Mohammad Rezwanul Haque Shawon[1], Xiaodong Liang[1], and Massimo Mitolo[2]**

[1]The University of Saskatchewan, Saskatoon, SK S7N 5A9, Canada
[2]Irvine Valley College, Irvine, CA 92618, USA
E-mail: shs054@mail.usask.ca; xil659@mail.usask.ca; mmitolo@ivc.edu

## Abstract

In this chapter, the planning of active distribution systems through microgrid formation is discussed. There are four critical steps involved: 1) defining the objectives, 2) defining the microgrid topology, 3) system modeling, and 4) network optimization. The objectives of such planning problems in the literature are extensively reviewed, dominant and popular objectives used are demonstrated. The modeling of microgrid components, such as dispatchable and nondispatchable DGs, load, and energy storage, is reviewed and summarized in the chapter. As critical tools, optimization techniques in planning are summarized. The switch placement in distribution networks is also discussed in this chapter.

**Keywords:** Optimal planning, active distribution systems, microgrid formation, objectives, microgrid topology, system modelling, optimization, switch placement.

## 6.1 Introduction

With the recent development of renewable power generation technology, the centralized bulk power systems with large-scale power generation connected to transmission networks have been gradually transformed to tomorrow's decentralized systems with multiple small-scale distributed generation (DG) units connected directly to distribution systems near consumers. In modern distribution networks with the increasing number of renewable energy based DGs, integration of nonconventional and nondispatchable DGs has transformed a passive distribution network into an active system.

The microgrid is the building block of smart grids and active distribution systems. The microgrid is defined by IEEE Standard 2030.7-2017 as "a group of interconnected loads and distributed energy resources with clearly defined electrical boundaries that acts as a single controllable entity with respect to the grid and can connect and disconnect from the grid to enable it to operate in both grid-connected or island modes". Based on IEEE Standard 1547.4-2011, distribution networks can be partitioned into microgrids to improve the system reliability. Therefore, to effectively integrate DGs, an existing distribution system can be divided into several microgrids by performing optimal planning, so that to improve voltage profile and system reliability, reduce losses and costs [1–6].

The objective of this chapter is to discuss the planning of active distribution systems using the microgrid formation concept to develop a reliable, efficient, and cost-effective distribution network, with improved power quality and customer satisfaction.

Optimal planning searches for the optimum result in a solution space, which must fulfill all constraints of the system. The economic, technical, and environmental performance must be considered as optimization objectives, and the economic perspective is to ensure the minimum deployment and operational costs [7]. Various system constraints used in the literature include: the node voltage [8–11]; the state of charge (SOC) of energy storage systems (ESS) [8, 12, 13]; the rated power of the ESS [8, 9, 12, 14]; the capacity of DGs [8, 12–14]; the power balance [12–14]; the line capacity limit [9, 14]; and load constraints [13, 14].

Allocation and sizing for DG units in the planning stage are very important to achieve optimal microgrid implementation [15], and the demand-response strategy may significantly benefit the optimal sizing of DGs [12]. A microgrid planning strategy can be realized by two steps [16]: 1) Step 1 - determine optimal sizes and locations of DGs in a distribution network; and 2) Step 2 - specify the electrical boundary of a microgrid by optimal allocation of switches. In the planning stage, distribution system operators may build a time bound investment plan to implement microgrids [17].

Accurate planning of microgrids is a challenging task due to uncertainties in load demand and the output of DGs [16]. The power output of solar

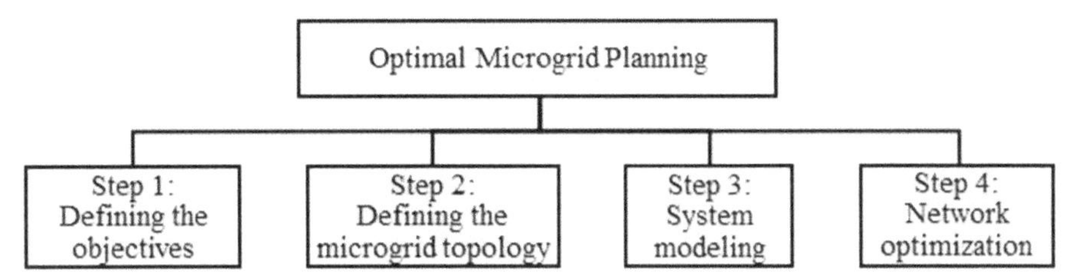


**Figure 6.1** Four critical steps in microgrid planning.

photovoltaic (PV) panels and wind turbines (WT) depends upon solar irradiation and wind speed, which are random and intermittent in nature. One way to mitigate the challenges of uncertain events is to accurately model uncertainties in the planning stage. The randomness and intermittency problems of renewable energy sources can also be mitigated by energy storage systems when the generation has a surplus [4].

In literature [8, 13], various planning strategies have been proposed, and developing an optimal planning strategy of a microgrid can be achieved by implementing four steps (Figure 6.1).

## 6.2 Step 1: Defining the Objectives

The first step of optimal microgrid planning is to define planning objectives. The distribution system operator, as the decision maker, determines specific objectives at the planning stage. Optimal deployment of DGs in the distribution network has been associated with various objectives, such as minimizing capital and operating costs, minimizing system losses, maximizing the profits, minimizing emissions, or a combination of multiple objectives [5, 9, 10, 18].

Reliability is one of the most promising aspects among all benefits when forming a microgrid [19]. Implementation of microgrids by allocating DGs in optimal locations offers increased reliability, efficiency, and flexibility of distribution systems in a cost-effective manner [9]. Certain standard indices are used to evaluate the system's reliability. For example, the system average interruption frequency index (SAIFI), the system average interruption duration index (SAIDI), and the momentary average interruption event frequency index (MAIFIe) are used in [2]; the energy index of reliability (EIR) is used in [19], which is defined as the expected energy served as a fraction of the total energy demand; energy not supplied (ENS) is used in [20], and load curtailment is used as a reliability index in [3]. A reliability index may be formulated as an objective or a constraint [21].

According to IEEE Standard 1366, reliability indices of power distribution networks can be divided into three main categories: time-based indices, frequency-based indices, and energy-based indices. Since energy can be converted to cost, the cost of energy not supplied (ENS) is used in [20] to determine the interruption cost as a measure of reliability.

When the penetration of DGs into the network is high, the DG generated power not only alters the power flow in distribution networks but also in transmission grids. Therefore, the connection of DGs to a distribution network may influence the overall power system stability (i.e., angle, frequency, and voltage stability) [22]. The deployment of DGs in terms of locations, sizes, numbers, and mix in a microgrid is one of the most important aspects of microgrid planning [5], [23], which is essential for minimizing losses, improving voltage profile (i.e., avoiding the possibility for the voltage swell), improving self-adequacy, and minimizing power imbalance, etc. Long transmission lines with central generation units lead to increased cost, increased energy losses, and reduced reliability; thus, the optimal allocation and sizing of DGs facilitates local supplying of loads in distribution systems, which in turn improves reliability, reduces cost and losses. Optimal allocation of ESS units can also be implemented along with optimal DG placement to improve the system performance [24]. System reconfiguration, capacitor bank installation, and load balancing are also options to reduce losses at distribution systems [5].

The high investment cost of implementing microgrids is a major concern [14]. Therefore, the objective function is generally formulated as a cost function by considering various costs associated with microgrid planning and operation [7, 12, 15]. The net cost of a microgrid is divided into two parts in [5]: 1) the net planning cost of DGs; and 2) all variable operational costs occurring in microgrids. In addition, uncertainty, losses reduction, and power quality improvement (i.e., voltage profile and power unbalance) are also important considerations [3, 25]. For example, the loss sensitivity factor (LSF) and the voltage sensitivity factor (VSF) are used to determine the optimal placement of microgrids to reduce the system loss and improve the system voltage profile; LSF and VSF are calculated after performing the load flow study of the distribution system [3].

Self-adequacy, a unique feature of a microgrid, refers to its capability to supply loads by DGs within the microgrid itself. The power balance within a microgrid is considered as an indication of its self-adequacy [24]. For networked microgrids (i.e., multiple interconnected microgrids) when the amount of transferred energy between the microgrids is low, or the generation-load imbalance present within them is low, the microgrids are considered self-sufficient [2, 26]. In the autonomous mode of operation (i.e., the island mode), not only more loads can be supplied within the microgrid, but also the energy losses on power lines connecting the microgrid are minimized [2]. A self-adequate microgrid can operate independently from the main power grid, or from each other for networked microgrids. Thus, minimizing the transferred energy between the microgrid and the main power grid, or among interconnected microgrids, is considered as an objective of optimal microgrid planning [6, 9, 11, 24]. During emergencies, splitting distribution networks into self-sufficient microgrids can prevent the propagation of a disturbance, and thus, avoid an unforeseen chain of events [9].

Maximizing self-adequacy during design will lead to maximizing the self-sufficiency and the supply-security during the operation of a microgrid; the supply-security of the microgrid is examined by defining a probabilistic adequacy index [2]. More loads can be supplied in its autonomous mode of operation when microgrids are created in a distribution system with the maximum supply security. Alternatively, the self-adequacy of a microgrid can be enhanced by introducing distributed reactive sources (DRS) to the distribution system, where the reactive power can be produced by DGs themselves or by capacitor banks [26]. The reactive sources can reduce system losses, improve voltage profile, and increase system capacity. To get the maximum benefits from distributed reactive sources, it is important to determine their optimal locations and sizes in the system.

The load factor (LF) is a measure of the electricity utilization rate, indicating the rate of energy usage vs. the maximum supply capacity [24]. Load factor is rarely considered in system planning and design studies. In [24], it is found that the system's load factor can be maximized by the optimal allocation of various distributed energy resources (DGs and energy storage).

The multi objective planning problems are investigated in [5, 9, 14, 27, 28]. In [5] and [27], a multiobjective problem has been converted to a single objective function by using the weighted sum method with equal weights to both objectives. In [9], two objectives are combined as a weighted summation to form a novel objective function. A fuzzy satisfaction-maximizing method is adopted in [28] to convert the original problem into a single objective optimization problem. The microgrid planning problem in [14] is decomposed into an investment master problem and an operation sub problem. Table 6.1 presents a summary of various objectives found in the literature for optimal microgrid planning.

**Table 6.1** Summary of objectives used for optimal microgrid planning in the literature.

| Objectives of Planning Strategy | Reference |
|---|---|
| Optimal number, location, and sizing of distributed generation, energy storage systems, distributed reactive sources, switches, and capacitor banks | [4], [5], [7], [14], [15], [16], [19], [20], [21], [23], [24], [25], [26], [29], [30], [31], [32], [33], [34], [35], [36] |
| Minimizing losses | [3], [4], [5], [24], [26], [32], [33] |
| Minimizing cost | [4], [5], [7], [8], [10], [12], [13], [14], [15], [16], [17], [18], [19], [20], [21], [23], [25], [27], [28], [29], [30], [32], [33], [34], [37], [38], [39] |
| Maximizing profits | [3], [23], [31], [35] |
| Maximizing self-adequacy | [6], [9], [11], [24], [26], [32], [40] |
| Maximizing customer satisfaction, or social welfare | [28], [37] |
| Maximizing load factor | [24] |
| Topology planning | [6], [40] |
| Minimizing emission | [4], [18], [27] |
| Improving voltage profile | [3] |
| Improving reliability | [2], [3] |

## 6.3 Step 2: Defining the Microgrid Topology

Transmission systems are often meshed networks, but most distribution networks have a tree-like radial topology [36]. The topology of the microgrid can be either radial or loop-based when implemented within a distribution network. To enhance the economy, resilience, and reliability of power delivery, the loop-based microgrid topology shows several advantages including 1) the reliable performance; 2) the coordination of protection devices; and 3) the microgrid operation strategy [6]. Although loop-based microgrids have high reliability in both islanded and grid-connected operations, the existing microgrid topology planning methods are mostly for radial structures [40].

Optimal planning of networked multi microgrids is investigated in [41], where each microgrid is autonomous by combining generation and load control to satisfy the local objectives. The benefits of networked microgrids includes [41]: 1) sharing reserves under critical conditions (e.g., loss of major generation), which minimizes emergency load shedding requirements, reduces the chance of the system collapse, and enhances the overall system reliability; 2) improving economic dispatch in either grid-connected or island mode (e.g., the excess power generation from a microgrid can be shared with a neighboring deficient microgrid); and 3) sharing energy storage and ancillary functions.

## 6.4 Step 3: System Modeling

System modeling is an important step of optimal microgrid planning. Major components of a microgrid include dispatchable and non-dispatchable DGs, the energy storage system, and the load.

### 6.4.1 Modeling of DGs

There are several types of DGs in microgrids, including wind turbines [2], photovoltaic (PV) modules [2], biomass generators [24], micro-turbines [32], and diesel engines [27]. All DGs can be categorized into two types: dispatchable and nondispatchable DGs (i.e., they cannot be turned on or off in response to the load demand).

Besides supplying electric power to the load, DGs in a microgrid play other important roles. DGs provide 1) voltage and frequency support in an islanded microgrid to satisfy reliability constraints [34]; and 2) the reactive power support to improve the system stability [5]. For reactive power support, capacitor banks are considered the most economical solution for volt/var control and power loss reduction for distribution systems [34]. Therefore, DGs and capacitor banks can be simultaneously placed and sized in microgrids to maximize their benefits.

#### 6.4.1.1 Dispatchable DG model

DGs with the controlled output that can be dispatched in response to the demand are termed as dispatchable DGs. Diesel engine [27, 42–44], micro-turbine [27, 32, 45], and fuel cell [45] are modeled dispatchable generators without considering uncertainty. It is assumed that a specified output power with a certain power factor can be generated during the planning horizon.

##### 6.4.1.1.1 Diesel engine model

Diesel engine generators generally serve as a backup power source. The fuel consumption (L/kWh) of the diesel generator is modeled as a linear function of their actual output power [27, 42, 43]. The fuel consumption is expressed as:

$$F = F_0 \times P_{rated} + F_1 \times P_{generated} \tag{6.1}$$

where $P_{rated}$ is the rated output power, and $P_{generated}$ is the actual output power. $F_0$ and $F_1$ are the fuel consumption curve fitting coefficients. In [27, 42–44], the values of $F_0 = 0.08415\ \mathrm{L/kWh}$ and $F_1 = 0.246\ \mathrm{L/kWh}$ are used.

#### 6.4.1.1.2 Micro turbine model

Micro turbines (MTs) are small high-speed gas turbines. The efficiency of MT increases with the increasing power generation of MT [45]. The cost function of the generated power of MT, $C_{MT}$ , is calculated as follows [27, 45]:

$$C_{MT}=\frac{C_{ng}}{L_h}\times\frac{P_{MT}}{\eta_{MT}} \tag{6.2}$$

where $C_{ng}$ is the price of natural gas in USD/m$^3$, $L_h$ is the low-hot value of natural gas in kWh/m$^3$, $P_{MT}$ is the generated power by each MT unit in kW, and $\eta_{MT}$ is the electrical efficiency of MT.

### 6.4.1.2 Nondispatchable DG model

DG with a random and uncertain output is defined as nondispatchable DG: PV and wind turbines are common types of non-dispatchable DGs.

Conventionally, the planning of microgrids is conducted under deterministic patterns, but due to the increasing number of DGs with uncertain parameters, it should be conducted under uncertain conditions [16]. Several major uncertainties include intermittent renewable power generation, load demand, and electricity price, which are significant obstacles to the economic and reliable design of microgrids [35]. Major renewable energy sources, such as wind and solar power, are dependent on meteorological factors, which are intermittent and volatile in nature, and highly unpredictable, causing considerable variability in power generation. To enable efficient integration of renewable energy sources, system planners traditionally considered the backup generation for smoothing out the variability [38]; however, the backup generation always introduces additional costs. Since the reliability of microgrids is greatly affected by the randomness of renewable energy (e.g., wind and solar energy) [8], the proper modeling of such uncertainties for DGs is essential for developing optimum microgrids in a distribution system [2].

The process of modeling uncertainty is based either on known statistical data or on the probability distribution function (PDF) of random variables, which is usually very challenging to obtain [31]. A straightforward approach is to use the historical meteorological data in a typical year.

#### 6.4.1.2.1 PV modeling

The power generated from PV modules depends on the characteristics of the modules, the amount of solar irradiance, and the ambient temperature [24, 26]. The amount of solar radiation is modelled by the beta PDF [2, 8, 9, 15, 24, 26]. The following equations are used in [24, 37] to model PVs:

$$f(s)=\frac{\Gamma(\alpha+\beta)}{\Gamma(\alpha)+\Gamma(\beta)}*s^{(\alpha-1)}*(1-s)^{(\beta-1)} \tag{6.3}$$

$$0\leq s\leq 1,\ \alpha\geq 0,\ \beta\geq 0$$

$$\alpha=\frac{\mu*\beta}{1-\mu},\ \ \beta=(1-\mu)*\left(\frac{\mu*(1+\mu)}{\sigma^2}\right)-1 \tag{6.4}$$

where $f(s)$ is the beta PDF of the solar irradiation, $s$ is the solar irradiation, $\alpha$ and $\beta$ are the shape parameters of the beta distribution, $\mu$ is the mean and $\sigma$ is the standard deviation of available historical data. The amount of solar radiation is modeled by a normal PDF in [11] by utilizing historical data. To obtain the PDF of solar irradiance, the study is conducted over one year, and then four days are selected as representatives of four seasons in [2, 26].

The power generated from the PV is calculated using the following equations [24]:

$$T_C=T_A+S_{AV}*\left(\frac{NOT-20}{0.8}\right) \tag{6.5}$$

$$I=S_{AV}\left[I_{SC}+K_i\left(T_C-25\right)\right] \tag{6.6}$$

$$V=V_{OC}-\left(K_v*T_C\right) \tag{6.7}$$

$$FF=\frac{V_{MPP}*I_{MPP}}{V_{OC}*I_{SC}} \tag{6.8}$$

$$P_S(S_{AV})=N*FF*V*I \tag{6.9}$$

where *NOT* is the nominal operating temperature, $T_C$ and $T_A$ are the cell and the ambient temperatures, and $S_{AV}$ is the average solar irradiation. $I_{SC}$ and $V_{OC}$ are the short circuit current and the open circuit voltage, respectively. $K_i$ and $K_v$ are the temperature coefficients of current and voltage, respectively. FF is the fill factor, $V_{MPP}$ and $I_{MPP}$ are the voltage and current at the maximum power point, respectively, and $P_S$ is the probabilistic PV generation of each module. However, the generated power from the PV is calculated in [11, 13, 37] using the following simplified formula:

$$P=\frac{P_{STC}*S_{AV}\left[1+K_p(T_C-25)\right]}{S_{STC}} \tag{6.10}$$

where $P_{STC}$ is the output of the PV module under standard test conditions (1000 W/m$^2$ at 25 °C), $S_{AV}$ is the average solar irradiation, $K_p$ is the temperature coefficient of power, $T_C$ is the working temperature of solar panels, and $S_{STC}$ is the solar irradiation under standard test conditions (1000 W/m$^2$).

#### 6.4.1.2.2 Wind turbine modeling

The wind speed, wind direction, and structure of wind turbines significantly affect wind power generation [24, 26]. To properly model the wind power generation, an appropriate pattern of wind speed is required. The Weibull PDF is used in [2, 8, 11, 24, 26], the Rayleigh PDF as a special case of the Weibull PDF is also used in [9] to describe the randomness of wind speed. Similar to solar irradiance, the study is conducted over 1 year period to obtain the PDF of wind speed, and 4 days are selected as representatives of the four seasons [2, 26]. The equations used in [24] to model wind turbines are:

$$V_m = \int_0^{\infty} vf(v)\,dv = \int_0^{\infty} v\left(\frac{2v}{c^2}\right) * \exp\left[-\left(\frac{v}{c}\right)^2\right] dv = \frac{\sqrt{\pi}}{2}c \tag{6.11}$$

$$f(v) = \left[\frac{2v}{c^2}\right] * \exp\left[-\left(\frac{v}{c}\right)^2\right] \tag{6.12}$$

where $V_m$ is the mean value of the wind speed calculated by utilizing the historical data, $c$ is the scale index, $v$ is the wind speed, and $f(v)$ is the Weibull PDF of the wind. The following equations are used in [13, 16, 24, 32, 33, 37] to calculate the output power of wind turbines as a function of the wind speed $v$.

$$P_{out} = \begin{cases} P_{rated} * \frac{v - v_c}{v_r - v_c}, & v_c \leq v \leq v_r \\ P_{rated}, & v_r \leq v \leq v_f \\ 0, & v < v_c \ or\ v > v_f \end{cases} \tag{6.13}$$

where $v_c$, $v_r$, and $v_f$ are the cut-in, rated and cut-out wind speed. $P_{out}$ and $P_{rated}$ are the output and rated power of each wind turbine.

In [21], clustering techniques serve as the basis for the methodology developed to capture the daily patterns of the wind speed, where representative scenarios are obtained along with their occurrence probabilities. In [34], 5 successive years of wind speed data have been collected and several PDFs of the wind speed including Weibull, Normal, and Gamma PDFs are evaluated. Using a goodness-of-fit test, the Weibull PDF has been selected to model the wind speed uncertainty.

### 6.4.2 Load Modeling

Load forecasting is another source of uncertainty, and therefore, the proper modeling of the load is essential during optimal microgrid planning. It is assumed in [11, 24] that the load has a standard Normal distribution with the mean value. The PDF of the active and reactive power of the load, respectively, $P_L$ and $Q_L$, are modeled using a Normal distribution in [8, 11, 16, 37]:

$$f(P_L) = \frac{1}{\sigma_P\sqrt{2\pi}}\exp\left(-\frac{(P_L-\mu_P)^2}{2\sigma_P{}^2}\right) \tag{6.14}$$

$$f(Q_L)= \frac{1}{\sigma_Q\sqrt{2\pi}}\exp\left(-\frac{(Q_L-\mu_Q)^2}{2\sigma_Q{}^2}\right) \tag{6.15}$$

where $\mu_P$ and $\mu_Q$ are the means of the normal distribution for active and reactive power, respectively; $\sigma_P$ and $\sigma_Q$ are the variances of the normal distribution for active and reactive power, respectively. The load is modeled using the beta PDF [15]. The load demand is modeled in [2, 9, 26, 37] by using the IEEE RTS model, where the hourly peak load is introduced as a percentage of the daily peak load for the four seasons, where each season is represented by 2 days, a weekday and a weekend. The historical data can also be used to accurately determine the load's PDF.

Figure 6.2 shows the summary of different methods used in literature to model the uncertainty of PV, WTs, and load.

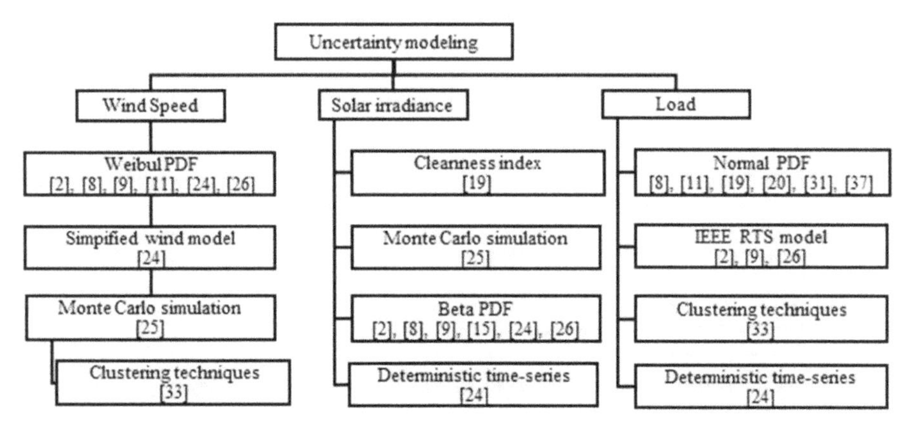


**Figure 6.2** Modeling of uncertainties for PV, WT, and load in optimal microgrid planning.

### 6.4.3 Energy Storage System Modeling

Increasing penetration of renewable energy resources such as WTs or PV modules in distribution systems may adversely affect the entire grid due to their intermittent characteristics. The energy storage system offers feasible solutions by storing the surplus energy, and facilitates the penetration of renewable generation in the system [26]. The energy storage system is modeled as a generator during on-peak hours (discharging period), and as a load during off-peak hours (charging period) [9, 24, 26]. From the point of view of the availability of the equipment, the energy storage system is characterized by two complementary states, available and unavailable [21]. The maximum capacity of the energy storage should be considered during the optimization procedure [24]. The energy storage technology can also be a mobile form, which can be transported using public transportation routes to locations where extra backup capacity is needed due to, for example, natural disasters, whereas these mobile energy storage units are operated as stationary resources under normal operations [17].

The integration of energy storage devices in modern distribution grids requires better coordinated distributed energy resources [25]. The energy storage system is modeled in [13] by considering the relationship among the state of charge (SOC) of a battery at the time $t$, the SOC of the battery at the previous time $t-1$, charge and discharge efficiency, and hourly attenuation. During discharge, $\mathrm{P_{SB}(t)} \geq 0$, and the SOC at the time $t$ is:

$$\boldsymbol{S(t)}=\boldsymbol{S(t-1)}\ (1-\sigma)-\frac{\boldsymbol{P_{SB}(t)}}{\eta_{\boldsymbol{d}}} \tag{6.16}$$

During charge, $\mathrm{P_{SB}(t)} \leq 0$, and the SOC at the time $t$ is:

$$\boldsymbol{S(t)}=\boldsymbol{S(t-1)}(1-\sigma)-\boldsymbol{P_{SB}(t)}\ \eta_{\boldsymbol{c}} \tag{6.17}$$

where $S(t)$ is the SOC at the time $t$, and $P_{SB}(t)$ is the charging or discharging power at the time $t$; $\eta_c$ is the charge efficiency; $\eta_d$ is the discharge efficiency; and $\sigma$ is the proportion of self-discharge.

## 6.5 Step 4: Network Optimization

The final step of optimal microgrid planning is to optimize the objective function, which is formulated based on the objectives of the planning, network constraints, and the system model. Various algorithms and methodologies are proposed in the literature to optimize the objective function.

Microgrid planning and operation optimization problems can be classified into three categories [27, 28]: 1) an optimal microgrid operational strategy with predefined capacities of distributed energy resources, where multiple objective functions, such as the minimization of operating costs, the maximization of reliability and the minimization of pollution, are involved; 2) the simultaneous optimization of the microgrid sizing and operational strategy, where planning and operation problems are divided into two stages (or sub problems), the first stage algorithm handles the planning problem, and the second stage algorithm optimizes the operational objectives; and 3) optimize the microgrid planning and operation problems jointly with the single optimization process instead of using a two-stage process in Category 2).

Identifying the optimal electrical boundaries for optimal microgrid planning is a complex problem because of the large search space. Heuristic algorithms may be employed for this level of complexity, such as backtracking search optimization (BTSO) [9]. The heuristic optimization algorithms are widely used in different research fields due to their ability to solve nonlinear, and nonconvex problems [24]. However, it is not possible to achieve definite optima using heuristic approaches. In the limited runtime process, the conventional group search optimization (GSO) may get stuck in local optima instead of achieving global ones. To overcome this issue, the adaptive group search optimization (AGSO) algorithm may be used for the optimal microgrid construction [24].

In [20], the Genetic algorithm (GA) and mixed-integer linear programming (MILP) are simultaneously applied to a model and solve a two-stage optimization considering utility's profit and customers' satisfaction. GA is applied for optimal sitting and sizing of DGs and placement of switches during the planning stage. The target switches are selected at the operational level by using MILP as boundaries of optimal microgrids. Unlike heuristic approaches, mixed integer programming methods can achieve optimal solutions more accurately.

The planning problem is formulated as a mixed integer nonlinear program (MINLP) in [7, 29] that specifies the microgrid topology, and optimally allocates and sizes distributed energy resources. Due to the computational complexity of the MINLP, the planning algorithm adopts a two-stage framework that presents a reduced computational complexity. The topology specification for the microgrid and the allocation and sizing of distributed energy resources are solved in the first stage and the technical feasibility evaluation through the power flow analysis for optimal scheduling of the installed equipment is investigated in the second stage. In [15], the objective function is formulated as a nonlinear convex stochastic optimization problem. Because the nonlinear problem is more difficult and complex, a sequential quadratic programming (SQP) algorithm is used to divide the overall nonlinear problem into several linear sub-problems by linearizing nonlinear constraints.

The stochastic behaviors of some uncertain parameters in the microgrid can affect optimal planning. A robust optimization (RO) approach is employed for integrating the load uncertainties, such as the variable renewable generation, market price forecasts, and microgrid islanding, to the operation sub-problem [14, 35]. In [16], the stochastic scenario-based approach is adapted to handle the uncertainty among different approaches, such as fuzzy mathematics, stochastic modeling, RO, and other techniques. In [23], a two-stage RO-based microgrid planning model is used that takes into account uncertainties of DG outputs and load consumptions

Compared with traditional stochastic optimization approaches, which rely on a probability distribution of the uncertainty data and sampled scenarios of the uncertainty realizations, RO has the following advantages: 1) RO only requires a few information of the uncertainty set, such as the mean and lower and upper bounds of the uncertain data, which are easier to obtain from the historical data, or estimated with certain confidence intervals in practice; and 2) RO calculates an optimal solution that is immune from all realizations of the uncertain data within a predetermined uncertainty set by considering the worst-case scenario; this is in contrast to the stochastic programming that provides probabilistic guarantees for constraint satisfaction.

These features make RO highly applicable in solving the microgrid planning problem because it may be difficult to obtain information of uncertain DG outputs and loads, while a large number of scenarios may be needed for the stochastic program approach to representing uncertainties during the overall planning period. On the other hand, RO also has certain disadvantages: 1) solutions of RO are often considered to be conservative; and 2) if the exact distribution of the uncertainty data is known, such information may not be fully used. To overcome this issue, the column-and-constraint generation (CCG) algorithm can be considered, it is a generic decomposition algorithm framework for a two-stage RO and is proven to be an efficient solution [36].

Based on the types of objective functions, various algorithms and methodologies have been applied to optimize them. Table 6.2 shows the summary of different optimization methodologies used in the literature. Among these methods, MINLP, MILP, GA, and particle swarm optimization (PSO) are the most commonly used.

**Table 6.2** Summary of different optimization methodologies used in literature.

| Optimization Algorithms | Reference |
|---|---|
| Particle swarm optimization (PSO) | [5], [7], [10], [31], [33] |
| Genetic algorithm (GA) | [5], [7], [8], [20], [21], [29], [34] |
| Backtracking search optimization (BTSO) | [9] |
| Tabu search optimization | [2], [26] |
| Mixed integer linear program (MILP) | [6], [12], [20], [27], [28], [40] |
| Mixed integer non-linear program (MINLP) | [3], [7], [29], [30], [34], [35], [39] |
| Multivariable nonlinear convex programming | [15], [17] |
| Decision theory-based approaches | [25] |
| Adaptive group search optimization (AGSO) | [24] |
| Imperialist competitive algorithm (ICA) | [11], [16] |
| Robust optimization | [14], [23], [36], [38] |
| NSGA-II | [18], [32] |
| Simulated annealing (SA) | [19] |
| Minimal cut-set (MCS) method | [41] |

### 6.5.1 Power Flow

The power flow analysis is an important tool for planning and operation studies in power systems. The adequacy of power supply at any load point in the system depends on generation, transmission, and distribution network parameters.

The forward-backward load flow algorithm is a well-known power flow method used for radial distribution systems, which is based on the direct application of Kirchhoff's Voltage Law (KVL) and Kirchhoff's Current Law (KCL) to overcome numerical issues associated with other load flow solutions algorithms (e.g., Newton-Raphson) when applied to sparse or weakly meshed distribution systems [26]. The forward-backward sweep algorithm exploits the radial nature of distribution systems, and it is computationally simpler and more efficient. As a consequence, this sweep technique has been modified to include DG units in the distribution system and microgrids [5]. However, the integration of DGs within conventional distribution networks changes the configuration of the network from radial to weakly meshed [11]. Therefore, the conventional load flow algorithm, such as Newton-Raphson, and backward-forward sweep techniques may not be suitable or efficient.

Nowadays, the high penetration of DGs (such as WTs and PV systems) with their stochastic properties into distribution networks and microgrids, together with ever-changing load patterns, make the power flow analysis challenging [11]. To address this problem, the probabilistic approach is used, such as the probabilistic power flow based on the Monte Carlo method [8].

In [11], a heuristic load flow method considering the effects of intermittent behavior of DGs and load is modeled in the probabilistic load flow algorithm. The method is suitable for both radial and weakly meshed distribution networks with DGs for the planning and operation of microgrids. A detailed AC optimal power flow algorithm is developed in [29], which can capture operational characteristics, and ensures that the optimal design satisfies all operational constraints of the network. DistFlow equations are often used to calculate the complex power flow and determine the voltage profile in a distribution system [36].

### 6.5.2 Demand Response

For optimal allocation and sizing of DGs in the microgrid planning problem, the demand-response can effectively reduce the required capacities of the energy storage system [28]. The demand side management (DSM) methods are usually preferred to minimize the imbalance between the load and the generation [12, 20, 28]. For the successful implementation of a DSM, the advanced metering infrastructure along with information, and communication technologies are required [27, 28].

Load management has rarely been considered in detail during the planning stage [12]. With the development of the advanced metering infrastructure, an integrated resource planning model based on the optimal direct load control is proposed in [13] to reduce the investment cost and minimize the total social cost. In [12], an integrated resource planning model is also proposed considering the impact of adjustable loads, which simultaneously deal with both supply-side and demand-side resources, and minimizes the overall planning cost of the microgrid. Therefore, the integrated resource planning method combined with demand response for microgrid planning is developed to determine the optimal sizing of DGs.

## 6.6 Switch Placement

The optimal placement of switches is an effective way to improve the reliability of active distribution networks [46]. Billinton and Jonnavithula [47] are among the earlier researchers who attempted to optimally place switches in distribution systems [48]. In a distribution system, sectionalizing switches are placed for different purposes, such as fault isolation, reliability improvement, network reconfiguration, etc. [47]. Therefore, the selection of optimal locations and the number of sectionalizing switches is an important consideration for microgrid planning in distribution systems. Installation of switches improves the reliability of the power supply, but it requires initial investments and increased operational costs over time [49]. Nevertheless, the optimal placement of switches is essential to improve the system's reliability and reduce power outages [50]. To justify the investment expenses, an optimization problem is required to be solved for the determination of the optimal number of switches and their locations [49, 51].

Power distribution companies presently consider distribution network automation an effective means to improve reliability and electric service quality [49]. To implement the distribution network automation, automation equipment, such as the remotely controlled switches [49, 52, 53], fault indicators [49, 53], and circuit breakers [54], are required in distribution networks. The cost of the interruption can be reduced, and the power system reliability improved due to the optimal placement of such equipment. Figure 6.3 shows major automation equipment used in active distribution systems.

Remotely controlled switches play an important role in an automated distribution network, which can reduce the time for isolating a faulted section, and perform a faster load restoration [51]. The system reliability can be effectively improved by installing remotely controlled switches, rather than manual switches [49]. Remote controlled switches, although more expensive than manual switches, can be employed when manual switches cannot meet relevant reliability requirements.

The remotely controlled switches placement is discussed in [55] through two scenarios: 1) the planning of a new distribution network with no switches in the system; and 2) automation of a distribution system equipped with existing manual switches, where the manual switches are optimally replaced with remotely controlled switches to meet the desired objectives. Table 6.3 shows a comparison between remotely controlled switches and manual switches.

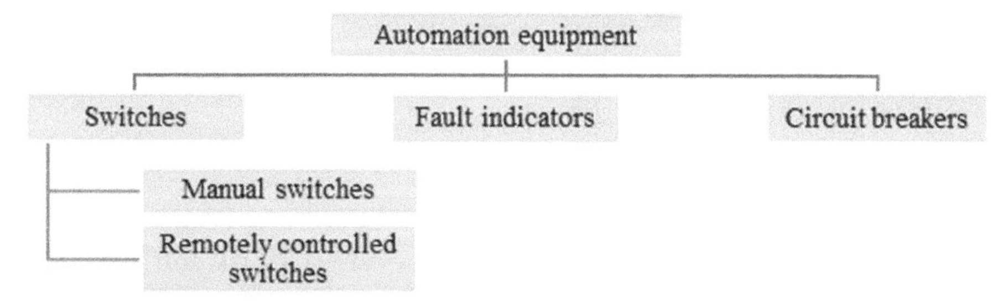


**Figure 6.3** Automation equipment used in active distribution systems.

**Table 6.3** A comparison between remotely controlled switches and manual switches.

| Performance | Remotely Controlled Switches | Manual Switches |
|---|---|---|
| Switching time | Less [50] | More [50] |
| Investment cost | Higher [49] | Lower [49] |
| Communication system failure's effect | Possibility of malfunction due to a communication system failure [60] | The communication system failure has no effect on its operation [60] |
| Fault isolation | Performed remotely | Repair crew is required to operate manually |
| Interruption cost | Lower [49] | Higher [49] |
| Reliability | Higher [49] | Lower [49] |

When a fault occurs in a distribution system, the fault management system should work promptly to bring the system back to a normal operating state. Three major tasks are performed by a typical fault management system: locating the fault, isolating the faulted section, and restoring services [49, 53]. After detecting the fault with the help of a fault indicator, sectionalizing switches isolate the healthy section from the faulted section in the network. The power supply to customers upstream the faulted area is restored using the original path, and customers downstream the faulted area are supplied through an alternative path temporarily created by the tie switches. These switching actions effectively reduce the cost of the interruption (e.g., loss of revenue), the number of interrupted customers, and the interruption duration, and thus, the power system reliability is improved. When the fault is repaired, the sectionalizing switches are closed to restore the power supply to the customers in the faulted section through the original path and the network operates in its normal operating status. The steps of the restoration procedures and switching operations are shown in Figure 6.4.

### 6.6.1 Objectives of Switch Placement

For the optimal placement of switches, improving the reliability and minimizing the cost are found to be the most used optimization objectives in literature.

The reliability indices, such as system average interruption frequency index (SAIFI), system average interruption duration index (SAIDI), and expected energy not supplied (EENS), are used to indicate the reliability of a distribution system [51, 54]. The objective function was formulated to improve the expected values of the reliability indices, SAIDI, SAIFI,

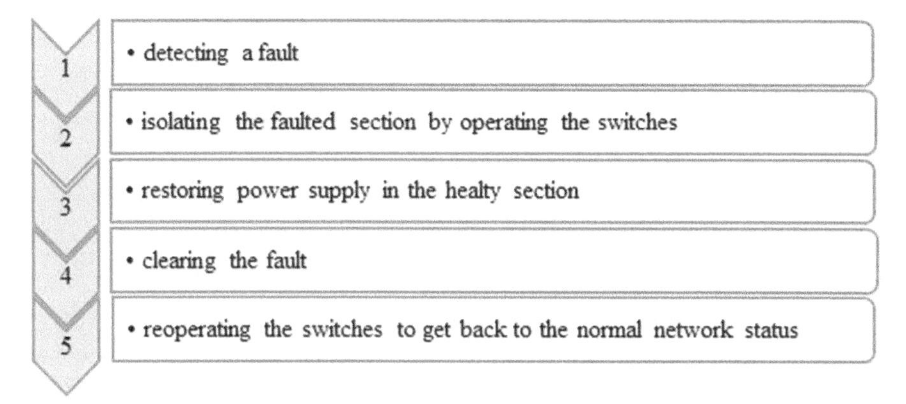


**Figure 6.4** The steps of the system restoration [56].

and EENS [56]. In [57], disconnect switches, including remotely controlled switches and manual switches and tie lines are deployed at the same time in the distribution network to improve the system reliability. It is found that the placement of disconnect switches and tie lines affects the duration of customer interruptions, SAIDI and EENS. However, the placement of tie lines at the end of feeders may not always be an optimal strategy to improve reliability.

The objective function is modeled to minimize the total costs of the distribution network, which includes the investment cost of switches, the investment cost of tie lines, the operational cost of the system, and the reliability-related cost [46]. The multiobjective switch placement algorithm in [50] has two member functions: the number of customers not supplied and the number of switches. The objective of the optimization model in [59] is to minimize the sum of DG curtailments in the distribution system. The objective function in [60] is formulated to maximize the revenue and minimize the operating cost. The multiobjective function in [49], [53] includes two single-objective functions to minimize the equipment and interruption cost. The objective in [47] is to select the optimum number of switches to minimize the outage cost, investment cost, and maintenance cost. The objective of the switch placement problem formulated in [51] is to minimize the total system cost in terms of customer outage cost.

Table 6.4 shows the summary of different objectives used for optimal placement of switches in literature. It is found that "minimizing cost" is the most dominant objective.

**Table 6.4** Summary of objectives used for optimal placement of switches in the literature.

| Objectives of Switch Placement | Reference |
|---|---|
| Improving reliability indices | [48], [54], [56], [58] |
| Minimizing cost | [46], [47], [48], [49], [51], [52], [53], [57], [58], [60], [61] |
| Minimizing the sum of DG curtailments | [59] |
| Maximizing revenue | [60] |

### 6.6.2 Methodology of Switch Placement

The selection of an effective methodology for allocating switches is very important because it is related to the restoration time and reliability index. The reliability oriented distribution switch optimization research has been carried out in [57, 58], which can be categorized into two groups based on the modeling techniques and problem-solving approaches [46]: heuristic based and mathematical programming based.

Generally, heuristic algorithms have a tendency of getting stuck at a local optima [46, 50, 51] and mathematical optimization methods, such as MILP, can reach the global optimum solution in a finite number of steps. To overcome the deficiency of heuristic approaches, the MILP based model for the reliability-oriented switch [46] and tie lines [57] placement in the distribution network is proposed. A MILP-based problem is formulated in [48] to minimize the normalized values for the total cost and the SAIDI. For the optimal placement of fault indicators and sectionalizing switches in the distribution network, the MILP-based optimization problems are solved in [49, 53]. A model is presented in [52], which incorporates the malfunction probability of sectionalizing switches into the optimal switch placement problem. The optimal placement of circuit breakers and switches is considered in [54], which proposes a new optimization model-based reliability assessment method for distribution networks.

In [56], an analytic hierarchical process (AHP) is used to develop a computer algorithm to solve the multiobjective remotely controlled switch allocation problem to improve the reliability index of distribution systems. In [50], a multiobjective optimization approach based on particle swarm optimization for switch placement is proposed. Instead of a single solution, the proposed algorithm provides a set of solutions. Evolutionary algorithms are largely applied to multiobjective optimization problems because generally more than one solution is obtained in such problems. A set of solutions can be best suited for several situations, and the decision making becomes simpler. In [59], a two-stage robust optimization model for the switch placement problem is formulated considering uncertainties of loads and DG outputs.

### 6.6.3 Discussion

The proposed method in [49] for optimal placement of fault indicators and manual switches was applied to Zhongshan (China) distribution network and the results are compared with the existing method, which shows that the energy not supplied and the cost of interruption are reduced: the energy not supplied is decreased by a total of 17.07 MWh, and the cost of interruption are reduced by 31.60%. The equipment configuration plan of the proposed method requires less equipment and the total cost decreases by 18.67% [49]. Reference [53] proposes a method to deploy remotely controlled switches, fault indicators, and manual switches to a typical Finnish 20 kV urban distribution network, and it is found that the interruption cost, SAIDI, and ENS of the network are decreased with the improved service reliability.

The results obtained in [48] show that both SAIDI and the total cost are decreased when automatic and manual switches are placed simultaneously, compared to the optimal placement of a single type of switch only. The result presented in [58] indicates that the customer interruption cost can be reduced effectively by allocating a small number of remotely controlled switches in a distribution system. The proposed method can also reduce the computation time due to the decreased number of integer variables in the optimization problem. However, the allocation of remotely controlled switches introduces the possibility of malfunction associated with the switches. The malfunction probability of sectionalizing switches is considered in the switch placement problem in [52]. The results show that even when malfunction probability is taken into consideration, the number of switches and the interruption cost are reduced and at the same time the reliability indices are improved.

## 6.7 Conclusion

This chapter provides a review of the planning strategy for optimal microgrid implementation in active distribution systems to improve the system reliability and the voltage profile, and reduce losses and costs. The challenging tasks in the planning stage include optimally sizing and allocating DGs, distributed reactive sources, energy storage systems, and switches considering the objectives of the planning. An additional challenge is how to properly model renewable energy sources considering their intermittent and random nature. This review covers critical elements of the planning study, including objectives, system modeling, microgrid topology, optimization algorithms, power flow studies, switch placement, etc. It provides an important guideline for planners to perform distribution system planning with microgrids.

## References

[1] IEEE Std 1547.4"IEEE Guide for Design, Operation, and Integration of Distributed Resource Island Systems with Electric Power Systems." 2011.

[2] S. A. Arefifar, Y. A. R. I. Mohamed, and T. H. M. EL-Fouly. "Optimum Microgrid Design for Enhancing Reliability and Supply-Security." *IEEE Transactions Smart Grid*, vol. 4, no. 3, pp. 1567–1575, Sep. 2013.

[3] M. H. S. Boloukat and A. A. Foroud. "Multiperiod Planning of Distribution Networks Under Competitive Electricity Market With Penetration of Several Microgrids, Part I: Modeling and Solution Methodology." *IEEE Transactions on Industrial Informatics.*, vol. 14, no. 11, pp. 4884–4894, Nov. 2018.

[4] M. Khalid, U. Akram, and S. Shafiq. "Optimal Planning of Multiple Distributed Generating Units and Storage in Active Distribution Networks." *IEEE Access*, vol. 6, pp. 55234–55244, 2018.

[5] M. V. Kirthiga, S. A. Daniel, and S. Gurunathan. "A Methodology for Transforming an Existing Distribution Network into a Sustainable Autonomous Micro-Grid." *IEEE Transactions Sustainable Energy*, vol. 4, no. 1, pp. 31–41, Jan. 2013.

[6] L. Che, X. Zhang, M. Shahidehpour, A. Alabdulwahab, and Y. Al-Turki. "Optimal Planning of Loop-Based Microgrid Topology." *IEEE Transactions Smart Grid*, vol. 8, no. 4, pp. 1771–1781, Jul. 2017.

[7] S. Mohamed, M. F. Shaaban, M. Ismail, E. Serpedin, and K. A. Qaraqe. "An Efficient Planning Algorithm for Hybrid Remote Microgrids." *IEEE Transactions Sustainable Energy*, vol. 10, no. 1, pp. 257–267, Jan. 2019.

[8] W. Fang, H. Liu, F. Chen, H. Zheng, G. Hua, and W. He. "DG Planning in Stand-Alone Microgrid Considering Stochastic Characteristic." *Journal of Engineering*, vol. 2017, no. 13, pp. 1181–1185, Jan. 2017.

[9] R. A. Osama, A. F. Zobaa, and A. Y. Abdelaziz. "A Planning Framework for Optimal Partitioning of Distribution Networks Into Microgrids." *IEEE System Journals*, vol. 14, no. 1, pp. 916–926, Mar. 2020.

[10] Z. Liu, X. Wang, R. Zhuo, and X. Cai. "Flexible Network Planning of Autonomy Microgrid." *IET Renewable Power Generation*, vol. 12, no. 16, pp. 1931–1940, Dec. 2018.

[11] N. Nikmehr and S. Najafi Ravadanegh. "Heuristic Probabilistic Power Flow Algorithm for Microgrids Operation and Planning." *IET Generation Transmission Distribution*, vol. 9, no. 11, pp. 985–995, Aug. 2015.

[12] L. Zhu, X. Zhou, X.-P. Zhang, Z. Yan, S. Guo, and L. Tang. "Integrated Resources Planning in Microgrids Considering Interruptible Loads and Shiftable Loads." *Journal of Modern Power System and Clean Energy*, vol. 6, no. 4, pp. 802–815, Jul. 2018.

[13] L. Zhu, Z. Yan, W.-J. Lee, X. Yang, Y. Fu, and W. Cao. "Direct Load Control in Microgrids to Enhance the Performance of Integrated Resources Planning." *IEEE Transactions Industry Applications*, vol. 51, no. 5, pp. 3553–3560, Sep. 2015.

[14] A. Khodaei, S. Bahramirad, and M. Shahidehpour. "Microgrid Planning Under Uncertainty." *IEEE Transactions Power Syststems*, vol. 30, no. 5, pp. 2417–2425, Sep. 2015.

[15] M. Kumar and B. Tyagi. "An Optimal Multivariable Constrained Non-linear (MVCNL) Stochastic Microgrid Planning and Operation Problem with Renewable Penetration." *IEEE System Journals*, vol. 14, no. 3, pp. 4143–4154, Sep. 2020.

[16] F. Samadi Gazijahani and J. Salehi. "Optimal Bilevel Model for Stochastic Risk-Based Planning of Microgrids Under Uncertainty." *IEEE Transactions Industrial Inform.*, vol. 14, no. 7, pp. 3054–3064, Jul. 2018.

[17] J. Kim and Y. Dvorkin. "Enhancing Distribution System Resilience With Mobile Energy Storage and Microgrids." *IEEE Transactions Smart Grid*, vol. 10, no. 5, pp. 4996–5006, Sep. 2019.

[18] L. Guo, W. Liu, B. Jiao, B. Hong, and C. Wang. "Multi-objective stochastic optimal planning method for stand-alone microgrid system." *IET Generation Transmission Distribution*, vol. 8, no. 7, pp. 1263–1273, Jul. 2014.

[19] J. Mitra, M. R. Vallem, and C. Singh. "Optimal Deployment of Distributed Generation Using a Reliability Criterion." *IEEE Transactions Industrial Applications*, vol. 52, no. 3, pp. 1989–1997, May 2016.

[20] A. Mohsenzadeh, C. Pang, and M.-R. Haghifam. "Determining Optimal Forming of Flexible Microgrids in the Presence of Demand Response in Smart Distribution Systems." *IEEE System Journals*, vol. 12, no. 4, pp. 3315–3323, Dec. 2018.

[21] C. Wang, B. Jiao, L. Guo, K. Yuan, and B. Sun. "Optimal Planning of Stand-Alone Microgrids Incorporating Reliability." *J. Modern Power System Clean Energy*, vol. 2, no. 3, pp. 195–205, Sep. 2014.

[22] H. Hedayati, S. A. Nabaviniaki, and A. Akbarimajd. "A Method for Placement of DG Units in Distribution Networks." *IEEE Transactions Power Delivery*, vol. 23, no. 3, pp. 1620–1628, Jul. 2008.

[23] Z. Wang, B. Chen, J. Wang, J. Kim, and M. M. Begovic. "Robust Optimization Based Optimal DG Placement in Microgrids." *IEEE Transactions Smart Grid*, vol. 5, no. 5, pp. 2173–2182, Sep. 2014.

[24] N. Daryani, K. Zare, and S. Tohidi. "Design for Independent and Self-Adequate Microgrids in Distribution Systems Considering Optimal Allocation of DG Units." *IET Generation Transmission Distribution*, vol. 14, no. 5, pp. 728–734, Mar. 2020.

[25] A. Andreotti, G. Carpinelli, F. Mottola, D. Proto, and A. Russo. "Decision Theory Criteria for the Planning of Distributed Energy Storage Systems in the Presence of Uncertainties." *IEEE Access*, vol. 6, pp. 62136–62151, 2018.

[26] S. A. Arefifar, Y. A.-R. I. Mohamed, and T. H. M. El-Fouly. "Supply-Adequacy-Based Optimal Construction of Microgrids in Smart Distribution Systems." *IEEE Transactions Smart Grid*, vol. 3, no. 3, pp. 1491–1502, Sep. 2012.

[27] L. Bhamidi and S. Sivasubramani. "Optimal Planning and Operational Strategy of a Residential Microgrid with Demand Side Management." *IEEE System Journals*, vol. 14, no. 2, pp. 2624–2632, Jun. 2020.

[28] J. Chen *et al.* "Optimal Sizing for Grid-Tied Microgrids with Consideration of Joint Optimization of Planning and Operation." *IEEE Transactions Sustainable Energy*, vol. 9, no. 1, pp. 237–248, Jan. 2018.

[29] A. O. Rousis, I. Konstantelos, and G. Strbac. "A Planning Model for a Hybrid AC–DC Microgrid Using a Novel GA/AC OPF Algorithm." *IEEE Transactions Power System*, vol. 35, no. 1, pp. 227–237, Jan. 2020.

[30] A. A. Hamad, M. E. Nassar, E. F. El-Saadany, and M. M. A. Salama. "Optimal Configuration of Isolated Hybrid AC/DC Microgrids." *IEEE Transactions Smart Grid*, vol. 10, no. 3, pp. 2789–2798, May 2019.

[31] N. Kanwar, N. Gupta, K. R. Niazi, and A. Swarnkar. "Optimal Distributed Resource Planning for Microgrids Under Uncertain Environment." *IET Renewable Power Generation*, vol. 12, no. 2, pp. 244–251, Feb. 2018.

[32] S. F. Contreras, C. A. Cortes, and J. M. A. Myrzik. "Optimal Microgrid Planning for Enhancing Ancillary Service Provision." *Journals Modern Power System Clean Energy*, vol. 7, no. 4, pp. 862–875, Jul. 2019.

[33] K. Karimizadeh, S. Soleymani, and F. Faghihi. "Optimal Placement of DG Units for the Enhancement of MG Networks Performance Using Coalition Game Theory." *IET Generation Transmission Distribution*, vol. 14, no. 5, pp. 853–862, Mar. 2020.

[34] A. H. Yazdavar, M. F. Shaaban, E. F. El-Saadany, M. M. A. Salama, and H. H. Zeineldin. "Optimal Planning of Distributed Generators and Shunt Capacitors in Isolated Microgrids With Nonlinear Loads." *IEEE Transactions Sustainable Energy*, vol. 11, no. 4, pp. 2732–2744, Oct. 2020.

[35] F. S. Gazijahani and J. Salehi. "Robust Design of Microgrids with Reconfigurable Topology under Severe Uncertainty." *IEEE Transactions Sustainable Energy*, vol. 9, no. 2, pp. 559–569, Apr. 2018.

[36] W. Yuan, J. Wang, F. Qiu, C. Chen, C. Kang, and B. Zeng. "Robust Optimization-Based Resilient Distribution Network Planning Against Natural Disasters." *IEEE Transactions Smart Grid*, vol. 7, no. 6, pp. 2817–2826, Nov. 2016.

[37] M. Baseer, G. Mokryani, R. H. A. Zubo, and S. Cox. "Planning of HMG with high penetration of renewable energy sources." *IET Renewable Power Generation*, vol. 13, no. 10, pp. 1724–1730, Jul. 2019.

[38] A. Khodaei. "Provisional Microgrid Planning." *IEEE Transactions Smart Grid*, vol. 8, no. 3, pp. 1096–1104, May 2017.

[39] S. D. Manshadi and M. E. Khodayar. "Expansion of Autonomous Microgrids in Active Distribution Networks." *IEEE Transactions Smart Grid*, pp. 1–1, 2016.

[40] C. A. Cortes, S. F. Contreras, and M. Shahidehpour. "Microgrid Topology Planning for Enhancing the Reliability of Active Distribution Networks." *IEEE Transactions Smart Grid*, vol. 9, no. 6, p. 9, 2018.

[41] L. Che, X. Zhang, M. Shahidehpour, A. Alabdulwahab, and A. Abusorrah. "Optimal Interconnection Planning of Community Microgrids with Renewable Energy Sources." *IEEE Transactions Smart Grid*, vol. 8, no. 3, pp. 1054–1063, May 2017.

[42] B. Zhao, X. Zhang, J. Chen, C. Wang, and L. Guo. "Operation Optimization of Standalone Microgrids Considering Lifetime Characteristics of Battery Energy Storage System." *IEEE Transactions Sustainable Energy*, vol. 4, no. 4, pp. 934–943, Oct. 2013.

[43] R. Dufo-López and J. L. Bernal-Agustín. "Multi-Objective Design of PV–Wind–Diesel–Hydrogen–Battery Systems." *Renewable Energy*, vol. 33, no. 12, pp. 2559–2572, Dec. 2008.

[44] Y. A. Katsigiannis, P. S. Georgilakis, and E. S. Karapidakis. "Hybrid Simulated Annealing–Tabu Search Method for Optimal Sizing of Autonomous Power Systems With Renewables." *IEEE Transactions Sustainable Energy*, vol. 3, no. 3, pp. 330–338, Jul. 2012.

[45] N. Nikmehr and S. Najafi-Ravadanegh. "Optimal Operation of Distributed Generations in Micro-Grids under Uncertainties in Load and Renewable Power Generation Using Heuristic Algorithm." *IET Renewable Power Generation*, vol. 9, no. 8, pp. 982–990, Nov. 2015.

[46] M. Jooshaki, S. Karimi-Arpanahi, M. Lehtonen, R. J. Millar, and M. Fotuhi-Firuzabad. "Electricity Distribution System Switch Optimization Under Incentive Reliability Scheme." *IEEE Access*, vol. 8, pp. 93455–93463, 2020.

[47] R. Billinton and S. Jonnavithula. "Optimal Switching Device Placement in Radial Distribution Systems." *IEEE Transactions Power Delivery*, vol. 11, no. 3, pp. 1646–1651, Jul. 1996.

[48] A. Shahbazian, A. Fereidunian, and S. D. Manshadi. "Optimal Switch Placement in Distribution Systems: A High-Accuracy MILP Formulation." *IEEE Transactions Smart Grid*, vol. 11, no. 6, pp. 5009–5018, Nov. 2020.

[49] B. Li, J. Wei, Y. Liang, and B. Chen. "Optimal Placement of Fault Indicator and Sectionalizing Switch in Distribution Networks." *IEEE Access*, vol. 8, pp. 17619–17631, 2020.

[50] J. R. Bezerra, G. C. Barroso, R. P. S. Leao, and R. F. Sampaio. "Multiobjective Optimization Algorithm for Switch Placement in Radial Power Distribution Networks." *IEEE Transactions Power Delivery*, vol. 30, no. 2, pp. 545–552, Apr. 2015.

[51] A. Abiri-Jahromi, M. Fotuhi-Firuzabad, M. Parvania, and M. Mosleh. "Optimized Sectionalizing Switch Placement Strategy in Distribution Systems." *IEEE Transactions Power Delivery*, vol. 27, no. 1, pp. 362–370, Jan. 2012.

[52] M. Farajollahi, M. Fotuhi-Firuzabad, and A. Safdarian. "Optimal Placement of Sectionalizing Switch Considering Switch Malfunction Probability." *IEEE Transactions Smart Grid*, vol. 10, no. 1, pp. 403–413, Jan. 2019.

[53] M. Farajollahi, M. Fotuhi-Firuzabad, and A. Safdarian. "Simultaneous Placement of Fault Indicator and Sectionalizing Switch in Distribution Networks." *IEEE Transactions Smart Grid*, vol. 10, no. 2, pp. 2278–2287, Mar. 2019.

[54] Z. Li, W. Wu, X. Tai, and B. Zhang. "Optimization Model-Based Reliability Assessment for Distribution Networks Considering Detailed Placement of Circuit Breakers and Switches." *IEEE Transactions Power System*, vol. 35, no. 5, pp. 3991–4004, Sep. 2020.

[55] Y. Xu, C.-C. Liu, K. P. Schneider, and D. T. Ton. "Placement of Remote-Controlled Switches to Enhance Distribution System Restoration Capability." *IEEE Transactions Power System*, vol. 31, no. 2, pp. 1139–1150, Mar. 2016.
[56] D. P. Bernardon, M. Sperandio, V. J. Garcia, L. N. Canha, A. da R. Abaide, and E. F. B. Daza. "AHP Decision-Making Algorithm to Allocate Remotely Controlled Switches in Distribution Networks." *IEEE Transactions Power Delivery*, vol. 26, no. 3, pp. 1884–1892, Jul. 2011.
[57] M. Jooshaki, S. Karimi-Arpanahi, M. Lehtonen, R. J. Millar, and M. Fotuhi-Firuzabad. "Reliability-Oriented Electricity Distribution System Switch and Tie Line Optimization." *IEEE Access*, vol. 8, pp. 130967–130978, 2020.
[58] S. Lei, J. Wang, and Y. Hou. "Remote-Controlled Switch Allocation Enabling Prompt Restoration of Distribution Systems." *IEEE Transactions Power System*, vol. 33, no. 3, pp. 3129–3142, May 2018.
[59] S. Lei, Y. Hou, F. Qiu, and J. Yan. "Identification of Critical Switches for Integrating Renewable Distributed Generation by Dynamic Network Reconfiguration." *IEEE Transactions Sustainable Energy*, vol. 9, no. 1, pp. 420–432, Jan. 2018.
[60] Z. Liu, Y. Liu, G. Qu, X. Wang, and X. Wang. "Intra-Day Dynamic Network Reconfiguration Based on Probability Analysis Considering the Deployment of Remote Control Switches." *IEEE Access*, vol. 7, pp. 145272–145281, 2019.
[61] A. Heidari, V. G. Agelidis, and M. Kia. "Considerations of Sectionalizing Switches in Distribution Networks with Distributed Generation." *IEEE Transactions Power Delivery*, vol. 30, no. 3, pp. 1401–1409, Jun. 2015.

# 7

# Overview on Reliability of PV Inverters in Grid-connected Applications

**Ranjith kumar Gatla[1], N.V Prasad K[2], P. Sridhar[1], Jianghua Lu[3], and Devineni Gireesh kumar[4]**

[1]Department of Electrical and Electronics Engineering, Institute of Aeronautical Engineering, Hyderabad, India
[2]School of Electrical and Automation Engineering, Hefei University of Technology, Hefei, PR China
[3]School of Information Science and Engineering, Wuhan University of Science and Technology, Wuhan, P R China
[4]Department of Electrical and Electronics Engineering, B.V Raju Institute of Technology, Narsapur, Telangana, India
E-mail: ranjith.gatla@gmail.com

## Abstract

With the quickly developing of grid-connected Photovoltaic (PV) systems, its reliability is being concentrated widely. The PV inverter is the weakest part of the PV system. Therefore, this paper presents an overview of the reliability of PV inverters in grid-connected applications. The discussion includes different PV inverter configurations for grid-connected systems, basic principles of reliability, and the importance of reliability evaluation in PV inverters. Finally, presents different lifetime estimation methods available to evaluate the reliability of the most frequent failure components in the PV inverter, which are power devices and dc-link capacitors. The main objective of this review is to provide the basics of reliability research in PV inverters.

**Keywords:** PV Inverter, Reliability, Thermal Stress, IGBT, and Dc-link Capacitor.

## 7.1 Introduction

In present days, the development in renewable energy sources (RES) to produce electric power has expeditiously grown. The various causes such as ever-increasing of electric power demand, the comprehensive nature of fossil fuel resources, and their increased prices. The installed capacity of the total renewable energy is approximately 2378 GW by the end of 2018 [1], and the capacity is expected to be even higher in the future. Mostly, the developing countries are increasing their capacity in the installation of renewable energy; China is the first on this list. For the past 10 years (2008–2018), China is in top ranking in developing and installing new renewable energy. Among the different renewable energy technologies, the electric power generated from PV energy has become the world's most popular RES, as per the global information demonstrating that more PV energy capacity is being installed than any other renewable energy. In 2018, The total freshly installed renewable power consisted of a major part of 55% PV energy, 28 % wind energy, and 11% hydropower [1].

PV energy has various benefits, such as supply chains, cost, and production scale that it is tough to spot any other renewable energy challenging than within the next decade or so [2]. From the last 10 years, the production of PV energy is rapidly increasing, PV global capacity and the annual addition during the last 10 years are shown in Figure 7.1 Mostly two types

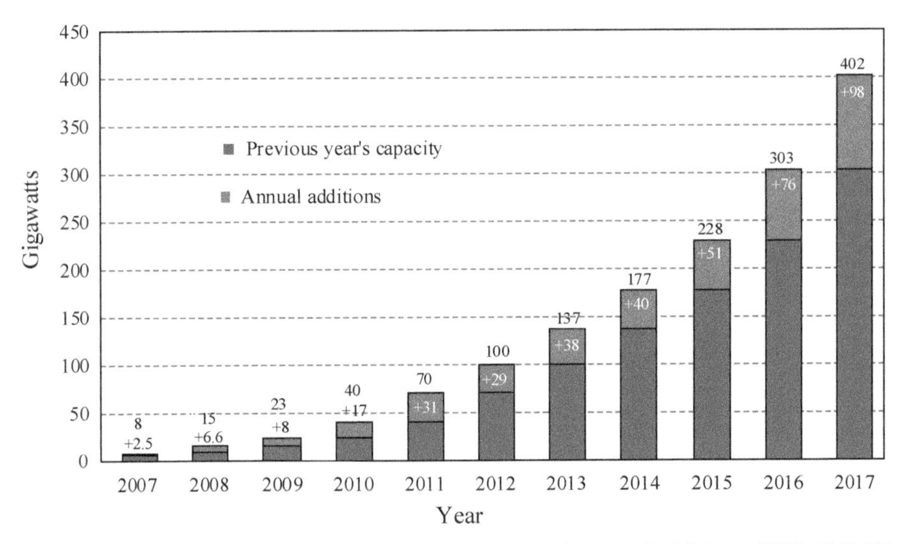


**Figure 7.1** Global capacity of the PV energy system and its annual additions, 2007–2017 [3]

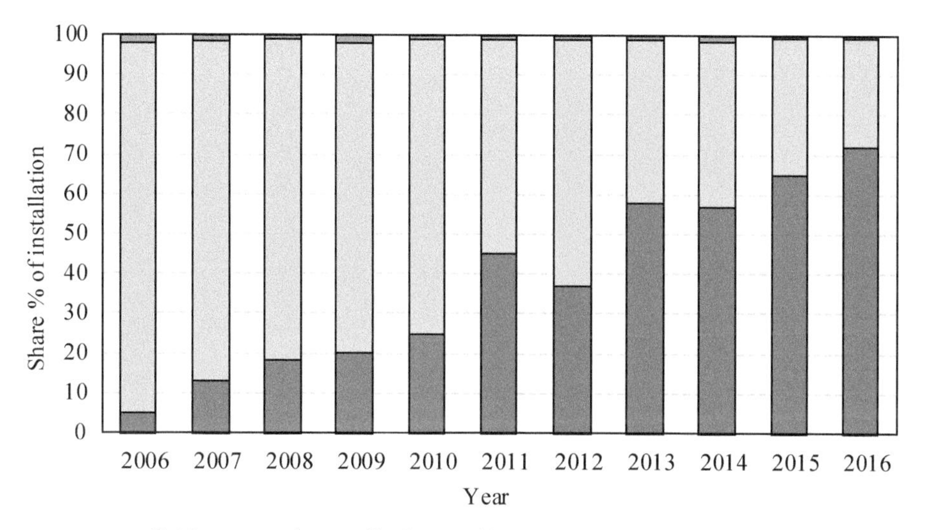


**Figure 7.2** A share of grid-connected, grid-connected decentralized and off-grid installations, 2006–2016 [3]

of PV systems are available: grid-connected and standalone PV systems. Currently, the grid-connected PV systems are the most evolving PV energy applications over standalone PV systems. The Share of the off-grid and grid-connected installations of the total PV installation is shown in Figure 7.2, it can be observed that almost more than 95% of the PV installation is grid-connected.

In recent days, remarkable attention has been devoted to the maintenance cost and reliability of PV inverters to satisfy the ambitious product requirements for the PV market-long operating hours under harsh environments. Inverters are employed in most modern life photovoltaic systems, nevertheless, they are considered as one of the weakest links in terms of reliability. According to a survey [4] as shown in Figure 7.3, the unscheduled maintenance that occurs due to PV inverter is about 37 % of the total occurrences. The unscheduled maintenance results in a loss of capacity in the generation that impacted one or more systems and required human treatment to restore the system to full operational capacity. Thus, as previously mentioned, the increasing reliability of PV inverters remains in immediate demand on top of the performance.

To understand the factors which, limit the reliability of PV inverters, each failure mechanism under critical stress conditions must be investigated.

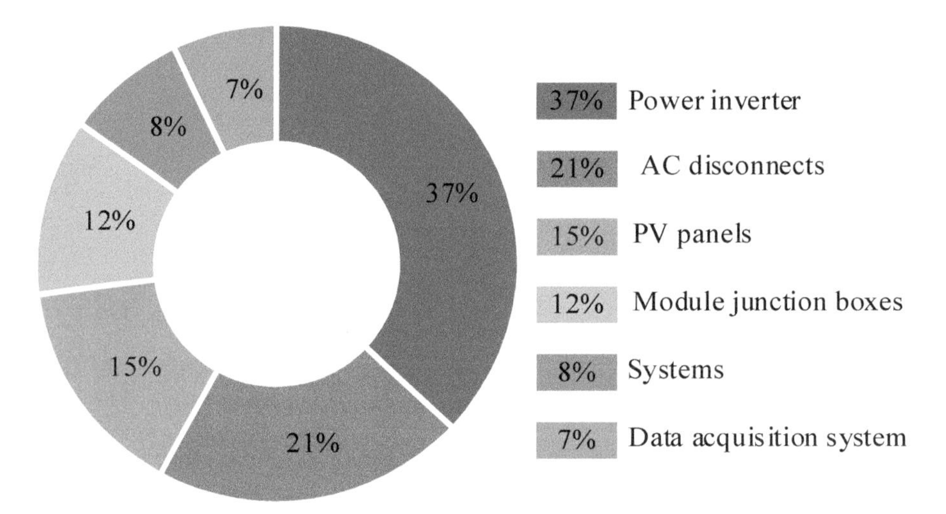


**Figure 7.3** Unscheduled maintenance occurrences by components [4]

A survey based on field experiences from power devices producers, customers in the motor drive, automation, aeronautics, and other manufacturing divisions [5], reveals that power semiconductors and dc-link capacitors are the most critical components in PV converters. According to the survey published in [5], it can be observed that 31% of the responders have chosen power semiconductors and 21% of responders have chosen dc-link capacitors are the main contributors to failure which is shown in Figure 7.4. The gate drives failure rate is 18%, and the connectors have a failure rate of 12%. The inductor, resistor, and other devices have a lower failure rate. Therefore, the Insulated Gate Bipolar Transistor (IGBT) and dc-link capacitors are the principal focus of this paper.

## 7.2 Power Converters for PV Systems

In PV systems, the use of all the available energy depends on the static converter topology that is used in it. Based on the topology of the inverter topology and configuration of the PV module, the grid-connected PV systems are categorized into different groups [6–7]: Centralized inverters; String inverters; multistring inverters; AC-module inverters; Modular cascaded inverters. The different PV system configurations are shown in Figure 7.5.

In the central inverter, the output voltage of the PV module is small, to increase the inverter output voltage, every PV module is connected in a

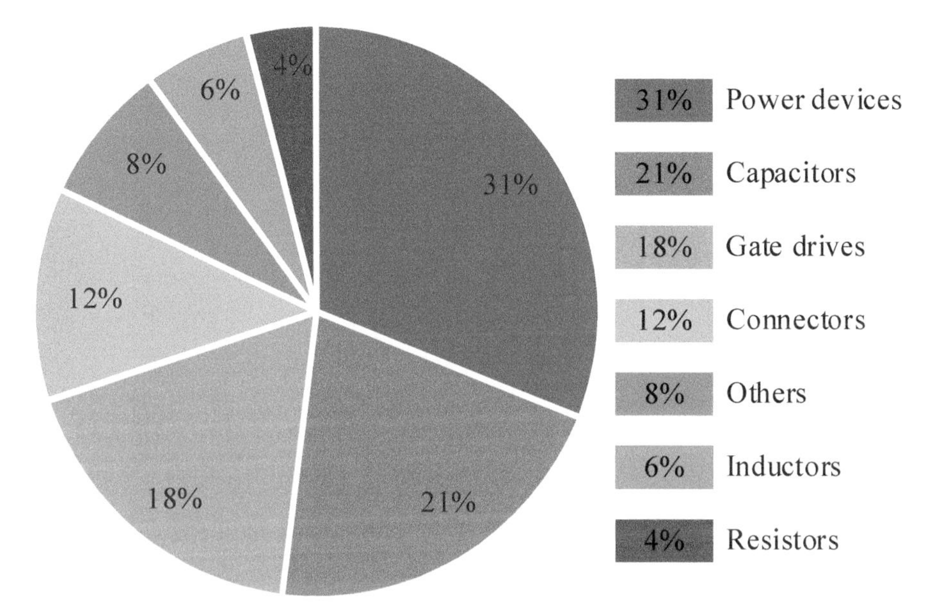


**Figure 7.4** Failure distribution of components in PV inverter [5]

series with each other so that further amplification can be avoided [6]. The series-coupled PV modules are known as PV strings. The string diodes are connected in a series with each string, to increase the inverter power level; these strings are connected in parallel, as shown in Figure 7.5 (a). The main drawback of the central inverter is the losses between the mismatching PV modules due to a centralized MPP tracking [8].

Smaller power level (1–5 kW) applications are used string inverse. In this configuration, the series-connected PV modules are used to boost the inverter output voltage. This configuration is only suitable for one PV string per inverter, as shown in Figure 7.5 (b). These string inverters are mainly used where the orientation of the modules is different.

In the multistring inverter, strings are connected in parallel, and every string is linked to its DC-DC boost converter. The parallel combinations of string-connected DC-DC converters are connected to a single main dc-ac inverter as shown in Figure 7.5 (c). In the multistring inverter, independent MPP tracking can be applied by using a DC-DC converter, this can improve the PV module utilization. One of the major disadvantages of this configuration is, it has two stages of power conversion, which causes a decrease in peak efficiency compared to other configurations [9–10].

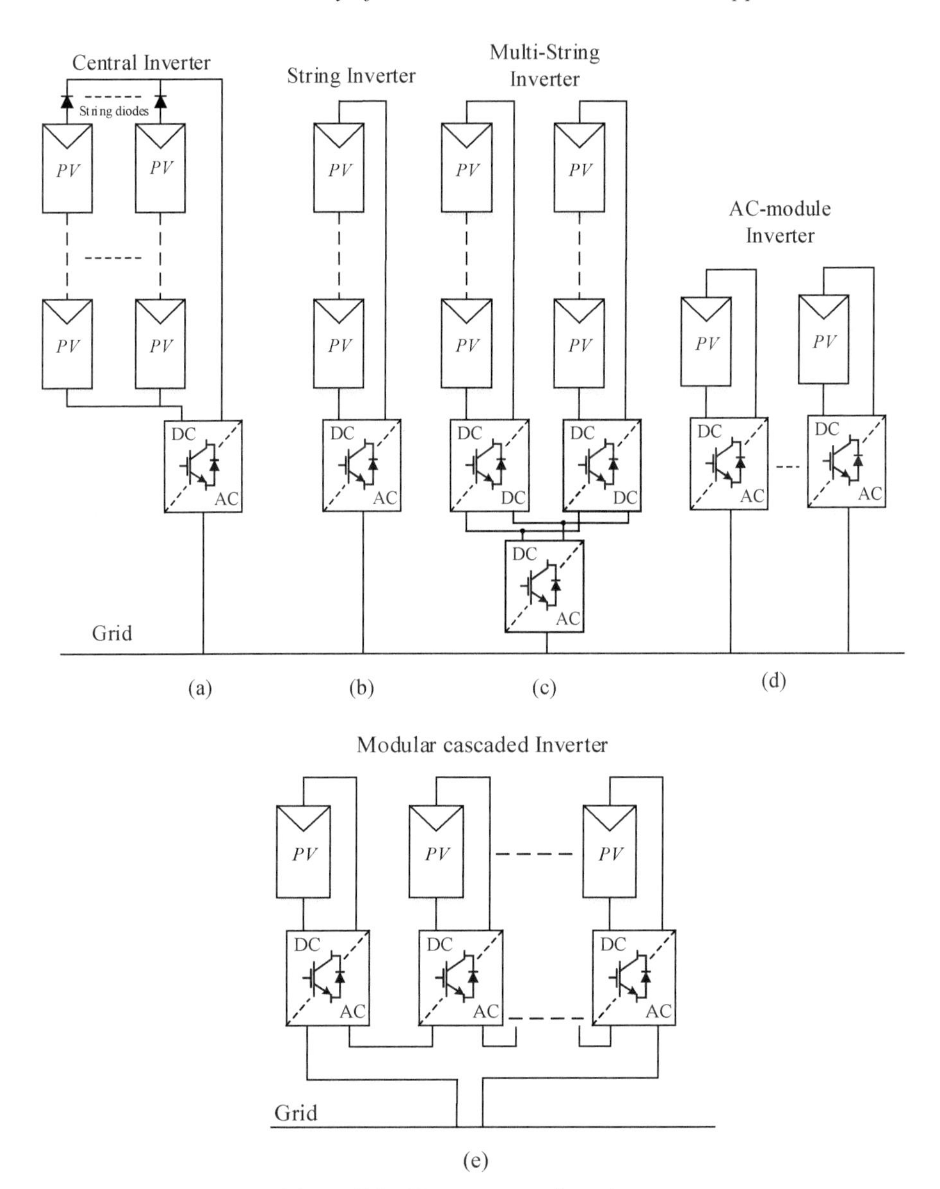


**Figure 7.5** PV system configurations

In the AC-module configuration, the dc-ac inverter is attached by only one PV panel as shown in Figure 7.5(d). In this configuration, the PV module mismatching losses will not exist. The PV system can be easily enlarged due to the modular configuration. The main disadvantage of the AC-module configuration is, it requires complex topologies to improve the amplification of the voltage, which increases the losses in the system.

In the modular cascaded inverter, every PV module is connected to its dc-ac converter, and the dc-ac converters are connected in series. The converter output voltage can be increased by increasing the number of series-connected dc-ac converters as shown in Figure 7.5(e). The cascaded inverters are two types; one is with two-stage power conversion configurations (one is DC/DC conversion and another one is DC/AC conversion), another one is a single-stage power conversion configuration [11–12].

## 7.3 Basic Principles of Reliability

Reliability is defined as “the ability of a system or device to work under specified conditions for a stated period,” also described as “the ability to work at a stated moment or period (Availability).” The theoretical definition of reliability is the frequency of failures, is the probability of success, is; a probability derived from maintainability, or in terms of availability and testability [19].

### 7.3.1 Failure Rate

Every product or device has a failure rate which is the number of units failing per unit time. It is the manufacturer’s aim to ensure that the device in the infant mortality period does not get to the customer; this leaves products with a useful life period during which failures occur randomly. The failure rate of a system normally relies on time, with the rate differing over the life cycle of the system. The reliability R(t) is the reliability at time t; it is expressed as

$$R(t) = e^{-\lambda t} \tag{7.3.1}$$

Where $\lambda$ represents the failure rate, which is expressed in failures/hour, $t$ is the period of operation. There is an assumption of random independent failures, which is generally useful for power electronic components. In practically the failure rate is not constant, it can be observed from a graph of failure rate versus component age often called a bathtub curve[20] which is represented in Figure 7.6 and the failure rate is constant only during the middle of the component life.

In the beginning, components have a high failure rate due to manufacturing defects that managed to pass tests but make the components wear out

**Figure 7.6** Typical failure rate curve as a function of time

very quickly, which is called as burn-in phase and can last depending on the system. After the burn-in phase, the useful product life does indeed provide a more or less constant failure rate. But we also need to take into consideration the end of product life where failure rate increases as components begin to wear-out due to use and age, which period is called a wear out phase. It is necessary to keep in mind that the device age axis is logarithmic, so the burn-in period is relatively short, while the end-of-life period is a very long slow ramp-up. But at some point, components wear out and need to be discarded or replaced.

### 7.3.2 Mean Time to Failure (MTTF)

The MTTF is the anticipated time before a failure takes place. Unlike reliability, MTTF does not rely upon a particular period. It provides the ordinary time in which a product works without fail. MTTF is an extensively priced estimate efficiency measure for the differentiation of numerous system designs. This index reflects the life circulation of a thing.

The following equation explains the MTTF

$$MTTF = \int_0^{+\infty} R(t)dt \tag{7.3.2}$$

The simplified equation for MTTF when $\lambda$ constant is expressed as

$$MTTF = \frac{1}{\lambda} \tag{7.3.3}$$

### 7.3.3 Mean Time to Repair (MTTR)

MTTR is a fundamental action of the maintainability of repairable items. It stands for the average time called to repair an unsuccessful part or device; it depends on maintenance conditions or contract [21]. Shared mathematically, it is the overall restorative upkeep time for failings separated by the complete variety of rehabilitative maintenance actions for failings throughout a provided time period.

The mathematical formulation is represented by the following equation

$$MTTR= \int_{0}^{+\infty} xN\left(x\right)dx = \frac{\beta}{\mu} \tag{7.3.4}$$

Where *N(x)* is the gamma distribution function with parameters $\mu$ and $\beta$. MTTR is usually evaluated by conducting experiments based on the previous repairs data[22].

### 7.3.4 Mean Time Between Failure (MTBF)

MTBF is a metric that worries the average time expired in between a failure and the following time it occurs. The MTBF is a key system specification in systems where failure rate requires to be handled, particularly for safety systems. The MTBF appears regularly in the engineering style requirements and also governs the regularity of needed system maintenance as well as examinations.

These lapses of time can be calculated by using a formula.

$$\text{MTBF} = \frac{\text{Operating time}}{\text{Number of failures}} \tag{7.3.5}$$

It can also be defined as the sum of MTTF and MTTR

$$MTBF = MTTF + MTTR \tag{7.3.6}$$

The random and emergency failures (which include a vast range of components and failures) can be predicted or measured by using *MTBF.* In [23], The accuracy measurement of the MTBF has been improved by the condition-based fault tree analysis (CBFTA). The results state that system-level reliability also improved by using CBFTA.

## 7.4 Power Module Reliability

Power electronics refer to the technologies behind switching power supplies, power converters, inverters, motors, and drives. The discrete components used within power electronic systems include diodes, power modules, capacitors, etc., The trends in the industry include conversion efficiencies which means less waste heat. However, as the package size becomes smaller and they increase the amount of power that they can carry combine this with harsh operating environments which affect the thermal behaviour of devices. The common thermal design pitfalls include inefficiently configured heat sinks. The consequences of improperly managing the thermal loads generated will cause inconsistent performance on these devices and shortened times of failure.

### 7.4.1 Reliability Analysis of IGBT Module

The main stressors to get a failure of the IGBT is thermal stress, electrical stress, mechanical stress, etc., Table 7.1 illustrates the effectiveness of the different stressors on the IGBT module, and it can be observed that the thermal stress is critical stressor in the IGBT module.

Steady-state temperature and cycling have one of the huge substantial influences on failure modes as well as power semiconductor failure mechanisms. Because of this, an exertion on investigating the most efficient power

**Table 7.1** Critical stressors for power devices (High to the low level of importance →***→**→*)

| Load | | | Power Device | | |
|---|---|---|---|---|---|
| Type of stress | Product design | Stressor | Die | Solder joints | Wire-bond |
| Temperature cycling and steady-state | • Thermal control<br>• Operating point.<br>• On/Off<br>• Operating power | Junction Temperature swing $\Delta T$ | *** | *** | *** |
| | | Mean junction temperature $T_m$ | ** | ** | ** |
| | | $dT/dt$ | * | * | * |
| Vibration/ chock | • Mechanical | Vibration /shock | * | | |
| Humidity/ Moisture | • Thermal control<br>• On/Off<br>• Breathing effect<br>• Operating power | Relative humidity | * | * | * |
| Contaminants and dust | • Enclosure design | Pollution | | | |

modules, control algorithms for bringing down the temperature, and low-cost heat sinks are the points of mainstream researchers.

Previous studies [24–28] revealed the failure mechanism of IGBTs - thermo-mechanical stresses in adjacent layers. As illustrated in Figure 7.1, several layers inside the IGBT are assembled with different thermal expansion coefficients. Therefore, during the normal operation, the uneven expansion and contraction due to thermal cycling generate shear strains and shear stresses until cracks or disconnections are triggered. Hence, the reduction of these stresses has become crucial to enlarge the reliability of the assembly.

The important cause of the power module failure is thermal cycling [29] because the structure of a power device is with different materials and these materials have different coefficients of thermal expansion in power semiconductors. For example, the power module consists of a silicon chip that is sitting on top of a ceramic layer, let consider, the silicon chip dimensions are 1 cm at 250C room temperature, and that masked the ceramic layer. When the temperature is at 1250C, the silicon might expand from 10mm to 11mm, and the ceramic might expand from 10 mm to 12mm, which means that two structures now have a mismatch and it is trying to develop a crack at the corners which cause failure in the power module.

Figure 7.7 show the internal structure of the IGBT module. The structure has the heatsink essentially at the bottom, the commonly used material for the heat sink is aluminum, and copper is also used where higher heating storage is needed. Thermal interface material or thermal grease is placed in between the base plate and heatsink which is a thin layer. The base plate is essentially the bottom of the module which tightens with screws to the heat sink, commonly used materials for the base plate are either copper or aluminum silicon carbide which has a better match with the thermal expansion coach. The Direct-Bond-Copper (DBC) is located above the base plate, which is a way of bonding a

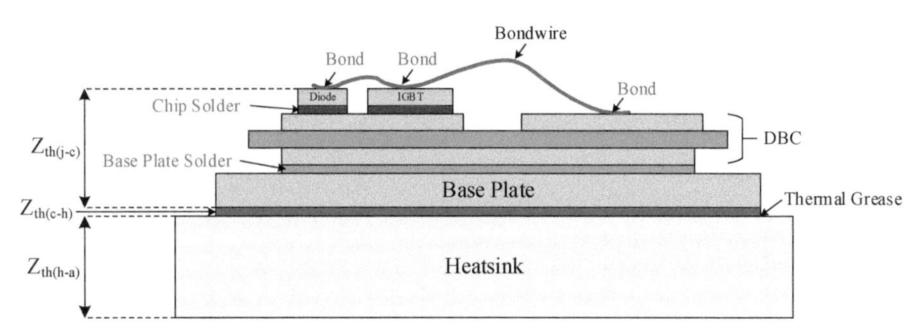


**Figure 7.7** The internal structure of an IGBT module [30]

ceramic to copper structure, and the upper side of the ceramic is also DBC which is used to connect the chips to the bottom of the module.

According to [30], solder joints cracking under the DBC, solder joints cracking under the chip (IGBT and diode), and the bond wire lift-off are the three main IGBT module wear-out mechanisms due to thermal cyclic stress

To establish the lifetime models for such failure mechanisms, accelerated temperature cycling tests are normally conducted. To ensure that the test will trigger both failure modes (solder degradation and wire bond), a test has been proposed based on the Bycycle in [31], which is nearer to genuine applications than singular tests. When the temperature variations are undergone in the IGBT's, they are normally presented to two kinds of cycling conditions - Thermal Cycling (TC) and Power Cycling (PC). PC is directed to decide the strength of the wire bonds; it raises and brings down the junction temperature of the chip at short intervals of time (sub-seconds to seconds system). TC is directed to determine the strength of the solder joints underneath the DBC and the chips, it raises and brings down the case temperature at long intervals (many seconds to several minutes). Once the ByCycle test has been applied, the number of cycles to failure ($N_f$) can be obtained via a mathematical approach. For extracting lifetime parameters, several analytical models have been proposed in [32, 33].

The previous study in the estimation of power electronic device lifetime has been generally concentrated on analytical analysis. throughout this time, MIL-HDBK-217F [34] handbook has been widely taken on to forecast the lifetime of electronic devices [35–40]. Primarily, the lifetime estimation can be accomplished in a statistic method, where the subsystem's reliability models (generally, constant failure rates) are specified. Various evaluation methods like Markov evaluation [41–44], fault tree analysis [45, 46], reliability obstruct layout analysis [47], and failure mode, and effect evaluation [48, 49] can then be used to evaluate the reliability from a system-level point of view.

Even though PV inverters represent a small percentage of the initial system cost (10–20%), as aforementioned they are one of the weakest points; therefore they may need to be replaced 3–5 times over the life of a PV system [50]. This fact introduces an extra cost to the overall PV system and thus limits Photovoltaic inverter's attractiveness.

To satisfy both stringent reliability requirements and constrained cost targets, the goal is to establish a reliability-cost model which can interrelate first the relationship between the varied lifetime and the cost as shown in Figure 7.8. It can be seen that when the cost is almost linearly increasing concerning the lifetime of the device,

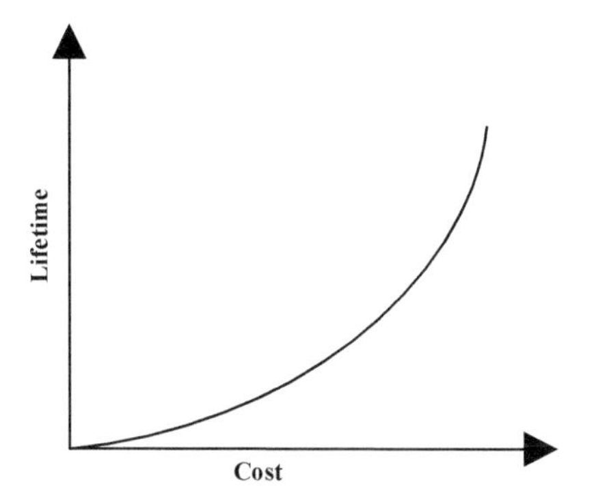


**Figure 7.8** Lifetime vs Cost

## 7.5 Capacitor Reliability

Capacitor life prediction methods can be classified into three categories. The first category is based on the constant failure rate model manuals such as Telcordia SR-332, IEC-TR-62380, and FIDES Guide 2009. These manuals assume the failure rate of the device during the stable failure period. It is constant, and the reliability R(t) satisfies the exponential distribution. One of the most famous manuals is the military manual MIL-HDBK-217F issued by the US Department of Defense, which is widely used to predict the life of devices. The first type of method was simple to use but did not consider the effects of device aging and temperature changes on failure rate. The MIL-HDBK-217F manual was canceled in 1995. The updated version of MIL-HDBK-217F, MIL-HDBK-217H, is further used [51], which takes into account the physical failure mechanism of the device. The reliability assessment method based on the reliability manual applies to devise failures associated with accidental failure and reliability evaluation of low power devices.

The second category is the life experience model, which is mainly derived from accelerated life experiments. It is currently the most widely used method for predicting device reliability [52, 53]. The main disadvantage of the life experience model is that it is mathematically derived from the available life data and does not directly describe the physical failure mechanism of the device.

The third category is the Physical Failure Life Model PoF (Physics of Failure), which is derived from the physical failure mechanism of the device. The Physical Failure Life Model can better describe the physical failure mechanism of the device, and the reliability evaluation accuracy is high.

However, in the practical application of the physical failure life model, the device material performance description and physical failure model are more difficult to obtain, and the failure model can only be evaluated under a single failure mechanism. Different failure mechanisms have different failure life models. [32, 55]. The different Failure modes, failure mechanisms, and the causes of the electrolytic capacitor are shown in Figure 7.9. In this research, the reliability of the device is evaluated using the second type of life experience model.

The causes of capacitor failure mainly include capacitor design defects, material aging, overvoltage and overcurrent, humidity, thermal stress, and mechanical stress. The failure of the capacitor can be caused by short-term overstress or accumulation of fatigue damage over a long period [56]. The

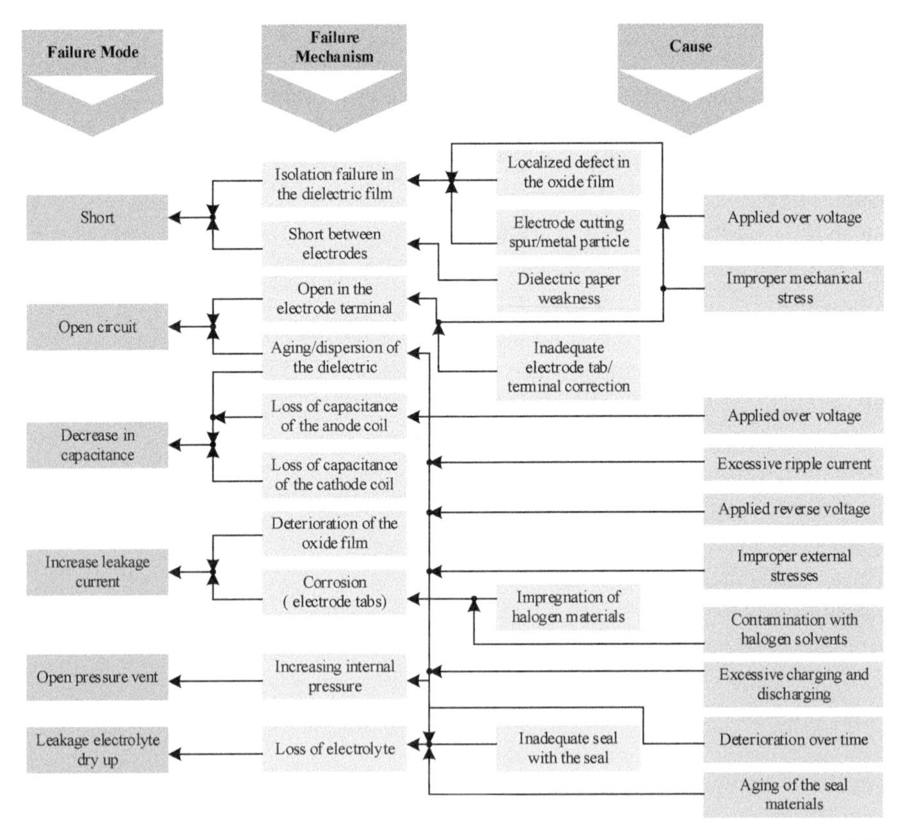


**Figure 7.9** Electrolytic capacitor failure modes, failure mechanism, and their causes [54]

capacitor lifetime is most closely related to the thermal relationship caused by the ambient temperature and the capacitance loss [57]. The capacitance loss is related to the capacitor current and the capacitor equivalent series resistance (ESR). Therefore, to accurately evaluate the capacitance reliability, it is necessary to analyze the capacitance current spectrum to obtain the refined capacitance loss and thermal model. In [58], based on the PWM modulation strategy, the three-phase inverter DC-Link capacitor current spectrum is calculated by the double Fourier analysis.

## 7.6 Lifetime Estimation Methods

### 7.6.1 Parts Stress Method

There are thoroughly utilized techniques for reliability assessment: Parts stress and part count techniques [59]. When there are no comprehensive system details, the part count technique is utilized. This technique is made use of to anticipate the failure rate under referral conditions. For that reason, the part count technique can be taken into consideration as an approximate technique. Unlike the part count technique, the part stress technique needs different stresses on every component, as well as exact environmental conditions are required. In the parts stress technique, the summing of the failure rate in each component gives the total failure rate in the system. In this thesis, the parts stress technique is used to have a more accurate failure rate based on the measurements

The total system failure rate ($\lambda_{system}$) is defined by the following mathematical formula

$$\lambda_{system} = \sum_{i=1}^{N} (\lambda_{part})_i \tag{7.6.1}$$

The failure rate of the capacitor, diode, switch, and inductor can be evaluated by the following equations

$$\lambda_{P,capacitor} = \lambda_{b,capacitor} \times \pi_{CV} \times \pi_Q \times \pi_E \tag{7.6.2}$$

$$\lambda_{P,diode} = \lambda_{b,diode} \times \pi_Q \times \pi_E \times \pi_C \times \pi_S \times \pi_T \tag{7.6.3}$$

$$\lambda_{P,switch} = \lambda_{b,switch} \times \pi_Q \times \pi_A \times \pi_E \times \pi_T \tag{7.6.4}$$

$$\lambda_{P,inductor} = \lambda_{b,inductor} \times \pi_C \times \pi_Q \times \pi_E \tag{7.6.5}$$

### 7.6.2 Lifetime Prediction Methods of Power Devices

There are different lifetime models presented in [32, 60, 61] such as Norris-Landzberg, Coffin-Manson-Arrhenius, Semikorn, Bayerer model, and Coffin-Manson. The lifetime models can be compared by the number of variables and model parameters considered in it. The lifetime model which considers more variables has more accuracy. The mathematical formulation of different lifetime models is described by the Equations (6.6) to (6.10).

#### 7.6.2.1 Coffin-Manson Lifetime Model

The Coffin-Manson Model [62, 63] is the simple analytical model to estimate the lifetime model of a semiconductor device, which considered only one variable parameter, i.e., $\triangle T_j$. It is expressed by the below equation

$$\mathrm{N_f} = \mathrm{A} \times (\triangle \mathrm{T_j})^{-\mathrm{n}} \tag{7.6.6}$$

Where $\triangle T_j$ represents the junction temperature swing, $\mathrm{N_f}$ represents the number of cycles to failure, *A* and *n* are the curve fitting parameters.

#### 7.6.2.2 Coffin-Manson-Arrhenius Lifetime Model

The Coffin-Manson-Arrhenius lifetime model [64] is an improved model of the Coffin-Manson model, which is considered $\mathrm{T_{jm}}$ besides $\triangle T_j$ considered in the Coffin-Manson model.

$$N_f = A \times (\triangle T_j)^{-n} \times e^{\left(\frac{E_a}{k_b \times T_{jm}}\right)} \tag{7.6.7}$$

Where $T_{jm}$ is the mean junction temperature, $k_b$ and $E_a$ are Boltzmann constant and activation energy, respectively.

#### 7.6.2.3 Norris-Landzberg Lifetime Model

A further improved model of the Coffin-Manson model described in (7.6.6) is the Norris-Landzberg model, which additionally considered the parameter *f*. the following equation describes the mathematical equation:

$$N_f = A \times f^{n_2} \times (\triangle T_j)^{-n_1} \times e^{\left(\frac{E_a}{k_b \times T_{jm}}\right)} \tag{7.6.8}$$

Where f is the frequency of temperature cycles [32].

**Table 7.2** Model parameters of the Semikorn model [68]

| Parameter | Value | Experimental Condition |
|---|---|---|
| $A$ | $3.436\times 10^{14}$ | |
| $\alpha$ | $-4.923$ | $64\,\mathrm{K} \le \Delta T_j \le 113\,\mathrm{K}$ |
| $\beta_0$ | $-1.942$ | $0.19 \le \mathrm{ar} \le 0.42$ |
| $\beta_1$ | $-9.012\times10^{-3}$ | $0.19 \le \mathrm{ar} \le 0.42$ |
| $\gamma$ | $-1.208$ | |
| $C$ | 1.434 | $0.07\,\mathrm{s} \le t_{ON} \le 63\,\mathrm{s}$ |
| $ar$ | 0.31 | |
| $k_b$ | $8.61733\times10^{-5}$ | |
| $E_a$ | 0.06606 | $32.5\,^0\mathrm{C} \le T_{jm} \le 122\,^0\mathrm{C}$ |
| $f_d$ | 0.6204 | |

#### 7.6.2.4 Semikorn Lifetime Model

The Semikorn lifetime model [65, 66] is relatively accurate; it is formulated by the following equation

$$N_f = A \times (\Delta T_j)^{\alpha} \times (ar)^{\beta_1 \Delta T_j + \beta_0} \times \left[\frac{C + (t_{ON})^{\gamma}}{C+1}\right] \times e^{\left(\frac{E_a}{k_b \times T_{jm}}\right)} \times f_d \tag{7.6.9}$$

Where ar the bond is wire aspect ratio, $f_d$ is the diode impact factor and $t_{ON}$ is the cycle period. $A$, $\alpha$, $\beta_0$, $\beta_1$, $\gamma$, and $C$ are model parameters as listed in Table 7.2.

#### 7.6.2.5 Bayerer Lifetime Model

The bayerer model [67] considers more parameters compared to all other lifetime models, and it also considers the power-on time and power module characteristics. It is defined as

$$N_f = A \times (\Delta T_j)^{\beta_1} \times e^{\overline{(T_{jm}+273)}^{\beta_2}} \times t_{on}^{\beta_3} \times I^{\beta_4} \times V^{\beta_5} \times D^{\beta_6} \tag{7.6.10}$$

Where $t_{on}$ is the on-time duration of the power cycle, $\beta_1$, $\beta_2$, $\beta_3$, $\beta_4$, $\beta_5$ and $\beta_6$ are the coefficients, which can be determined by conducting reliability experiments. The values of the coefficients and other parameters are listed in Table 7.3. *I, V,* and *D* are the current per bond wire, blocking voltage and diameter of the bonding wire, respectively.

The summary and overview of the comparative analysis of different lifetime models are presented in Table 7.4. It can be seen that the lifetime models are compared by the number of variables and model parameters considered in them. The lifetime prediction of a system has different values

**Table 7.3** Bayerers model parameters [67]

| Parameter | Value |
|---|---|
| Technology factor ($A$) | $9.34\times10^{14}$ |
| Coefficient ($\beta_1$) | $-4.416$ |
| Coefficient ($\beta_2$) | 1285 |
| Coefficient ($\beta_3$) | $-0.463$ |
| Coefficient ($\beta_4$) | $-0.716$ |
| Coefficient ($\beta_5$) | $-0.761$ |
| Coefficient ($\beta_6$) | $-0.5$ |

**Table 7.4** Summary and comparative analysis of different lifetime models

| Lifetime model | Variables considered | Model parameters |
|---|---|---|
| Coffin-Manson | $\Delta \mathrm{T_j}$ | A n |
| Coffin-Manson-Arrhenius | $\Delta \mathrm{T_j}$, $\mathrm{T_{jm}}$ | A, n, $\mathrm{k_b}$, $\mathrm{E_a}$ |
| Norris-Landzberg | $\Delta \mathrm{T_j}$, $\mathrm{T_{jm}}$, f | A, $\mathrm{n_1}, \mathrm{n_2}$, $\mathrm{k_b}$, $\mathrm{E_a}$ |
| Semikorn | $\Delta \mathrm{T_j}$, $\mathrm{T_{jm}}$, $\mathrm{t_{ON}}$, $\mathrm{f_d}$ | A, $\alpha$, $\beta_0$, $\beta_1$, $\gamma$, C, $\mathrm{k_b}$ |
| Bayerer | $\Delta \mathrm{T_j}$, $\mathrm{T_{jm}}$, $\mathrm{t_{on}}$, I, V, D | A, $\beta_1$, $\beta_2$, $\beta_3$, $\beta_4$, $\beta_5$, $\beta_6$, $\mathrm{k_b}$ |

when we use various lifetime models. Mainly, the number of variables and model parameters considered will affect the lifetime prediction models of the system. The lifetime prediction method which considered a high number of parameters will give a more accurate lifetime prediction. The Bayerers model considered a greater number of parameters when compared to all other lifetime prediction models. So, which has its relative accuracy due to the consideration of power on time and characteristics of a power module.

### 7.6.3 Lifetime Prediction Methods of DC-Link Capacitors

Lifetime designs are essential for predicting lifetime, criteria of various capacitor services, and online condition monitoring. One of the most commonly utilized empirical designs for a capacitor is presented in (1) which defines the impact of voltage stress and temperature.

$$L = L_0 \times \left(\frac{V}{V_0}\right)^{-n} \times exp\left[\left(\frac{E_a}{K_b}\right)\left(\frac{1}{T} - \frac{1}{T_0}\right)\right] \tag{7.6.11}$$

here L is the lifetime at the operating condition voltage ($V$) and operating condition temperature ($T$). $\mathrm{L_0}$ is the lifetime at the test condition voltage ($V_0$) and test condition temperature ($T_0$). n is the exponent of the voltage stress, $\mathrm{E_a}$ reparents activation energy, and $K_b$ is Boltzmann's constant.

For film capacitor and Al-cap capacitors, a streamlined design from (7.6.11) is widely used as follows

$$L = L_0 \left( \frac{V}{V_0} \right)^{-n} 2^{\frac{T_0 - T}{10}} \tag{7.6.12}$$

In [69], rather than a power regulation relationship, a linear equation is discovered to be preferable to define the effect of voltage stress. Furthermore, the lifetime reliance on temperature provided in (7.6.12) is an estimation only [70]. A lifetime design of the electrolytic capacitor is recommended based upon ESR drift because of electrolyte loss and dissipation. The evaluation of ESR is based upon the electrolyte volume reduction and pressure of the electrolyte. The forecast results in fir well with the temperature-lifetime relationship shown in (7.6.11). To get the physical description of the lifetime design variations from various capacitor suppliers, a common model is obtained in [19] as follows:

$$\frac{L}{L_0} = \begin{cases} exp[a_1(V_0 - V)] \times exp[\frac{E_{a0} - a_0\xi}{K_B T} - \frac{E_{a0} - a_0\xi_0}{K_B T_0}] \\ (\frac{V_0}{V})^{-n} \times exp[(\frac{E_a}{K_B})(\frac{1}{T} - \frac{1}{T_0})] \\ (\frac{V_0}{V}) \times exp[(\frac{E_a}{K_B})(\frac{1}{T} - \frac{1}{T_0})] \end{cases} \tag{7.6.13}$$

Where $\xi$ and $\xi_0$ are the temperature/voltage stress variables underused and test conditions, respectively. $a_1$ and $a_0$ represents the constants relating to the temperature and voltage colony of $E_a$. $E_{a0}$ is the activation energy at test condition. It can be kept in mind that the impact of voltage stress is designed as exponential equations, power law, and linear, respectively for high voltage stress, medium voltage stress, and low voltage stress. Currently, the lifetime prediction method presented in (7.6.12) is the most widely used method for predicting capacitor reliability which is used in this research.

## 7.7 Conclusion

In this paper, the review is mainly focused on the reliability of PV inverters in grid-connected applications. Initially, various PV converters for the PV system have been discussed based on the configuration of the PV module and the inverter topology. Then after the weakest components in the PV inverter have been identified which relate to the lifetime of the PV inverter. Further, the discussion mainly focused on different effects to get failure of the IGBT modules and dc-link capacitors. Finally, different lifetime evaluation methodologies for IGBT modules and dc-link capacitors have been discussed. It was observed that there are different model parameters and variables are considered in different lifetime models and therefore the lifetime prediction method which considered a high number of parameters will give a more accurate lifetime prediction.

## References

[1] R.E.P.N.f.t.s.C. (REN21), "Renewables 2018 Global Status Report." Paris: REN21 Secretariat, 2018.

[2] D.M.I.F. Blaabjerg, Y. Yang, and H. Wang. "Renewable Energy Systems Technology Overview and Perspectives." in *Renewable Energy Devices and Systems with Simulations in MATLAB and ANSYS, F. B. a. D. M. Ionel, Ed., ed: CRC Press LLC*, 2017.

[3] REN21. "Renewables 2017 Global Status Report [Online]." Available: www.ren21.net/wp-content/uploads/2017/06/17-8399_GSR_2017_Full_Report_0621_Opt.pdf

[4] L.M. Moore and H.N. Post. "Five Years of Operating Experience at a Large, Utility-Scale Photovoltaic Generating Plant" *Progress in Photovoltaics: Research and Applications*, vol. 16, pp. 249–259, 2008.

[5] S. Yang, A. Bryant, P. Mawby, D. Xiang, L. Ran, and P. Tavner, "An Industry-Based Survey of Reliability in Power Electronic Converters." *IEEE Transactions on Industry Applications*, vol. 47, pp. 1441–1451, 2011.

[6] M. Calais, J. Myrzik, T. Spooner, and V.G. Agelidis. "Inverters for Single-Phase Grid Connected Photovoltaic Systems-an Overview." In 2002 *IEEE 33rd Annual IEEE Power Electronics Specialists Conference*. Proceedings (Cat. No.02CH37289), 2002, pp. 1995–2000.

[7] S.B. Kjaer, J.K. Pedersen, and F. Blaabjerg. "A Review of Single-Phase Grid-Connected Inverters for Photovoltaic Modules. " *IEEE Transactions on Industry Applications*, vol. 41, pp. 1292–1306, 2005.

[8] F. Blaabjerg, C. Zhe, and S.B. Kjaer. "Power Electronics as Efficient Interface in Dispersed Power Generation Systems" *IEEE Transactions on Power Electronics*, vol. 19, pp. 1184–1194, 2004.

[9] M. Meinhardt, G. Cramer, B. Burger, and P. Zacharias. "Multi-String-Converter with Reduced Specific Costs and Enhanced Functionality " *Solar Energy*, vol. 69, pp. 217–227, 2001/07/01/ 2001.

[10] J. Schonberger. "A Single Phase Multi-String PV Inverter with Minimal Bus Capacitance." In 2009 *13th European Conference on Power Electronics and Applications*, 2009, pp. 1–10.

[11] G.R. Walker and P.C. Sernia. "Cascaded DC-DC Converter Connection of Photovoltaic Modules." *IEEE Transactions on Power Electronics*, vol. 19, pp. 1130–1139, 2004.

[12] E. Roman, R. Alonso, P. Ibanez, S. Elorduizapatarietxe, and D. Goitia. "Intelligent PV Module for Grid-Connected PV Systems." *IEEE Transactions on Industrial Electronics*, vol. 53, pp. 1066–1073, 2006.

[13] Y. Lei, C. Barth, S. Qin, W.C. Liu, I. Moon, A. Stillwell, et al. "A 2-kW Single-Phase Seven-Level Flying Capacitor Multilevel Inverter With an Active Energy Buffer." *IEEE Transactions on Power Electronics*, vol. 32, pp. 8570–8581, 2017.

[14] P. Sochor and H. Akagi. "Theoretical Comparison in Energy-Balancing Capability between Star- and Delta-Configured Modular Multilevel Cascade Inverters for Utility-Scale Photovoltaic Systems." *IEEE Transactions on Power Electronics*, vol. 31, pp. 1980–1992, 2016.

[15] S.P. Gautam, S. Gupta, and L. Kumar. "Reliability Improvement of Transistor Clamped H-Bridge-Based Cascaded Multilevel Inverter" *IET Power Electronics*, vol. 10, pp. 770–781, 2017.

[16] E. Ozdemir, S. Ozdemir, and L.M. Tolbert. "Fundamental-Frequency-Modulated Six-Level Diode-Clamped Multilevel Inverter for Three-Phase Stand-Alone Photovoltaic System." *IEEE Transactions on Industrial Electronics*, vol. 56, pp. 4407–4415, 2009.

[17] R. Selvamuthukumaran, A. Garg, and R. Gupta. "Hybrid Multicarrier Modulation to Reduce Leakage Current in a Transformerless Cascaded Multilevel Inverter for Photovoltaic Systems." *IEEE Transactions on Power Electronics*, vol. 30, pp. 1779–1783, 2015.

[18] B. Xiao, L. Hang, J. Mei, C. Riley, L.M. Tolbert, and B. Ozpineci. "Modular Cascaded H-Bridge Multilevel PV Inverter With Distributed MPPT for Grid-Connected Applications." *IEEE Transactions on Industry Applications*, vol. 51, pp. 1722–1731, 2015.

[19] H. Wang, K. Ma, and F. Blaabjerg. "Design for Reliability of Power Electronic Systems." *In IECON 2012 - 38th Annual Conference on IEEE Industrial Electronics Society*, 2012, pp. 33–44.

[20] R.E. Giuntini. "Mathematical characterization of human reliability for multi-task system operations." in Smc 2000 conference proceedings. 2000 *IEEE international conference on systems,*

*man and cybernetics. 'cybernetics evolving to systems, humans, organizations, and their complex interactions'* (cat. no.0, 2000, pp. 1325–1329 vol.2.

[21] Y.H. Lee, D. "A Study on the Techniques of Estimating the Probability of Failure." Journal of the Chungcheong Mathematical Society, pp. 573–583, 2008.

[22] A. Goel. "A New Approach to Electronic Systems Reliability Assessment." *Ph.D. Thesis, Rensselaer Polytechnic Institute*, Troy, NY, USA, 2007.

[23] D.M. Shalev and J. Tiran. "Condition-Based Fault Tree Analysis (CBFTA): A New Method for Improved Fault Tree Analysis (FTA), Reliability and Safety Calculations." *Reliability Engineering & System Safety*, vol. 92, pp. 1231–1241, 2007/09/01/ 2007.

[24] P. Gromala, J. Reichelt, and S. Rzepka. "Accurate Thermal Cycle Lifetime Estimation for BGA Memory Components with Lead-free Solder Joints." *Proceedings 10th EuroSimE*; April 27–29, 2009; Delft, Netherlands

[25] N. Pan et al. "An Acceleration Model for SnAg-Cu Solder Joint Reliability under various Thermal Cycle Conditions." *Proceedings SMTAI* 2006, pp. 876–883

[26] J. Reichelt, P. Gromala, and S. Rzepka. "Accelerating the Temperature Cycling Tests of FBGA Memory Components with Lead-Free Solder Joints without Changing the Damage Mechanism." *In 2009 European Microelectronics and Packaging Conference*, 2009, pp. 1–8.

[27] K. Upadhyayula and A. Dasgupta. "Guidelines for Physics-of-Failure Based Accelerated Stress Testing." in Annual Reliability and Maintainability Symposium. 1998 Proceedings. *International Symposium on Product Quality and Integrity*, 1998, pp. 345–357.

[28] K.C. Norris and A.H. Landzberg. "Reliability of Controlled Collapse Interconnections." *IBM Journal of Research and Development*, vol. 13, pp. 266–271, 1969.

[29] M. Andresen, K. Ma, G. Buticchi, J. Falck, F. Blaabjerg, and M. Liserre. "Junction Temperature Control for More Reliable Power Electronics." *IEEE Transactions on Power Electronics*, vol. 33, pp. 765–776, 2018.

[30] H. Wang, M. Liserre, F. Blaabjerg, P.d.P. Rimmen, J.B. Jacobsen, T. Kvisgaard, et al. "Transitioning to Physics-of-Failure as a Reliability Driver in Power Electronics." *IEEE Journal of Emerging and Selected Topics in Power Electronics*, vol. 2, pp. 97–114, 2014.

[31] G.J. Riedel and M. Valov. "Simultaneous Testing of Wirebond and Solder Fatigue in IGBT Modules." in CIPS 2014; 8th *International Conference on Integrated Power Electronics Systems*, 2014, pp. 1–5.
[32] I.F. Kovacevic, U. Drofenik, and J.W. Kolar. "New Physical Model for Lifetime Estimation of Power Modules." in The 2010 *International Power Electronics Conference* - ECCE ASIA -, 2010, pp. 2106–2114.
[33] C. Busca, R. Teodorescu, F. Blaabjerg, S. Munk-Nielsen, L. Helle, T. Abeyasekera, et al. "An Overview of the Reliability Prediction Related Aspects of High Power IGBTs in Wind Power Applications." *Microelectronics Reliability*, vol. 51, pp. 1903–1907, 2011/09/01/ 2011.
[34] U.S.D.o. Defense, MIL-HDBK-217F – Military Handbook for Reliability Prediction of Electronic Equipment. Washington DC, USA: Department of Defense, 1991.
[35] X. Shi and A.M. Bazzi. "Solar Photovoltaic Power Electronic Systems: Design for Reliability Approach." in 2015 17th *European Conference on Power Electronics and Applications* (EPE'15 ECCE-Europe), 2015, pp. 1–8.
[36] C. Jais, B. Werner, and D. Das. "Reliability Predictions - Continued Reliance on a Misleading Approach." in 2013 *Proceedings Annual Reliability and Maintainability Symposium (RAMS)*, 2013, pp. 1–6.
[37] M. Aten, G. Towers, C. Whitley, P. Wheeler, J. Clare, and K. Bradley. "Reliability Comparison of Matrix and Other Converter Topologies." *IEEE Transactions on Aerospace and Electronic Systems*, vol. 42, pp. 867–875, 2006.
[38] S.E.D. León-Aldaco, H. Calleja, and J.A. Alquicira. "Reliability and Mission Profiles of Photovoltaic Systems: A FIDES Approach." *IEEE Transactions on Power Electronics*, vol. 30, pp. 2578–2586, 2015.
[39] R. Burgos, G. Chen, F. Wang, D. Boroyevich, W.G. Odendaal, and J.D.V. Wyk. "Reliability-Oriented Design of Three-Phase Power Converters for Aircraft Applications." *IEEE Transactions on Aerospace and Electronic Systems*, vol. 48, pp. 1249–1263, 2012.
[40] D. Hirschmann, D. Tissen, S. Schrder, and R.W.D. Doncker. "Reliability Prediction for Inverters in Hybrid Electrical Vehicles." in 2006 37th *IEEE Power Electronics Specialists Conference*, 2006, pp. 1–6.
[41] A.M. Bazzi, A. Dominguez-Garcia, and P.T. Krein. "Markov Reliability Modeling for Induction Motor Drives Under Field-Oriented Control." *IEEE Transactions on Power Electronics*, vol. 27, pp. 534–546, 2012.
[42] H. Chen, H. Yang, Y. Chen, and H.H.C. Iu. "Reliability Assessment of the Switched Reluctance Motor Drive Under Single Switch

Chopping Strategy," *IEEE Transactions on Power Electronics*, vol. 31, pp. 2395–2408, 2016.
[43] J. Yuan, C.H. Lin, S.J. Chang, and S.H. Lai. "Reliability Modeling & Evaluation for Networks under Multiple & Fluctuating Operational Conditions," *IEEE Transactions on Reliability*, vol. R-36, pp. 557–564, 1987.
[44] E.C.D.J. Rogers. "A Comparison of Grid-Connected Battery Energy Storage System Designs." *IEEE Transactions on Power Electronics*, vol. 32, pp. 6913–6923, 2017.
[45] W.S. Lee, D.L. Grosh, F.A. Tillman, and C.H. Lie. "Fault Tree Analysis, Methods, and Applications & #2013; A Review." *IEEE Transactions on Reliability*, vol. R-34, pp. 194–203, 1985.
[46] G.R. Biswal, R.P. Maheshwari, and M.L. Dewal. "Cool the Generators: System Reliability and Fault Tree Analysis of Hydrogen Cooling Systems," *IEEE Industrial Electronics Magazine*, vol. 7, pp. 30–40, 2013.
[47] M. Èepin, "Reliability Block Diagram" in *Assessment of Power System Reliability: Methods and Applications*, M. Èepin, Ed., ed London: Springer London, 2011, pp. 119–123.
[48] H.C. Liu, J.X. You, P. Li, and Q. Su. "Failure Mode and Effect Analysis Under Uncertainty: An Integrated Multiple Criteria Decision Making Approach," *IEEE Transactions on Reliability*, vol. 65, pp. 1380–1392, 2016.
[49] S. Haghbin. "Electrical Failure Mode and Effect Analysis of a 3.3 kW Onboard Vehicle Battery Charger." in 2016 18th *European Conference on Power Electronics and Applications* (EPE'16 ECCE Europe), 2016, pp. 1–10.
[50] G.J.T. McMahon, and R. Hulstrom. "Module 30 year life: What does it Mean and is it Predictable/Achievable?." in eliability *Physics Symposium*, 2008, pp. 172–177.
[51] J.G. McLeish, "Enhancing MIL-HDBK-217 Reliability Predictions with Physics of Failure Methods." *in 2010 Proceedings - Annual Reliability and Maintainability Symposium (RAMS)*, 2010, pp. 1–6.
[52] A. Morozumi, K. Yamada, T. Miyasaka, and Y. Seki. "Reliability of Power Cycling for IGBT Power Semiconductor Modules." in *Conference Record of the 2001 IEEE Industry Applications Conference*. 36th IAS Annual Meeting (Cat. No.01CH37248), 2001, vol. 3, pp. 1912–1918.

[53] D. Hirschmann, D. Tissen, S. Schroder, and R.W.D. Doncker. "Reliability Prediction for Inverters in Hybrid Electrical Vehicles." *IEEE Transactions on Power Electronics*, vol. 22, pp. 2511–2517, 2007.

[54] Mouser. "Reliability of Aluminum Electrolytic Capacitors." ed America: ELNA Co., Ltd, 2011, pp. 1–17.

[55] L. Yang, P.A. Agyakwa, and C.M. Johnson. "Physics-of-Failure Lifetime Prediction Models for Wire Bond Interconnects in Power Electronic Modules." *IEEE Transactions on Device and Materials Reliability*, vol. 13, pp. 9–17, 2013.

[56] H. Wen, W. Xiao, X. Wen, and P. Armstrong. "Analysis and Evaluation of DC-Link Capacitors for High-Power-Density Electric Vehicle Drive Systems." *IEEE Transactions on Vehicular Technology*, vol. 61, pp. 2950–2964, 2012.

[57] J.W. Kolar and S.D. Round. "Analytical Calculation of the RMS Current Stress on the DC-link Capacitor of Voltage-PWM Converter Systems." *IEE Proceedings - Electric Power Applications*, vol. 153, pp. 535–543, 2006.

[58] B.P. McGrath and D.G. Holmes. "A General Analytical Method for Calculating Inverter DC-Link Current Harmonics" *IEEE Transactions on Industry Applications*, vol. 45, pp. 1851–1859, 2009.

[59] A. Khosroshahi, M. Abapour, and M. Sabahi. "Reliability Evaluation of Conventional and Interleaved DC–DC Boost Converters." *IEEE Transactions on Power Electronics*, vol. 30, pp. 5821–5828, 2015.

[60] A. Syed. "Limitations of Norris-Landzberg Equation and Application of Damage Accumulation Based Methodology for Estimating Acceleration Factors for Pb Free Solders." in 2010 *11th International Thermal, Mechanical & Multi-Physics Simulation, and Experiments in Microelectronics and Microsystems (EuroSimE),* 2010, pp. 1–11.

[61] Y. Zhang, H. Wang, Z. Wang, Y. Yang, and F. Blaabjerg. "Impact of Lifetime Model Selections on the Reliability Prediction of IGBT Modules in Modular Multilevel Converters." in 2017 *IEEE Energy Conversion Congress and Exposition (ECCE)*, 2017, pp. 4202–4207.

[62] K.E. Horton, J.M. Hallander, and D.D. Foley. "Thermal-Stress and Low-Cycle Fatigue Data on Typical Materials " p. V001T01A013, 1965.

[63] H. Cui. "Accelerated Temperature Cycle Test and Coffin-Manson Model for Electronic Packaging. " in *Annual Reliability and Maintainability Symposium*, 2005. Proceedings, 2005, pp. 556–560.

[64] N.U.A. Wintrich, T. Werner, and T. Reimann. "Application Manual Power Semiconductors. " ed. *Nuremberg*, Germany: Semikron Int.GmbH, 2015.

[65] N. Sintamarean, F. Blaabjerg, H. Wang, F. Iannuzzo, and P.d.P. Rimmen. "Reliability Oriented Design Tool for the New Generation of Grid Connected PV-Inverters. " *IEEE Transactions on Power Electronics*, vol. 30, pp. 2635–2644, 2015.

[66] U. Scheuermann. "Pragmatic Bond Wire Model." in *ECPE Workshop Lifetime Modeling Simulation*, DÃijsseldorf, Germany, July 2013.

[67] R. Bayerer, T. Herrmann, T. Licht, J. Lutz, and M. Feller; "Model for Power Cycling lifetime of IGBT Modules - Various Factors Influencing Lifetime." in 5th *International Conference on Integrated Power Electronics Systems*, 2008, pp. 1–6.

[68] U.S.a.U. Hecht. "Power Cycling Lifetime of Advanced Power Modules for Different Temperature Swings." in PCIM, *Nuremberg*, Germany, 2002, pp. 59–64.

[69] Y. Ko and J. Seo. "Text Classification from Unlabeled Documents with Bootstrapping and Feature Projection Techniques." *Information Processing & Management*, vol. 45, pp. 70–83, 2009/01/01/ 2009.

[70] M.L. Gasperi. "Life Prediction Model for Aluminum Electrolytic Capacitors." in IAS '96. *Conference Record of the 1996 IEEE Industry Applications Conference Thirty-First IAS Annual Meeting*, 1996, pp. 1347–1351 vol.3.

[71] R. Kaplar, R. Brock, S. DasGupta, M. Marinella, A. Starbuck, A. Fresquez, S. Gonzalez, J. Granata, M. Quintana, M. Smith, and S. Atcitty. "PV Inverter Performance and Reliability: What is the Role of the IGBT?." *Photovoltaic Specialists Conference (PVSC), 37th IEEE*, pp. 1842–1847, June 2011.

# 8

# Energy Storage

**Sanjeevikumar Padmanaban[1], Mohammad Zand[2], Morteza Azimi Nasab[3], Mohamadmahdi Shahbazi[4], and Heshmatallah Nourizadeh[5]**

[1,2,3,4]CTIF Global Capsule, Department of Business Development and Technology, Herning 7400, Denmark
[5]Department of Electrical Engineering, Shahid Beheshti University of Tehran, Iran
E-mail: Sanjeev@btech.au.dk; Dr.zand.mohammad@gmail.com; Morteza1368morteza@gmail.com; M.shahbazi@basu.ac.ir; H_nourizadeh@sbu.ac.ir

## Abstract

Due to the inefficiency and increasing reduction of fossil fuel resources and the advancement of technology and smart grid design, in order to produce optimal power for power shocks, special threats are needed. In the meantime, the optimal use of production capacity and storage of excess power can improve the performance of power networks and reduce the amount of production capacity. For this purpose, energy storage systems (ESS) are considered as one of the important elements in power network design. In this chapter, the updated equipment of several energy storage technologies is reviewed and their various features are analyzed. These studies include their use, classification, storage capacity, the current state of the industry, and future installation. This chapter also examines the main features of energy-saving technologies suitable for renewable energy systems.

## 8.1 Introduction

The future development of modern electrical systems towards smart grids is one of the issues in power systems today that is highly regarded by planners and system operators, many changes, and non-uniformity of the load curve at different hours of the day. This has caused all the installed production capacity to be used only during peak hours, and a large amount of installed capacity is out of the circuit during low and medium load hours, which means capital sleep.

On the other hand, due to fluctuations in fuel prices and declining fossil fuel resources, environmental pollution, and growing energy demand, all of which have led to the emergence of new alternative energy called renewable energy [1, 2]. However, the uncertainty of renewable resources and their high investment cost has raised serious concerns [3–6]. Renewable sources cannot produce energy continuously due to changes in energy production rates over different periods of time. Therefore, in order to convert renewable energy sources into the main sources of the network, there is an urgent need to use energy storage devices [7, 8]. These problems are more or less seen in world power networks that have highly variable load curves. This has led researchers to think about storing electrical energy by looking at human experiences and the background of storage for a long time.

Energy storage systems have been proposed as a solution to the above challenge. Energy storage systems are usually under independent control, with the help of which the energy produced in the power system can be stored and, if necessary, it can be stored in a used electrical system. Because of the cost of electricity generation and its selling price at different hours of the day and night, due to the start of the electricity market, there are significant differences, and also, on the other hand, energy from renewable sources must be stored when generating surplus and then used when necessary. Therefore, the idea of energy storage was proposed. There are several reasons why energy storage should be used, some of which are as follows:

Facilitate the increase in the penetration of renewable energy sources

1. Load curve leveling
2. Help control frequency
3. Delaying the development of transmission lines
4. Reduce voltage fluctuations
5. Increase power quality and reliability
6. Eliminate short-term and random demand fluctuations

Today, the development of electrical energy storage technology to generate power in an emergency has found a special place in the power system [13–8]. Also, since the power system is undergoing major changes, energy storage in the power system is a crucial choice to cover issues such as restructuring the electricity market, bringing in renewable sources, and helping to increase distributed generation. Improving power quality and contributing to network performance under environmental protection laws.

The purpose of this chapter is to review the types of energy storage systems and an overview of the most important energy storage technologies available or under development, which we will discuss in detail in the following operating principles and the most important specifications of each technology. This chapter will also discuss the important criteria for energy-saving technologies suitable for renewable energy applications.

## 8.2 Installed Capacity in the World

At least 160 gigawatts (GW) of large-scale energy storage is currently installed in power grids around the world. The vast majority (99%) of this capacity includes storage pump (PSH) technologies. 1. The other includes a combination of battery technologies, compressed air storage (CAES), flywheels, and hydrogen storage [9]. Most of the technology is thermal energy storage, which is related to home hot water tanks. Other technologies, such as ice and cold water storage in several countries, such as Australia, the United States, China, and Japan, have played an important role in picking up

**Table 8.1** Some energy storage technologies and examples of related projects [9]

| Primary Application | Initial Investment Cost (USD/kW) | Efficiency (%) | Output | Location | Technology |
|---|---|---|---|---|---|
| Long-term storage | 500–4600 | 50–85 | Electricity | Supply | PSH |
| Long-term storage | 3400–4500 | 50–90 | Thermal | Supply | UTES |
| Low,medium& high-temp applications | 1000–3000 | 80–99 | Thermal | Supply, deman | Thermochemical |
| Short-term Storage | 130–500 | 90–95 | Electricity | T&D | Flywheels |
| Short-term Storage | 130–515 | 90–95 | Electricity | T&D | Super capacitors |
| Medium temperature applications | 500–3000 | 50–90 | Thermal | Demand | Solid media |
| Low-temp applications | 300–600 | 50–90 | Thermal | Demand | Cold storage-water |

and reducing electricity subscribers' electricity bills. Underground thermal storage systems (UTES) are found in abundance in Canada, Germany, and other European countries.

## 8.3 Application of Energy Storage Devices

The value of energy storage technology lies in the services they provide in different locations of the energy system. These technologies can be used throughout the power grid, in heating and cooling systems, as well as distribution systems and off-grid applications. In addition, they can provide infrastructure support services throughout the supply, transmission, and distribution sectors. Provide energy system consumption.

Large-scale electrical energy storage technologies have many applications. These applications range from fast power quality to improve reliability to long-term energy management applications to improve efficiency. Some of the key applications of storage can be categorized as follows:

Seasonal storage: The ability to store energy for days, weeks, months to compensate for long-term power supply shortages or seasonal variability in the supply and demand side of the energy system (e.g. summer heat storage for Use in winter through underground thermal energy storage systems).

Energy trading (Storage trades): In this case, cheap energy is stored during periods when demand is low and then sold when electricity demand is high. This type of energy trade is generally referred to as two energy markets.

Frequency regulation: Frequency regulation is the continuous balancing of supply and demand shifts under normal conditions, which is done automatically alternately at minute-minute or even shorter intervals.

1. Voltage support: Supports voltage, injection, or absorption of reactive power to maintain the voltage levels of the transmission and distribution system under normal conditions.
2. Black start: In rare cases where the power system collapses and not all solutions are workable, the restart feature allows the power supply to be restarted without the use of mains electricity. Be launched.

Demand shifting and peak reduction: In order to match the demand and supply of electricity and to help integrate different sources of supply, energy demand time can be shifted and shifted. This movement is facilitated by changing the timing of certain activities (for example, heating water, or space).

Variable supply resource integration: The use of variable supply resource integration, the use of energy storage to change and optimize the output of variable supply sources (e.g. wind and solar), reduce rapid and seasonal output changes, the geographical and temporal distance between both supply and demand to increase the quality and quantity of supply.

The important point here is the difference between storage devices for energy and electricity applications. In electrical applications, the short time period is about a few seconds to a few minutes, but in energy applications, the time period is about a few minutes to a few hours.

## 8.4 Classification of Energy Storage Devices

Extensive problems in electricity storage have shaped the technology of power systems today. There are various ways to store energy on a large scale. However, any type of storage is expensive and requires economic calculations. Electricity storage is done through four methods: chemical, thermal, mechanical, and electrical. Figure 8.1 shows the classification of methods and technologies specific to each method.

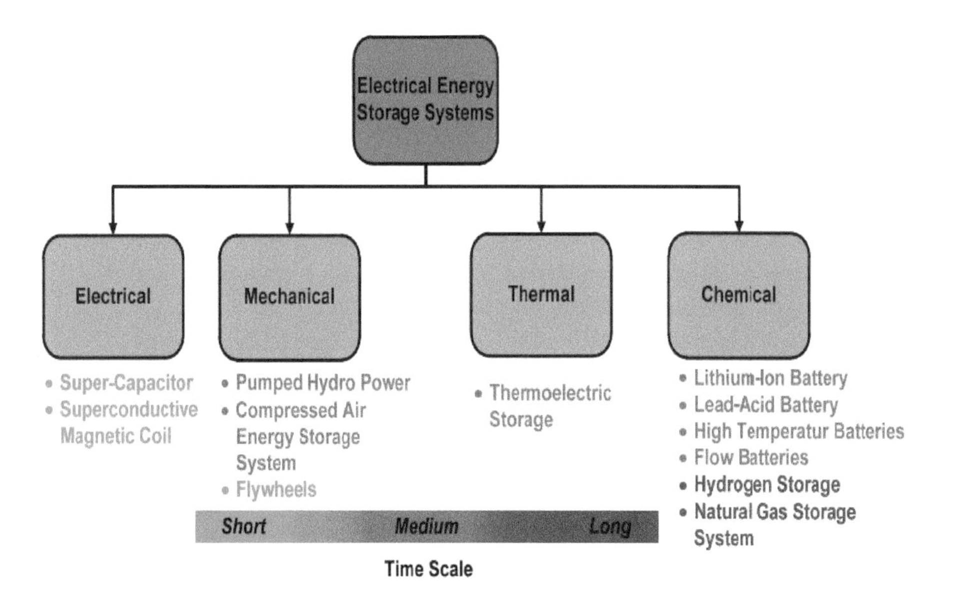


**Figure 8.1** Classification of EES energy storage systems [11, 10].

**Table 8.2** Types of advanced electrical energy storage technologies [12].

| Technology | Primary Application | Challenges |
|---|---|---|
| CAES | Energy Management - Backup and Periodic Storage - Renewable Integration | Geographical constraints - Lower productivity due to reciprocity conversion - Environmental impacts |
| The wheels of the planes | Load leveling - frequency adjustment - pixelation and non-peak storage | Rotor Tensile Strength Limitations - Limited Energy Storage Time Due to High Losses |
| Advanced lead acid | Load leveling and adjustment - network stability | Discharge limit - lower energy density |
| NaS | Power quality - density elimination | Liquid contamination issues (electrodes) |
| Li-ion | Power quality - frequency adjustment | High production cost - scalability - inability to deep discharge |
| SMES | Power quality - frequency adjustment | Lower energy density - high cost of raw materials and fabrication |
| Electrochemical capacitor | Power quality - frequency adjustment | Heavy current cost |
| Thermal energy saver | Load adjustment and leveling - network stability | Heavy current cost |

### 8.4.1 A Variety of Storage Technologies in the Supply Chain to Consume Electricity

An electrical energy storage system is a system in which electrical energy is converted into a type of chemical, thermal, magnetic, etc. energy that can be stored to be converted into electrical energy again if needed. Table 8.2 shows the types of advanced electrical energy storage technologies, initial applications, available information, and their challenges.

To choose an electrical energy storage system, you must pay attention to the required power and energy rates, charge and discharge speed, temperature, volume, and weight. The different types of technologies specific to each method of electrical energy storage (chemical, thermal, mechanical, and electrical) are introduced below.

### 8.4.2 Electrical storage technologies

**Capacitor cloud storage**

A supercapacitor is a type of electrochemical capacitor that has a much higher storage capacity than conventional capacitors. This type of capacitor stores energy in the electric field between two dielectrics. A supercapacitor can be thousands of times larger in volume than an electric capacitor. Larger supercapacitors with a capacity of more than 5,000 farads are used as batteries and have a higher energy density than conventional capacitors. The largest supercapacitor has a power density of 30 watts per kilogram, which is less than the density of lithium-titanium (fast-charging) batteries. These capacitors have high permeability and electrodes are very close to each other. Common types of supercapacitors include two-layer EDLC electric capacitors that have a very high power density and efficiency of more than 90% and are also very expensive. Some of the applications of different types of supercapacitors are:

1. Military equipment and military lasers
2. Wind turbines
3. Medical Equipment
4. Solenoid valve starter
5. Backup power supplies
6. Actuator of electromagnetic motors
7. Solar remote control units
8. Solar systems

Air capacitors have a longer life than conventional rechargeable batteries with advantages such as charging time of about a few seconds compared to

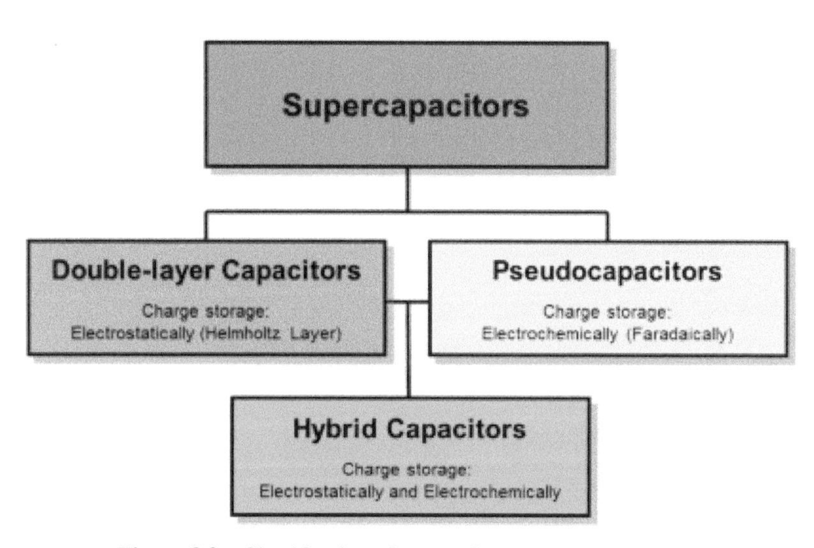


**Figure 8.2** Classification of types of supercapacitors [10–12]

charging for about a few hours and also millions of charge and discharge cycles than 1,000 times charge and discharge. This is because in supercapacitors, no chemical changes normally occur on the electrodes and the energy is physically stored. Also, supercapacitors have very low internal resistance and high efficiency.

Recent research into the use of graphene (Graphene) promises supercapacitors that can be charged in 16 seconds and charged and discharged more than 10,000 times. It should be noted that the best supercapacitors can have an energy density equivalent to conventional lithium-ion batteries. Supercapacitors are used on a small scale to store energy, and if progress is made in increasing their energy density, they can be expected to replace electrochemical batteries.

In supercapacitors, the electrodes are not bonded by chemical reactions, indicating that they can be fully discharged. This, as a function of the load state, leads to more voltage fluctuations. For this reason, supercapacitors can not completely replace batteries, and only supercapacitor battery systems can increase storage efficiency and increase battery life [14].

This energy storage system can be useful for electrical appliances and wind energy applications. Other applications include maintaining a constant frequency and modulating voltage drops in decentralized power plants. The extremely high energy storage capacity of SCs has made any device that uses

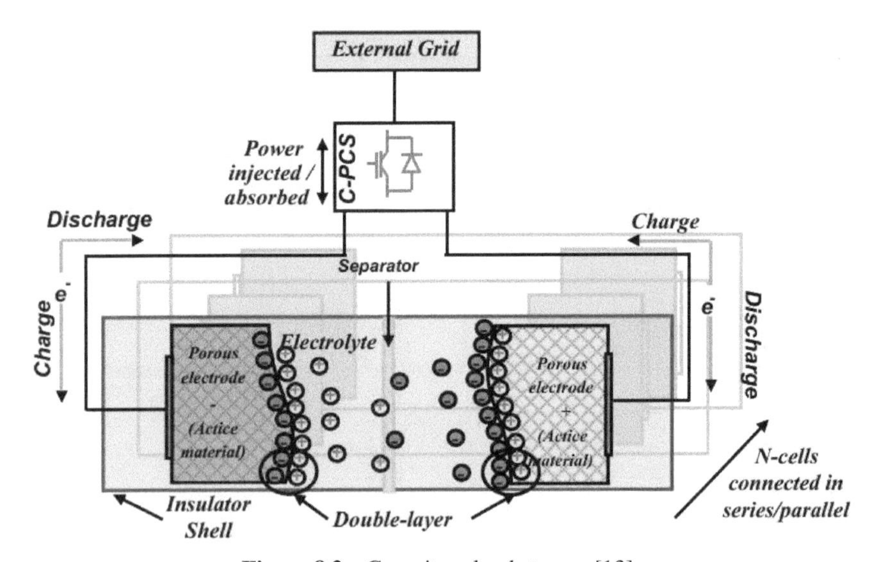


**Figure 8.3** Capacitor cloud storage [13]

a battery as a power source a candidate for the use of a supercapacitor. Early capacitors, unlike batteries, were able to store high energy with low power.

They could store high-energy energy, but that energy would soon run out. Supercapacitors have solved this problem to some extent by combining high-power energy storage as well as large amounts of energy. This energy storage source is currently one of the most popular options for use in electric or dual-fuel vehicles, and its advantages over batteries have led experts to consider it suitable for battery replacement [15].

## 8.5 Superconducting Magnetic Storage (SMES)

Superconducting objects increase their storage capacity and do not withstand current flow at low temperatures. The use of superconductors is limited by factors such as temperature change, critical magnetic field, and critical current density. The superconducting magnetic storage system consists of three basic parts: the superconducting coil, the power correction system, and the cooling system.

The superconducting coil is used as an inductor and during non-peak hours electrical energy is stored in the form of magnetic energy through a direct current (DC) in the said inductor field. If the coil is made of materials such as copper, a lot of magnetic energy is lost in the wire due to resistance, and if the wire is superconducting, the energy is stored in a stable state for as long as necessary. The superconductivity of the coil causes very little system loss and its current remains almost unchanged.

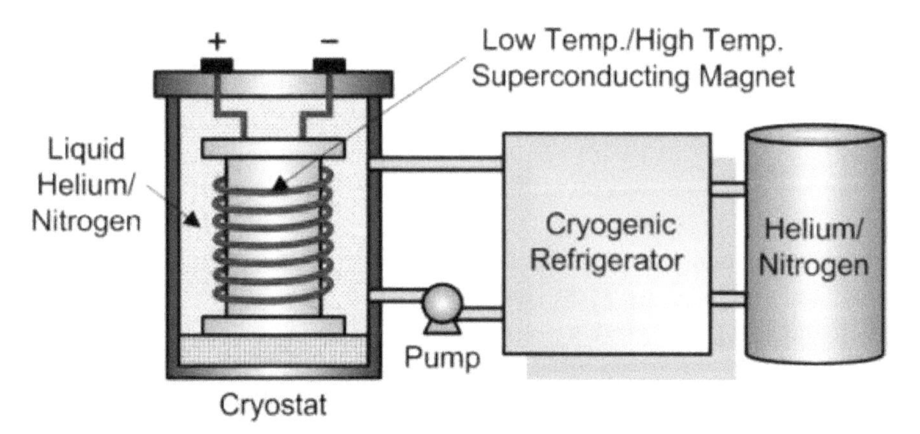


**Figure 8.4** How a superconducting magnetic storage system works [15].

Superconductors are not resistant to DC, but they have losses in the application of AC electric current, which can be reduced with proper design. For both AC and DC operating modes, a lot of energy is stored. It has superconducting properties, the best temperature for devices is 50–77 degrees Kelvin. The stored energy can be released in the network through the electrical discharge of the coil. Application of power management system, which is the same as power correction system, converts alternating current to direct and direct to alternating current in the SMES charge and discharge routine, respectively, using an inverter rectifier converter consisting of two AC / DC converters with six thyristors and a three-power transformer. The reducing coil is for converting alternating current (AC) to direct current or vice versa. The two ends of the superconducting coil can be controlled continuously by controlling the firing angle of thyristor's DC voltage in a wide range of positive and negative voltage values. [15].

The SMES system is very efficient, with a charge and discharge efficiency of more than 90% and less waste than other storage devices during the energy storage process, and in fact, the inverter rectifier converter causes a loss of 2–3% of energy in each path. However, due to the high energy required for cooling and the high cost of superconducting wires, SMESs are currently only used for short-term energy storage and power quality improvement. At present, the cost per unit capacity of this storage is about $ 50,000 per kilowatt, which is expected to reach $ 31,000 per kilowatt by 2022. The prospect of becoming more competitive is down to $ 1,000 per kilowatt.

## 8.6 Mechanical Storage Method

### 8.6.1 Storage Pump

The pumping power plant is a storage facility to store the potential of the water in the reservoir of the dams to generate hydropower energy during peak hours. In this method, water is pumped from downstream tanks to upstream tanks, and by doing so, electrical energy is stored as potential energy in pumped water at higher altitudes. This method gives the ability to increase water capacity by pumping water from a lower reservoir to a higher reservoir when demand for electricity is low (for example, at night. This water is stored when demand is high or electricity is needed). It goes back to the bottom tank to turn the turbines and finally turn the generator.

In the 1930s, a reversible hydropower turbine was built and put into operation. These turbines were able to operate as turbines - generators and in the opposite direction as pumps equipped with electric motors. The latest

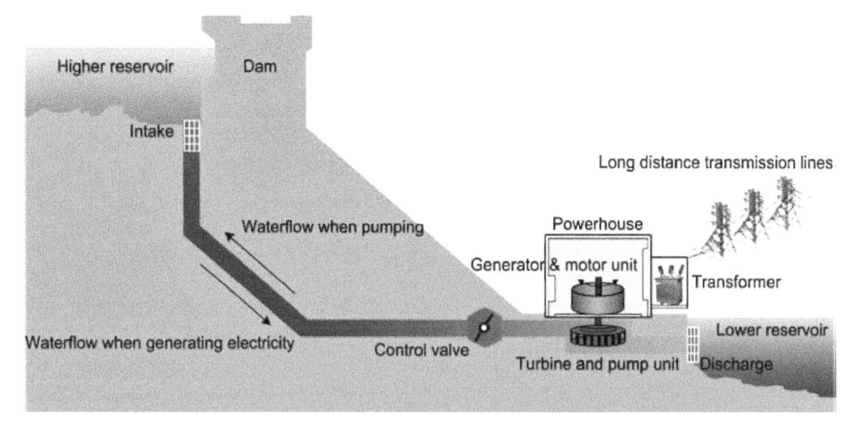


**Figure 8.5** How the storage pump system works [16]

large-scale turbines are the engineering technology of variable speed turbines that have a higher efficiency of about 80%. These turbines generate electricity at the same time as the grid frequency and operate as a pump motor independent of the grid frequency. Due to the fact that storage pump power plants smooth the load fluctuations in the electricity network and allow the thermal power plants that produce base load electricity to continue to operate at maximum peak efficiency. As the most cost-effective means of storing large amounts of electrical energy based on exploitation.

The amount of power produced by the turbine depends on the amount of upstream reservoir water as well as the height of the waterfall. The planning period in systems is 24 hours and is divided into different time scales [16]. This system has an advantage over thermal storage that also provides fast PHS response as important components for grid frequency control. The lifespan of this system is about 30–50 years and its efficiency is 613–75% [6].

## 8.6.2 Compressed Air Storage

The main components of a CAES system include: engine, compressor, air storage chamber, combustion chamber, turbine, and generator. The operation of this system is that in non-peak hours, it takes electricity from the network and compresses, the air to about 75 bars by a compressor and puts it inside an underground chamber. Of course, for optimal use of space, the air is cooled before being injected into the cavity. To create an underground air storage chamber, it can either be made artificially, which is very expensive, or use groundwater aquifers or various mines. Compressed air can be stored in the

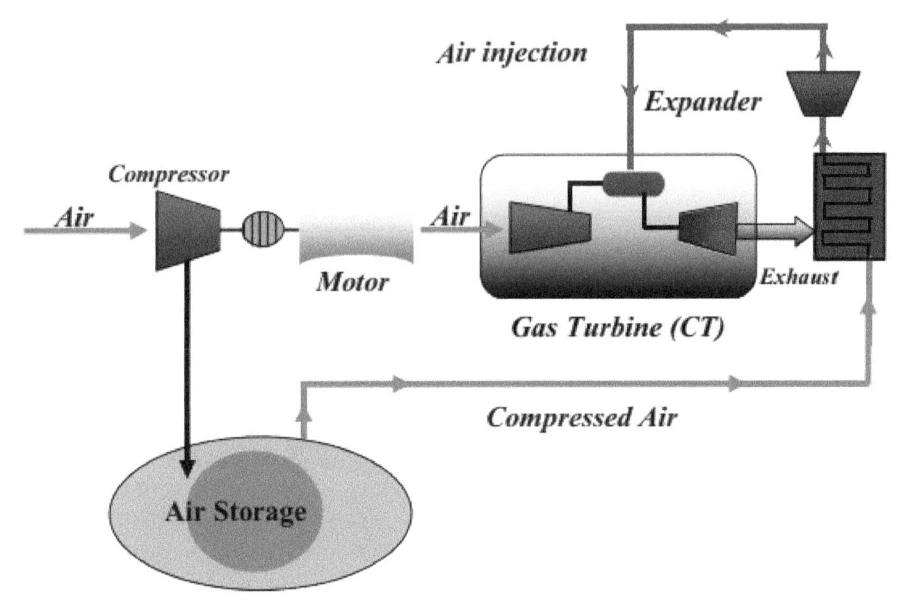


**Figure 8.6** Scheme of operation of a CAES storage system [17]

chamber with very little loss, and when necessary, after leaving, it is mixed with some fuel in the combustion chamber and then transferred to the gas turbine and finally generated from it by generating electricity.

CAES storage method using natural containers that are economically viable and practical equivalent to using a compressor in a gas power plant without spending more than half of the production capacity of gas turbines to rotate the compressor, led to the idea of building power plants gas has been provided in places where underground chambers can be used.

In this case, during non-peak hours, the power plant compressor can be done using the CAES system, and during peak hours, the power plant compressor can be taken out of the circuit, thus almost doubling the production capacity. After the storage pump method, the CAES system has the largest capacity among storage devices.

The typical capacity of CAES systems is around 50 to 300 MW, and due to the small losses of this system, the storage period is up to one year. The commissioning time of the CAES system is about 10 minutes, which is less than the same time for gas power plants (20–30 minutes). The typical energy density of compressed air is about 108 joules per gram. In the meantime, storing natural aquifers is very cost-effective in terms of investment. In the charging mode, the generator rotates in reverse mode, in other words, it is working in the motor mode, and its purpose is to create mechanical power for the compressor and transfer air to the combustion tank. When the tank is empty, compressed air is used to control the turbine's combustion. Currently, only two compressed air energy storage systems have been built in the world, one of which belongs to the United States with a capacity of 110 MW and Germany with a capacity of 290 MW [17]. Also, the investment cost in this project, depending on the storage conditions, is between $ 400 and $ 800 per kilowatt-hour [18].

### 8.6.3 Flight Wheel Storage

The flywheel stores electrical energy by converting it into kinetic energy. This is done by increasing the speed of a rotor and storing energy in the form of rotational energy. The rotor speed decreases with energy and increases with energy. The main components of a flywheel storage system include: generator motor, flywheel, bearings, vacuum chamber, and control system. This system takes energy from the network during non-peak hours and uses its engine to rotate the aircraft wheel, and during peak hours, the network uses this kinetic energy. Since the energy stored inside the aircraft is directly related to the square of its rotational speed, in order to increase the energy stored in the aircraft, its rotational speed must be increased.

The stored energy is lost in friction after a while and the rotor stops moving and practically does not allow the use of FES. The source of this problem is the tension in the wheel due to rotational forces and inertia. To solve it, new rotors made of carbon-composite filaments with a rotation speed

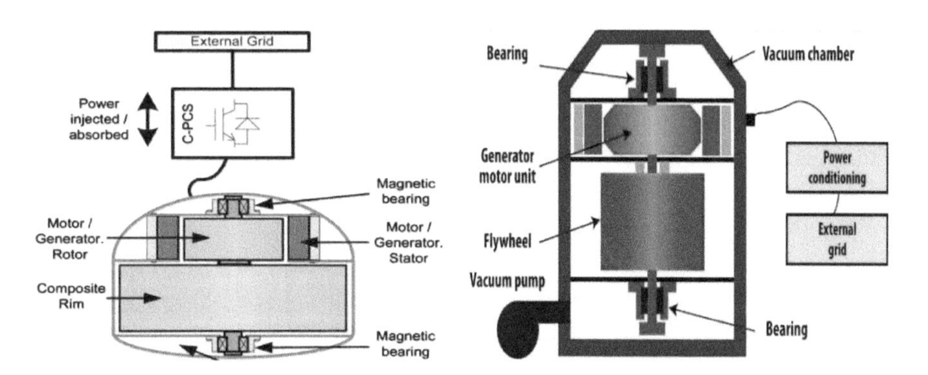


**Figure 8.7** How the flywheel storage system works [15] and [19]

of about 10,000 to 20,000 rpm replaced the usual metal rotors with a smoke rate of 4,000 rpm, and due to the fact that composite wheels have less weight. At a certain rotational speed, less stress was created in them. In addition, new composite materials are more durable than old materials. In general, the new aircraft wheels have less weight, higher resistance, and the ability to rotate at much higher speeds compared to the old type [19].

To solve the problem of friction in the aircraft wheels, a vacuum chamber and magnetic bearings are used, the losses of which are almost negligible. In this method, the magnetic bearings hold the rotor using a magnetic field to prevent mechanical contact, which in ordinary bearings causes high losses. The presence of a vacuum also prevents friction losses with the air. The system is used to improve power quality and has some peak capability. The advantages and disadvantages of new aircraft wheels include the following:

Advantages: compactness and low volume - high efficiency - low and insignificant maintenance - no noise

Disadvantages: Safety considerations - High cost of raw materials - High price of magnetic bearings

One of the applications of flywheel technology is the use of flywheel energy storage systems to create virtual inertia in independent networks with low inertia. In general, the user of this technology can be expressed as follows:

1. Creating stability and damping in the system (System stability and damping)
2. Power quality
3. Load leveling

Compared to lead-acid batteries, flywheels have higher discharge speeds, longer life, less risk, and very little maintenance, and are suitable for advanced applications. But at the moment they are relatively expensive and their storage capacity is low compared to other storage devices. With the commercial production of new types and the development of manufacturing techniques, it is expected that these storage devices will be able to compete more with other devices in the future [20].

## 8.7 Thermal Storage Method

Thermal energy storage (TES) is created to combine renewable energy with power generation on the birth side, but this technology can be used on the customer side or on demand. Using TES systems, the load can be moved by transferring electrical charges from peak to off-peak hours. Therefore, it can become a powerful tool in demand–side management.

This technology has energy stability and good quality compared to other storage modes. These types of storage can also temporarily store thermal energy at high or low temperatures, which increases the efficient use of thermal energy equipment. TES systems are used to balance production and demand.

The way TES systems work is the same in all structures and applications. First, the energy is injected into the TES system, and then this energy is stored and discharged in different applications for the use of the TES system [23, 24]. There are several criteria for classifying TES systems. If the amount and level of temperature stored in the TES system are based on the desired criterion, it can be divided into two categories: "Low-temperature thermal energy storage (LTTES)" and "High-temperature thermal energy storage (HTTES)". The operating temperature of the systems in LTTES mode is below 200 degrees Celsius and its application is in systems such as heating and cooling of residential houses, solar greenhouses, and solar water boilers [25, 26]. In the article, systems in HTTES mode use heat recovery and renewable energy. This method is mostly used in construction mines and the metallurgical industry.

However, if the measurement criterion is based on the heating time of the TES system, it can be divided into two categories: "short-term" and "long-term". Also, if the scale and measurement criterion is based on the amount of energy storage materials, it can be divided into three categories: "reasonable heat storage", "latent heat storage" and "chemical heating heat storage" [12].

### 8.7.1 Reasonable Thermal Energy Storage Systems

In a logic TES, energy is stored by changing the temperature of the storage device. The amount of heat stored is proportional to the density, specific temperature, volume, and temperature change in the storage material. The performance of a storage system mainly depends on the specific density and heat consumption, which affects the required volume [20]. There are two drawbacks to most logical storage systems. These are usually large and are caused by temperature fluctuations caused by the accumulation and rational extraction of energy [14]. If a large temperature fluctuation is allowed, the storage size is reduced in order. However, compared to other storage methods,

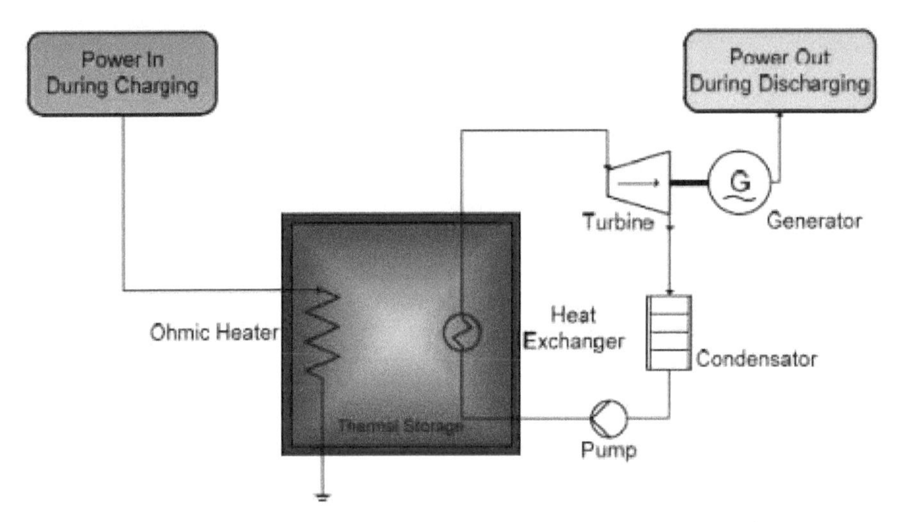


**Figure 8.9** How high temperature thermal energy storage system works [19].

the size will still be large. Size can also be reduced by using storage materials that have a large heat capacity. Large reserves, in addition to taking up a lot of space and increasing costs, also cause a lot of heat loss [18].

### 8.7.2 Sensible thermal energy storage systems

In this method, which is known as one of the ETS methods, energy is stored

The operation of TEES systems is similar to that of storage pump systems and CAES, except that it is suitable for medium-term applications. This technology is currently being researched and will be introduced to the market within the next 5 to 10 years. Information about the parameters and their range is not available for this technology, but it is expected to be cost-effective in megawatt capacities.

## 8.8 Chemical Storage Method

### 8.8.1 Chemical Storage Systems with Internal Storage

#### 8.8.1.1 Hydrogen storage system (HES)

Liquid hydrogen has a lower energy density in terms of volume and capacity compared to hydrocarbon fuel. The charging process is done by electricity through electrolysis. The hydrogen produced is compressed and stored, for

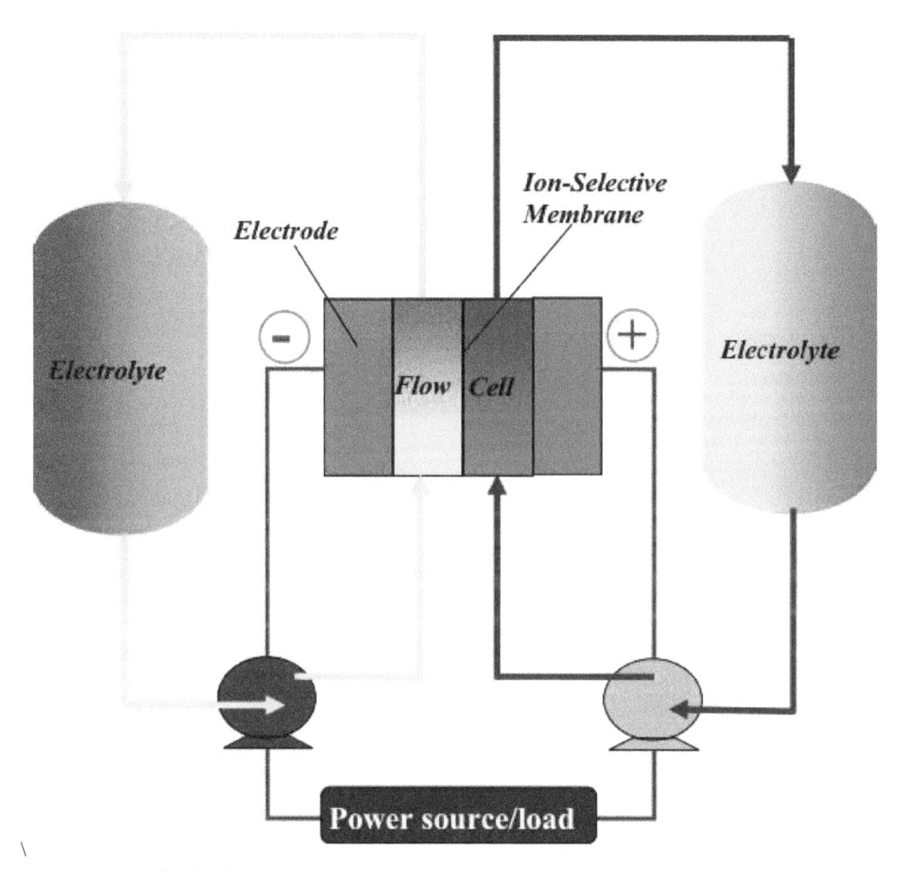


\

**Figure 8.10** How high temperature thermal energy storage system works [8].

example, in salt caves or special reservoirs. Hydrogen is used in expansion turbines or fuel cells after the discharge process. In addition, electricity generated from hydrogen is used directly in hydrogen cars that run on fuel cells, internal combustion engines, or heat generators.

The low density of hydrogen in the gaseous state makes it difficult to use hydrogen as an energy carrier. This means that compared to liquid fuels such as gasoline with methanol, it has a low energy content per unit volume (about 12.07 kJ / kg), so it is used as rocket fuel. Liquid hydrogen has the highest energy density of all fuels. Therefore, achieving the desired heat requires high energy consumption. In order to prevent excessive boiling, suitable tanks must be installed, but adding insulation leads to higher costs.

#### 8.8.1.2 GAT power system: Artificial natural gas methanation

Artificial natural gas storage is an alternative to hydrogen storage technology. Methanation, a thermogenic reaction known as the Fisher-Tropez process, is the process that produces artificial natural gas from hydrogen and carbon dioxide. Its final product is methane, which is the main constituent of natural gas and is therefore fully compatible with natural gas infrastructure. Therefore, it can be injected into the natural gas network without any restrictions.

Its storage capacity is about 400 terawatt-hours (like the German gas network) which can be used for medium-term and long-term storage purposes. The main advantage of methanation is the direct use of hydrogen, which is exactly compatible with the existing value chain of natural gas. The main drawbacks are the reduction of efficiency and increase in the cost. In addition, this process requires an external source of CO2 and heat loss. If this heat is not used for heating residential units or industrial processes, then the overall efficiency will be further reduced. Conventional power plants or biogas can be used as a source of CO2. The non-synchronization of the period of extra electricity for metatation and the period of CO2 production by the power plant causes the synchronization of the carbon burning process for energy production and the use of electricity for methane production. In other words, storing carbon dioxide imposes an additional cost on the system.

#### 8.8.1.3 Current batteries

Using current-powered batteries is a low-cost yet efficient way to store energy produced from sources such as solar and wind power plants. These batteries have a structure similar to acidic batteries, except that the electrolyte material is stored in an external tank. Current batteries are a type of rechargeable battery in which two liquids with different electrical charges (called electrolytes) exchange ions with each other and from this ion exchange electrical energy is generated. The electrolytes are housed separately in two large chambers by the retaining material, and the electrolyte is pumped into these chambers as needed. In this way, by changing the amount of electrolyte available, the electrical energy produced can be controlled and the output of these batteries can be changed from a few kilowatts to several megawatts. The newly designed current battery can generate more than 795 MW per square centimeter of electrical energy, which is three times more than other methods of generating power without a buffer and 10 times more than lithium-ion batteries. These types of batteries are very powerful in terms of current and their current capacity is directly related to the volume of the electrolyte tank.

### 8.8.2 Chemical Storage Systems with External Storage

Chemical storage systems with external storage mainly include batteries. They are electrodynamic devices used to store energy from DC or AC sources for future use. This technology, which is more than 150 years old, is one of the most common storage devices that is widely used throughout the supply chain for the consumption of electricity and is especially suitable for renewable energy combined power plants. Batteries are used when the load demand is low with microgrid sources not being able to meet the load demand.

There are different types of batteries, the most common of which are Acid-Lead and Lithium-ion. These types of batteries are relatively inexpensive and have reached maturity, and their limitations, defects, and technical defects, as well as their need for repair and maintenance, are well known. These batteries can be used for most applications. Batteries are significant from three perspectives:

Reliability: Batteries come to the aid of production resources in the event of a system malfunction and compensate for the lack of microgrid power until it returns to normal. In this situation, the batteries are responsible for maintaining the microgrid frequency within the allowable range.

1. Load management: Batteries will prevent the pressure of microgrid units by storing power during low load hours and providing energy during peak load hours.
2. Economical: Batteries can generate significant profits by purchasing energy from the mains during off-peak hours and selling stored energy during off-peak hours.

Due to the fact that the batteries are located next to other sources, including wind turbines, they use the energy of their nearby generators for charging most of the time, and during discharge, they also feed their adjacent loads. This slightly increases the loss rate, but due to the storage of energy during the cheap hours and its discharge during the peak hours of the network when the energy price is high, the batteries significantly reduce the total operating costs.

#### 8.8.2.1 Lithium-ion battery

In general, each battery consists of three main parts: a positive electrode, negative electrode, and electrolyte. In lithium-ion batteries, the positive electrode or cathode is made of a combination of lithium, such as lithium cobalt oxide, and the negative electrode or anode is made of carbon with a separating

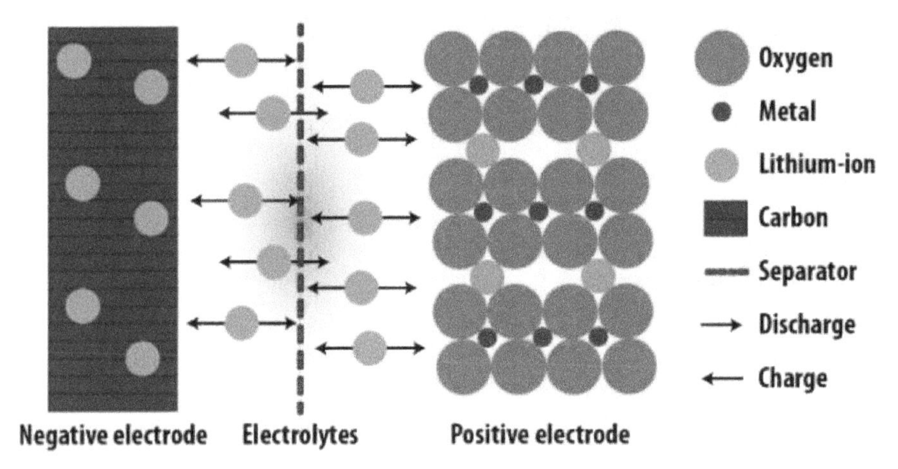


**Figure 8.11** How lithium-ion battery works [19]

layer between them. The electrolyte in lithium batteries is also made from lithium salts in an organic solvent. The use of organic solvents in the role of electrolytes due to their flammability requires some safety measures.

The variety of storage technologies in the supply chain to the consumption of electrolyte electrical energy in these batteries consists of a set of materials, each of which has its own function. The defective performance of any of the electrolyte components will result in the defective performance of the entire battery. Lithium batteries are equipped with electronic protection circuits and fuses to prevent polarization, overvoltage, overheating, and other safety issues. Since lithium-ion batteries play a key role in the development of portable electronics, extensive research has been done to develop them. This type of rechargeable battery, like other types of batteries, consists of electrochemical shovels. Each cell in turn consists of two electrodes that are connected to each other by an electrolyte. Each of these components has a special and unique structure. It should be noted that nano–materials have made dramatic changes in the efficiency and longevity of lithium batteries.

### 8.8.2.2 Lead-acid battery

Lead Acid Battery is the most mature and cheapest energy storage device of all available battery technologies. If we put two metals of the same name (electrode) in the acid (electrolyte) liquid, a current is established between the two electrodes. The positive electrode absorbs the ions and combines with the positive electron and the same lead oxide (2Pbo), creating an electric potential at both the negative and positive pole ends of the battery.

When charged at the positive pole, lead sulfate combines with water to produce lead oxide and hydrogen ions, and hydrogen produced by the positive electrode reacts with lead sulfate to produce HS04 ions. The opposite is done. Lead acid batteries have low investment costs ($ 250 to $ 550 / kWh) and high reliability and efficiency (713–90%) [18]. Among the problems of these batteries can be their dysfunction at low temperatures with high ambient. Also, the life of these batteries is very short.

#### 8.8.2.3 High-temperature batteries (sulfur - sodium)

These batteries, in addition to being able to meet energy demand at peak times and deliver electricity to areas that previously had shortages due to transmission and distribution restrictions, are a ready source for protection. They also act against electrical disturbances. Their energy density is three times that of ordinary lead and acid batteries, and they can recharge and store stored electrical energy in the same time in 8 hours, and store indefinitely at a temperature of about 900 degrees Fahrenheit. The use of NAS batteries stabilizes the demand for electricity, which has significant fluctuations between day and night, one of the results of which will be to reduce the cost of electricity to customers.

By charging the battery during off-peak hours and using the electricity stored during peak hours, power plants provide energy more efficiently. This process will be economical for a consumer customer when there is a large

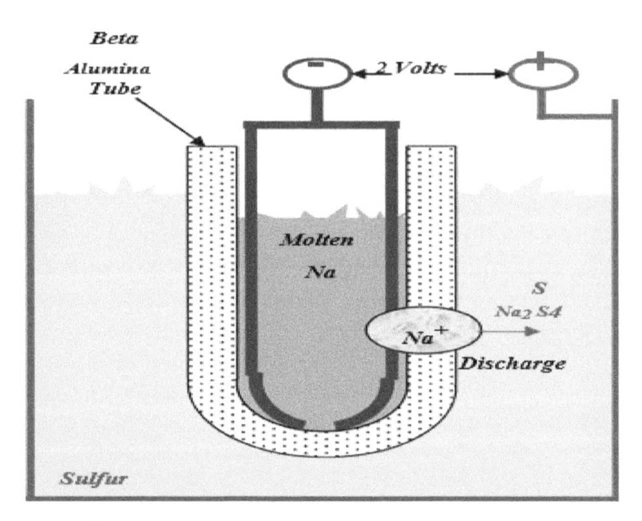


**Figure 8.12** High temperature battery operation (sulfur-sodium) [15].

difference between peak and off-peak electricity costs. (Double tariff meters). These batteries have a feature called Nas Pulse Factor, according to which it can inject up to 1 to 5 times its nominal power to the power grid with the help of appropriate electronic equipment. According to this characteristic, it is charged with its nominal power in seven hours, and with the same time and power, it can inject power into the network. The batteries installed in the world so far have been about 319 MW, most of which are in Japan and have the capacity to produce 1899 MWh. Compared to lead-acid batteries, this type of battery has a higher density of about 4 four times that of lead-acid batteries, as well as a very fast response in milliseconds for both full charge and discharge modes. The most important feature of this battery is its 90% recycling [2].

#### 8.8.2.4 Nickel (nickel-cadmium) battery

Nickel-based batteries are mainly nickel-cadmium (NICD) composed of nitrogen. There are two types of nickel-metal batteries and nickel-zinc batteries. All three use similar materials for the positive electrode and the electrolyte, with nickel hydroxide and an aqueous solution of potassium hydroxide with some lithium hydroxide, respectively [17]. As for the negative electrode, the NiCD type uses cadmium hydroxide, the NIMH uses a metal alloy, and the NIZN uses zinc hydroxide. Among these, NICD batteries can be compared with lead-acid batteries because they have a longer lifespan than lead-acid batteries and a higher energy density, as well as the cost and duration of repairs and maintenance of these batteries, are low. Is wet.

#### 8.8.2.5 Status of energy storage technologies

Status of Energy Storage Technologies Electric and thermal storage technologies (cold or heat) have different levels of early research and development (R&D), full maturity and development in terms of development. In the figure below, some of the key technologies according to the initial investment need and technology risk in relation to its current situation (for example, being in the stages of research and development, demonstration and deployment, or commercialization). it has been shown.

## 8.9 Storage Cost

To calculate the cost of energy storage devices, three parameters of efficiency, investment costs, and life cycle should be considered [12–16]. The costs of energy storage systems are of particular concern to power companies. In

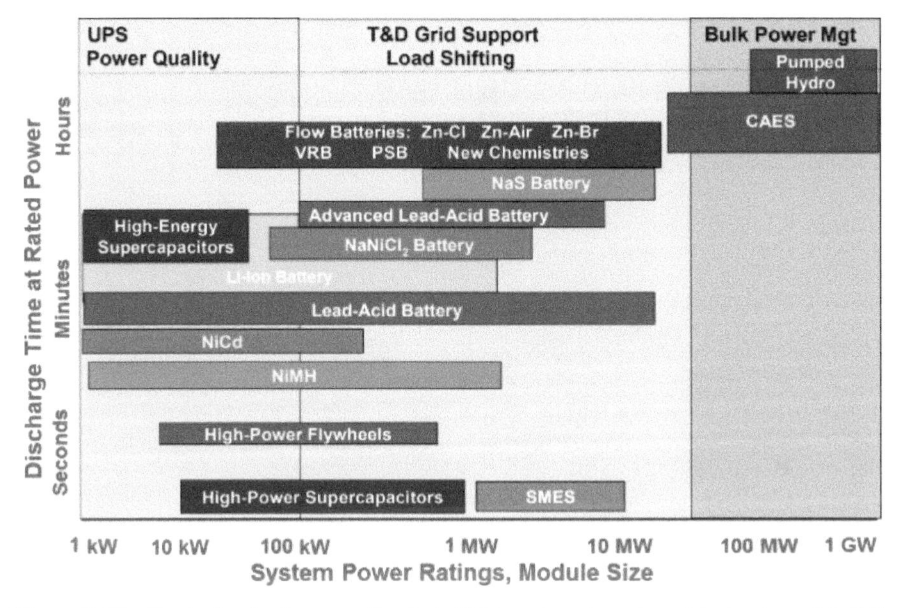


**Figure 8.13** Key technologies according to initial investment needs and technology risk [15]

energy storage systems, these costs vary with the interval between which the energy storage system is emptied. This time interval is a function of the program load profile. It is clear that different energy storage systems may have economic advantages or disadvantages, depending on the discharge time for the program.

## 8.10 Criteria for Determining Appropriate Energy Storage Technologies

Energy storage can be used across parts of the energy system including supply, transmission and distribution, and demand (end consumer). The best place to deploy storage technology depends on the capability that is expected in the energy system process from supply to consumption. In addition, developments in smart grids and other new technologies will determine the optimal location of these storage devices.

Before choosing an energy storage technology, a set of clear criteria is needed to help make the right decisions. Each of these criteria should be prioritized depending on their application. The following figure shows the range of storage used during the energy system process from supply to consumption.

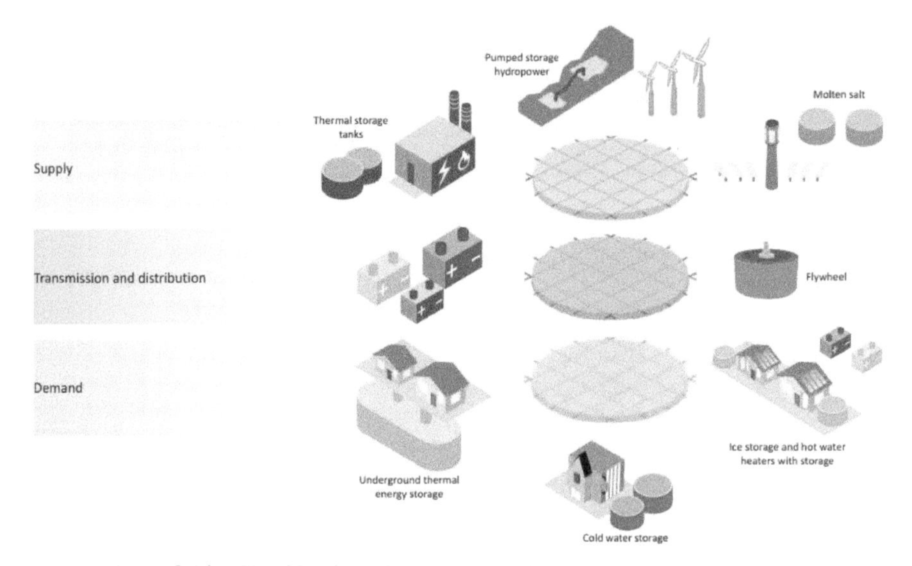


**Figure 8.14** Classification of storage usage in energy system process [13].

As can be seen from the figure, the appropriate technology for the electrical energy supply sector is a storage pump, thermal storage tanks and molten salt in the transmission and distribution sector of the appropriate technologies are batteries and flywheels. Of course, there are various types of batteries that are also used for storage in the consumption sector. In the consumption sector, in addition to batteries, underground thermal energy storage technologies, cold water storage, ice storage, and hot water storage with heaters are used.

## References

[1] Dincer, Ibrahim, and Marc A. Rosen. Thermal energy storage systems and applications. John Wiley & Sons, 2021.

[2] Olabi, A. G., et al. "Application of graphene in energy storage device–A review." Renewable and Sustainable Energy Reviews 135 (2021): 110026.

[3] M. Zand, M. A. Nasab, A. Hatami, M. Kargar and H. R. Chamorro, "Using Adaptive Fuzzy Logic for Intelligent Energy Management in Hybrid Vehicles," 2020 28th ICEE, pp. 1-7, doi: 10.1109/ICEE50131.2020.9260941. IEEE Index

[4] Hamed Ahmadi-Nezamabad, et al.. "Multi-objective optimization based robust scheduling of electric vehicles aggregator." Sustainable Cities and Society vol. 47, 101494, 2019.

[5] Zand, Mohammad; Nasab, Morteza Azimi; Sanjeevikumar, Padmanaban; Maroti, Pandav Kiran; Holm-Nielsen, Jens Bo: 'Energy management strategy for solid-state transformer-based solar charging station for electric vehicles in smart grids', IET Renewable Power Generation, 2020, DOI: 10.1049/iet-rpg.2020.0399 IET Digital Library, https://digital-library.theiet.org/content/journals/10.1049/iet-rpg.2020.0399

[6] Ghasemi M, et al. (2020). An Efficient Modified HPSO-TVAC-Based Dynamic Economic Dispatch of Generating Units, Electric Power Components and Systems doi.org/10.1080/15325008.2020.1731876

[7] Nasri, Shohreh, et al, Maximum Power Point Tracking of Photovoltaic Renewable Energy System Using a New Method Based on Turbulent Flow of Water-based Optimization (TFWO) Under Partial Shading Conditions. 978-981-336-456-1

[8] Rohani A, et al, "Three-phase amplitude adaptive notch filter control design of DSTATCOM under unbalanced/distorted utility voltage conditions," Journal of Intelligent & Fuzzy Systems, , 2020, 10.3233/JIFS-201667

[9] M. Zand, M. A. Nasab, O. Neghabi, M. Khalili and A. Goli, "Fault locating transmission lines with thyristor-controlled series capacitors By fuzzy logic method," 2020 14th International Conference on Protection and Automation of Power Systems (IPAPS), Tehran, Iran, 2019, pp. 62-70, doi: 10.1109/IPAPS49326.2019.9069389.

[10] Z. Zand, M. Hayati and G. Karimi, "Short-Channel Effects Improvement of Carbon Nanotube Field Effect Transistors," 2020 28th Iranian Conference on Electrical Engineering (ICEE), Tabriz, Iran, 2020, pp. 1-6, doi: 10.1109/ICEE50131.2020.9260850.

[11] Lilia Tightiz et all, An intelligent system based on optimized ANFIS and association rules for power transformer fault diagnosis, ISA Transactions, Volume 103, 2020, Pages 63-74, ISSN 0019-0578, https://doi.org/10.1016/j.isatra.2020.03.022.

[12] Zand M, et al "A Hybrid Scheme for Fault Locating in Transmission Lines Compensated by the TCSC," 2020 15th International Conference on Protection and Automation of Power Systems (IPAPS), 2020, pp. 130-135, doi: 10.1109/IPAPS52181.2020.9375626

[13] M. Zand, M. Azimi Nasab, M. Khoobani, A. Jahangiri, S. Hossein Hosseinian and A. Hossein Kimiai, "Robust Speed Control for Induction

Motor Drives Using STSM Control," 2021 12th (PEDSTC), 2021, pp. 1-6, doi: 10.1109/PEDSTC52094.2021.9405912.

[14] P. Sanjeevikumar, M. Zand, M. A. Nasab, M. A. Hanif and M. S. Bhaskar, "Spider Community Optimization Algorithm to Determine UPFC Optimal Size and Location for Improve Dynamic Stability," 2021 IEEE 12th Energy Conversion Congress & Exposition - Asia (ECCE-Asia), 2021, pp. 2318-2323, doi: 10.1109/ECCE-Asia49820.2021.9479149

[15] Azimi Nasab, M.; Zand, M.; Eskandari, M.; Sanjeevikumar, P.; Siano, P. Optimal Planning of Electrical Appliance of Residential Units in a Smart Home Network Using Cloud Services. Smart Cities 2021, 4, 1173–1195. https://doi.org/10.3390/smartcities4030063

[16] Azimi Nasab, M.; Zand, M. Sanjeevikumar, P. et, al. An efficient robust optimization model for the unit commitment considering of renewables uncertainty and Pumped-Storage Hydropower. Computers and Electrical Engineering 2022,

[17] Azimi Nasab, Mortez, Zand, Mohammad, Padmanaban, Sanjeevikumar,Dragicevic, Tomislav, Khan, Baseem, "Simultaneous Long-Term Planning of Flexible Electric Vehicle Photovoltaic Charging Stations in Terms of Load Response and Technical and Economic Indicators" World Electric Vehicle Journal ,2021, P 190, doi:10.3390/wevj12040190

[18] Zand, Mohammad, et al. "Big Data for SMART Sensor and Intelligent Electronic Devices–Building Application." Smart Buildings Digitalization. CRC Press 11-28. https://doi.org/10.1201/9781003201069, 2022

[19] Wang, Xiaowei, et al. "Insights of Heteroatoms Doping−Enhanced Bifunctionalities on Carbon Based Energy Storage and Conversion." Advanced Functional Materials 31.11 (2021): 2009109.

[20] Boretti, Alberto. "Integration of solar thermal and photovoltaic, wind, and battery energy storage through AI in NEOM city." Energy and AI 3 (2021): 100038.

# 9

# A Comprehensive Review of Techniques for Enhancing Lifetime of Wireless Sensor Network

**Raj Gaurang Tiwari[1], Alok Misra[2], Ambuj Kumar Agarwal[3], and Vikas Khullar[4]**

[1]Chitkara University Institute of Engineering and Technology, Chitkara University, Punjab, India
[2]Institute of Engineering and Technology, Lucknow
[3]Chitkara University Institute of Engineering and Technology, Chitkara University, Punjab, India
[4]Chitkara University Institute of Engineering and Technology, Chitkara University, Punjab, India
E-mail: India rajgaurang@gmail.com; alokalokmm@gmail.com; ambuj4u@gmail.com; vikas.khullar@gmail.com

## Abstract

A sensor is a device that detects an event and turns it into an electrical, mechanical, or another form. In diverse application areas like phones, electronic and electronic devices, mechanical devices, industries, etc., sensors can be utilized. There are diverse varieties of sensors such as seismic, optical, magnetic, infrared, thermal, acoustic, and radar, which can examine an outsized variety of ambient circumstances. There are several resource constraints on a sensor, like storage, energy, communication, and computation capabilities. A sensor network can be constructed by coalescing identical or diverse sensors. In the wireless sensor network, physical information is sensed by a set of independent sensors and also transmitted to main locations through the network. Armed forces applications, such as the supervision of the battlefield, are motivated by the development of sensor networks.

As of today, we can find their applications in process control, device status monitoring, health monitoring through body area networks, civil and disaster management, environmental, and commercial applications, etc. These kinds of networks can also be utilized to supervise and analyze tornado/storm movement, monitor the warmth of the surrounding volcano, and monitor wild animal behavior. In this chapter, we intend to present an inclusive analysis of the recent literature on prolonging the lifetime and coverage area of sensor networks.

**Keywords:** WSN, Sensor, Coverage Area, Lifespan of WSN.

## 9.1 Introduction

In recent times Ad hoc networks have become a popular area of study. Several issues are still to be resolved. Some of such issues are as follows [1]:

1. Security
2. Scalability
3. Energy conservation
4. Routing
5. Interoperation
6. Quality of service
7. Client-server model shift
8. Node cooperation

Ad hoc networks, by their very nature, have limited capacity to expand. If the available capacities are such as bandwidth, the antenna radiation pattern specifies some communication limits [2].

### 9.1.1 Scalability

Ad hoc networks, by their very nature, have limited capacity to expand. If the available capacities are such as bandwidth, the antenna radiation pattern specifies some communication limits [2].

### 9.1.2 Routing

Routing in dedicated wireless networks is unnecessary because of an exceedingly vibrant environment. Ad Hoc networks comprise a set of wireless mobile nodes, construct a vigorously momentary network without using any pre-existing network infrastructure or central direction [3].

### 9.1.3 Quality of Service

As the conception of networking was introduced to provide only the best possible service, there is a big challenge for network designers because of the heterogeneity of existing Internet applications. Direct audio, video, and information transfer are only a few examples of applications with a wide range of needs.

### 9.1.4 Safety Measures

As ad hoc networks are particularly vulnerable to detrimental activities, one of the very imperative issues of ad hoc networks is their security. A very sky-scraping extent of security against enemies and an active/passive listening attacker is expected when such networks are utilized in armed forces operations or clandestine meetings. As centralized network management or certification authority is deficient in such networks, thus these enthusiastically changing wireless structures are vulnerable to hacking, listening, intrusion, etc. Safety measures are often the foremost "obstacle" in business applications.

### 9.1.5 Energy/Power Preservation

In traditional Ad hoc networking research, energy/power preserving networks have emerged as extremely accepted networks. There are two main themes of nearly identical research: maximizing one battery life and maximizing the life of the entire network [4]. The previous relates to business applications and collaboration issues related to nodes whereas the second is more substantial, for example, in environments where the armed forces are supposed to cooperate with the nodes. Up to a great extent, research has been done on the routing layers; however, there are still many investigations to be carried out [5].

### 9.1.6 Node Collaboration

Node collaboration occurs when a node requests services from other nodes. Nodes rely on other nodes for services; hence there is no substitute except to rely on other people's data. However, when changes in data amount and priority are considered, the situation becomes significantly more complex.

### 9.1.7 Interoperation

One of the unexplored research topics of ad hoc networks is the implications that arise when two separately created networks come physically close to each other. The intervention with each other becomes inevitable when ad hoc networks move into the same vicinity. Idyllically, such networks that are in the same vicinity should be amalgamated but it is not a very straightforward task to join two networks it is just because the networks may be utilizing diverse synchronization, or even dissimilar MAC or routing protocols. One of the major concerns is security also. Thus the basic idea of interoperation is how networks can be designed so that they can be attuned with any 3G or 4G cellular network.

## 9.2 Intricacy while Deployment of Manet

Here are some key routing issues to consider when deploying MANETs:

1. Fickleness of surroundings
2. Impulsiveness of Medium
3. Nodes with limited recourses
4. Communication faults
5. Node malfunction
6. Jam-packed Links or Nodes
7. Connection collapse
8. Varying Topology
9. Route Breakages.

## 9.3 Wireless Sensor Networks

A sensor network has 100–1000 nodes, each connected to one or more others. Antenna, transceiver, microcontroller, battery, and interface electronic circuit make up each node. The data can be routed or flooded among nodes. A WSN sensor node is made up of one or more sensors, an embedded processor, a low-power wireless transmitter, and a battery. A sensor node is a device that senses environmental factors such as temperature, humidity, and pressure. Following sensing, the embedded CPU, microcontroller, or processor processes the signal and delivers data packets to neighboring nodes. Each sensor node serves as both a transmitter and a receiver. The drain node makes a decision based on data from other sensor nodes [6–9].

Because most sensors are powered by batteries, it is vital to develop energy-efficient routing solutions. Furthermore, because all sensors in the same network must share the restricted channel bandwidth, packet jamming and loss occur [10]. Despite all its advantages, some precincts affect the performance of such networks. Such factors are power constrictions, hardware constrictions, error forbearance, and scalability. Many researchers are currently trying to conquer these precincts and have been able to rise somewhat above them [11].

## 9.3.1 Sensor Network Communication Architecture

According to Antonio [12], a sensor network that comprises several sensor devices such as every node has the competence to congregate information and then transmit this valuable information to the sink node and end-users. With the aid of multi-hop infrastructure, the congregated information is routed back to the final user through the sink as shown in Figure 9.1.

Here, in this network, the sink sends commands or queries to other sensor nodes in the sensor area, and on the other hand, the sensor node operates in the group to achieve the sensor task and send sensitive information to the sink. Meanwhile, the sink acts as a gateway to external networks. Furthermore, it collects information from sensor nodes, performs simple processing on this collected information, and then finally sends the appropriate data to the end-user over the Internet. Each node of the network sensor uses a single-stage remote transmission to transmit information to the sink [13]. Both the sink and the nodes are used as a stack of protocols where they combine power and routing awareness, integrate information with network protocols, efficiently transmit via wireless media power efficiently and enhance the joint effort of sensors. The protocol stack encompasses the application, transport, network, data link, physical layers, and power, mobility, and job management layers as shown in the Figure 9.2.

However, this technique is expensive in terms of power utilization for long-distance transmission [14].

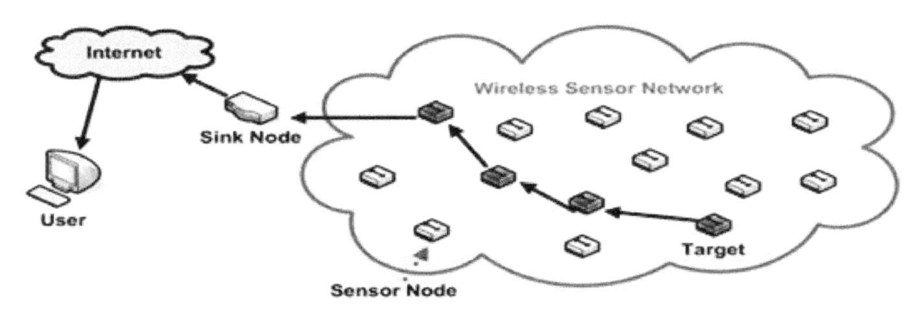


**Figure 9.1** Wireless network

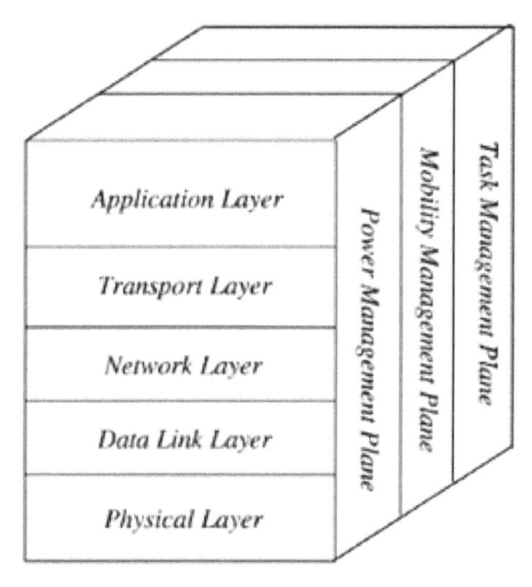


**Figure 9.2** Sensor networks protocol stack

## 9.4 Coverage Problem in Sensor Network

Numerous scholars are now researching an optimization technique for covering and locating nodes in wireless sensor networks. The TianD algorithm developed by M Cardei [15], the CCP algorithm developed by Wang [16], and the Huang algorithm developed by Liang [17] are all viable approaches in this context.

To increase coverage, Xia Bang Chen et al., [18] introduced Hy-MFCO. They determined through actual experiments and computer calculations that Hy-MFCO is capable of boosting detection coverage while maintaining energy economy.

"Maximum separation set coverage problem" has been solved by Mihila Cardi et. al., [19]. There is a maximum coverage area for coupled nodes, while all other nodes are dormant. Efficiencies in energy and network lifespan are increased as a result. Genetic restriction algorithms were used to build the weighed genetic algorithm and method for maximizing coverage proposed by Jia Jie et al., [20]. For the evolutionary algorithm to work properly, a complete coverage interrupt is required to forecast ideal node combinations and absolute configurations for each working node, hence prolonging the network's endurance. To find the optimal coverage range and extend the network's life, Yoon-hee Han [21] presented a strategy based on genetic algorithm coverage programming and evolutionary comprehensive research methodologies for monitoring all targets.

## 9.5 Lifetime Maximization of Wireless Sensor Networks

Since the sensors are battery-powered, the power consumption of sensors to be deployed in WSN should be minimized. To conserve energy, sensors should have their power supplies, which should get turned off when not in use [22]. Due to this limitation, prolonging the lifetime of WSNs by assuring the quality of coverage is questionable. Therefore extensive research and investigation about this issue are important.

The use of Energy Harvesting Wireless Sensor Networks (EHWSNs) in place of conventional battery-powered wireless sensor networks (WSNs) may result in a longer network lifetime [23–25]. However, due to the expensive expense of energy harvesting equipment, complete WSNs cannot rely only on them. Figure 9.3 depicts a classification of network lifespan maximization strategies.

The network is affected by routing[26]. To maximize battery life, use a dynamic path specified by sensors. In part, this is due to data collection causing routing congestion. In addition, sensor node batteries are depleted [27]. Controlled movement using mobile sensors or sinks has solved this problem thus ensuring a well-balanced network traffic load. Coverage is the act of assessing the quality of individual events that occurred within a target area, a sensing point, or a sensor-covered barrier field. The network must maintain coverage of the target area and expected communication between

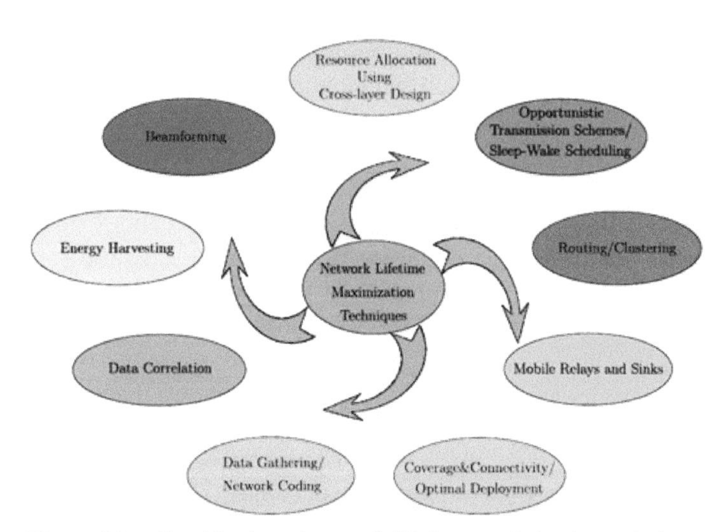


**Figure 9.3** Classification of network lifetime maximization techniques

subsets of sensors and the sink node [28]. Assume that each node has sensors and has been assigned specific responsibilities. The sensors' activity has been scheduled in such a way that the coverage is reliable. By ensuring coverage, the likelihood of maximizing the network lifetime is increased [29].

Energy conservation can be accomplished by reducing tele-traffic, which eliminates data redundancy and thus maximizes network lifetime. This can be accomplished by utilizing an energy-efficient cross-layer design for optimal data collection from sensors, thereby minimizing energy waste due to redundant data transmission. Ultimately, through conserving energy, the network's lifetime has been increased [30–31].

Multiple single-antenna transmitters were constructed using a distributed or collaborative beam forming technique, which increased transmission distance. As a result, the antenna's transmission power has decreased [32]. Because the power dissipation is spread across numerous transmitters, each emitter can reduce it's transmit power. As a result, the network's lifetime has been increased.

Sensor nodes in a WSN may fail to owe energy depletion or device failure. As a result, fault tolerance [33] is a critical concern in WSNs. The energy efficiency of a system is determined by the amount of energy dissipated and the rate of data loss. Additionally, energy efficiency can be increased by properly deploying nodes in a cluster, selecting an appropriate topology, and employing appropriate transmission mechanisms. The energy consumption of the nodes in a system can be used to detect node problems. To ensure nodes have a long period of autonomy, protocols and procedures strive to optimize power consumption, which is essential to extend both the nodes' and the network's lifetime [34].

By implementing appropriate sleeping techniques, the life of a network can be extended. Sleeping techniques have been introduced to conserve energy and extend the network's lifetime. Additionally, energy efficiency can be increased by turning on the node by a wakeup schedule [3]. The sensor node sleeps normally or aggressively, depending on the duty cycle assigned. The WSN can assess the effect of adopting sleeping mechanisms on data transmissions from the industrial area to the base station by comparing the number of transmissions caused by aggressive sleeping behavior to the number caused by regular sleeping behavior.

The lifespan of WSNs is determined by the capacity of the battery. Each node in a multihop WSN serves as both a data originator and a data router. Node failures can result in changes to the network's topology, requiring the network to be reorganized. Power conservation and management are critical elements to address, which have resulted in the development of numerous technologies such as energy-aware protocol development and hardware optimizations. These research and development activities resulted in the commercialization of many motes or wireless radio modules.

The implementation of WSN for real-time applications is hampered by the sensor's inadequate energy storage capacity. Energy backup can be enhanced by integrating renewable energy sources with sensor nodes to extend the life of the WSN. The amount of power required by the sensor node during transmission establishment may vary depending on the distance between the sensor node and the sink. The sensor node's power consumption is directly proportional to the distance between the sensor node and the Cluster Head (CH), which also impacts the network longevity. In relay or multihop transmission models, intermediate routers are employed to expand the network's length, which affects both power consumption and network longevity.

Battery backup is required to power the sensor node, and solar panels may also be used to power the sensor node, depending on the environment in where the sensor is situated. Design constraints may vary according to the application in which the monitoring function is performed. The environment has a significant impact on the network's size, deployment method, and topology. If the network is small, fewer nodes are required to construct it in an indoor setting, whereas if the network is large, more nodes are required to cover a wider area in an outdoor environment. Obstacles in the surroundings may obstruct communication between sensor nodes.

WSN is a network of tiny low-power devices capable of sensing, processing, and communicating on a single chip. Two forms of WSN exist. The first is made up of hundreds or thousands of nodes that operate across enormous geographical areas. This network communication technology is employed in military and environmental surveillance applications. The second is made up of tens of sensors utilized in remote measurement applications. The majority of these applications utilize WSNs just for data collecting and transmission, leaving feature extraction and fault detection to a central computer. Sensor feature extraction and problem detection is a viable alternative to raw data transmission that can significantly reduce the amount of data transmitted, conserve energy, and extend the node's lifetime.

Typically, in a WSN, power optimization is concentrated on the networking and sensing subsystems. Even when node components are idle, they consume a significant amount of energy. As a result, power management strategies are utilized to temporarily disable node components. Energy is consumed more efficiently by the communication subsystem than by the calculation subsystem. According to analysis, transmitting a single bit can take as much energy as running a few thousand instructions. As a result, communication should be sacrificed in favor of computation. Generally, systems based on networks and nodes are regarded to be more energy efficient. The network-based approach processes sensor data at the base station, whereas the node-based approach processes sensor data at the sensor nodes.

## 9.6 Overview of Optimization Techniques Employed to Maximize the Lifetime of WSN

Prolonging the life of WSNs by ensuring coverage quality is dubious. As a result, considerable inquiry and analysis into this subject are critical. The preceding section discussed the conventional approaches for extending the life of WSNs. This section discusses the evolutionary strategy for resolving the coverage issue.

An evolutionary approach has been proposed for determining the maximum number of sensor disjoint sets capable of thoroughly covering the target area [36].

To address the aforementioned challenge, a mimetic algorithm has been created that employs both Darwinian evolution and Lamarckian local improvement to find the ideal solution while taking local exploitation and global exploration into account.

A Hybrid approach genetic algorithm based on a Schedule Transition Operations with Genetic Algorithm (STHGA) is utilized to solve target coverage problems based on a forward encoding scheme [37]. The uniqueness of this method is that the chromosomal maximum gene value is increased gradually with the quality of the solution obtained that is related to the number of disjoint complete cover sets. Apart from sensing ranges and the number of sensors, the influence idleness of sensors has also been correlated with the performance of STHGA.

Kuhn-Munkres Parallel Genetic Algorithm is employed to overcome the set cover problem, which leads to prolonging the network lifetime of WSNs in a large-scale environment. This algorithm schedules the sensors into several complete cover sets called disjoints. Then the sensors are triggered for in batch of conservation of energy. It utilizes the divide-and-conquer technique for reduction of dimensionality and adopts the polynomial Kuhn-Munkres algorithm to splice the feasible solutions obtained in each sub-area to improve the efficiency of search significantly.

With the Connected Set Covers (CSC) issue, the goal is to link as many sensor nodes as possible to the base station. Three optimization strategies lowered the computational complexity of the CSC problem: IP-based, greedy, distributed and localized heuristics. Target Coverage (TC) problems are recast as Maximal Set Cover (MSC) problems. To prolong the life of a network, more set covers are required. When a set cover is activated, its sensors detect all targets at once, while the other sensors slumber. As a result, the MSC problem is NP-hard complete and can be solved centrally using linear programming and greedy techniques.

The power-saving issue is posed as a Set Covering Problem (SCP). An energy-efficient sensor network can be created by adding Modified Ant Colony Optimization (MACO), which solves for the lowest set cover for a given collection of nodes [38].

The suggested research utilizes Ant Colony Algorithms (ACO), Intelligent Water Drop Algorithms (IWD), and Flower Pollination Algorithms (FPA). The following sections provide overviews of certain algorithms.

The ACO algorithm is a natural metaphor algorithm inspired by ant behavior. While travelling, the ants discover food deposits and follow these deposited pheromones to their nests; the other ants then follow these deposited pheromones. Although pheromones degrade with time, they provide up new possibilities when ants agree to choose a heavy pheromone path [39]. Thus, ants can determine the shortest route between their colony and a food source using only pheromone information. The ACO algorithm's fundamental mathematical concept was initially used for the Travelling Salesman Problem. The TSP challenge is to discover the shortest path between two cities that passes through each city precisely once and then returns to the beginning city. There are N cities in the ACO algorithm for the TSP, and the total number of ants is M. Each ant selects the next place with a probability proportional to the distance between the current and next locations and the pheromone's intensity.

Two types of ants are used in ACO algorithms: Forward Ants (FA) and Backward Ants (BA). FAs investigate and collect information about the pathways between source and destination nodes [40]. While the ants walk, they build a tree as they merge or reach their goal, and data is conveyed along the tree paths. While returning from the destination node to the source nodes, the BAs fulfill their primary duty of updating the nodes they pass through. The two significant downsides of ACO-based algorithms are that the ant becomes blind in search once it deviates from PoIs during the solution building process and that a large number of redundant sensors (RSs) are included in the final solution, which increases the coverage cost unnecessarily [41].

Shah-Hosseini invented IWD [42]. The IWD algorithm is inspired by natural water movement. Water always takes the path of least resistance. This is a great way to build a swarm relay system. The IWD method has two main parameters: velocity and soil. The IWD algorithm's lifetime changes these two parameters. The IWD influences the soil it travels through, increasing or lowering its speed [43–44]. The IWD's velocity increases nonlinearly and inversely proportionate to the amount of soil between two sites. As a result, IWDs with less dirt gain speed. The IWD also transports soil, the amount of which is governed by the IWD's trip duration. The network's memory is soil exchanged between nodes. IWDs like to choose the easiest route (or soil). Thus, the IWD algorithm is an optimization solution.

## 9.7 Conclusion

In general, the performance of WSN can be increased by deploying the sensor in an effective manner that is done by covering each point in a region of interest by employing at least one sensor node. Hence there arises a need to find maximum coverage area through the deployment of the minimum number of sensor nodes. Further factors like technological advancement and increased network size also create the necessity to increase the coverage area for improved connectivity and lifetime of the network. This research work focuses on constructing a connected and energy-efficient sensor network for applications that expects optimal sensor placement so that the lifetime of the network is improved. The coverage area of a sensor node can be maximized using widely studied heuristic optimization techniques like Genetic Algorithms and Particle Swarm Optimization. Even though these algorithms give good results with lesser computational overhead, the iterative nature, as well as the memory requirements, limits the algorithm for high speed and resource-rich applications. The objective of the chapter is to focus on improving the network lifetime by maximizing the coverage area with respect to energy and point of interest using problem-specific intelligent optimization techniques like ACO, IWD.

## References

[1] M. Conti and S. Giordano. "Mobile Ad Hoc Networking: Milestones, Challenges, and New Research Directions." *IEEE Communications Magazine*, 52(1), pp. 85–96, 2014.

[2] S. Sharafeddine and O. Farhat. "A Proactive Scalable Approach for Reliable Cluster Formation in Wireless Networks with D2D Offloading." *Ad Hoc Networks*, 77, pp. 42–53, 2018.

[3] A. M. Desai and R. H. Jhaveri. Secure Routing in Mobile Ad Hoc Networks: A Predictive Approach." *International Journal of Information Technology*, 11(2), pp. 345–356, 2019.

[4] N. A. Pantazis, S. A. Nikolidakis, and D. D. Vergados. Energy-Efficient Routing Protocols n Wireless Sensor Networks: A Survey." *IEEE Communications surveys & tutorials*, 15(2), pp. 551–591, 2012.

[5] Y. Yao, Q. Cao, and A. V. Vasilakos. "EDAL: An Energy-Efficient, Delay-Aware, and Lifetime-balancing data collection protocol for heterogeneous Wireless Sensor Networks." *IEEE/ACM Transactions on Networking (TON)*, 23(3), pp. 810–823, 2015.

[6] O. I. Khalaf, and B. M. Sabbar. "An Overview on Wireless Sensor Networks and Finding Optimal Location of Nodes." *Periodicals of Engineering and Natural Sciences (PEN)*, 7(3), pp. 1096–1101, 2019.

[7] M. Pule, A. Yahya, and J. Chuma. "Wireless Sensor Networks: A Survey on Monitoring Water Quality." *Journal of applied research and technology*, 15(6), pp. 562–570, 2017.

[8] D. Kandris, C. Nakas, D. Vomvas, and G. Koulouras. "Applications of Wireless Sensor Networks: An Up-to-Date Survey." *Applied System Innovation*, 3(1), p. 14 2020.

[9] H. Agarwal, P. Tiwari, and R. G. Tiwari,. "Exploiting Sensor Fusion for Mobile Robot Localization." In 2019 *Third International conference on I-SMAC (IoT in Social, Mobile, Analytics and Cloud)(I-SMAC)* (pp. 463–466). IEEE, 2019 December.

[10] R. Priyadarshi, R., B. Gupta, B. and A. Anurag. "Deployment Techniques in Wireless Sensor Networks: A Survey, Classification, Challenges, and Future Research Issues." *The Journal of Supercomputing*, pp. 1–41, 2020.

[11] L. Hamami and B. Nassereddine. "Application of Wireless Sensor Networks in the Field of Irrigation: A Review." *Computers and Electronics in Agriculture*, 179, p. 105782, 2020.

[12] P. Antonio, F. Grimaccia, and M. Mussetta. "Architecture and Methods for Innovative Heterogeneous Wireless Sensor Network Applications." *Remote Sensing*, 4(5), pp. 1146–1161, 2012.

[13] V. C. Gungor, B. Lu, and G. P. Hancke. "Opportunities and Challenges of Wireless Sensor Networks n Smart Grid." *IEEE transactions on ndustrial electronics*, 57(10), pp. 3557–3564, 2010.

[14] A. Dunkels, J. Alonso, T. Voigt, H. Ritter, and J. Schiller. "Connecting Wireless Sensornets with TCP/IP Networks." *In International Conference on Wired/Wireless Internet Communications* (pp. 143–152). Springer, Berlin, Heidelberg, 2004 February.

[15] Y. Yang, and M. Cardei. "Adaptive Energy Efficient Sensor Scheduling for Wireless Sensor Networks." *Optimization letters*, 4(3), pp. 359–369,2010.

[16] X. R. Wang, G. L. Xing, Y. F. Zhang, C. Y. Lu, R. Pless, and C. D. Gill. "Integrated Coverage and Connectivity Configuration n Wireless Sensor Networks." *Proceedings of the 1st ACM Conference on Embedded Networked Sensor Systems (SenSys'03)*, Los Angeles, CA, 2003.

[17] Y. Liang, P. Zeng, and H. Yu, "Energy Adaptive Cluster-Head Selection for Wireless Sensor Networks." *Information Control*. 35, pp. 141–146, 2006.

[18] Chen, Chia-Pang, S. C. Mukhopadhyay, C. L. Chuang, T. S. Lin, M. S. Liao, Y. C. Wang, and J. A. Jiang. "A Hybrid Memetic Framework for Coverage Optimization n Wireless Sensor Networks." *IEEE transactions on cybernetics* 45, no. 10, pp. 2309–2322, 2015.

[19] M. Cardei, and D. D. Zhu, "Improving Wireless Sensor Network Lifetime Through Power Aware Organization." *Wireless Network* 11, pp.'333–340, 2005.

[20] J. Jia, J. Chen, J. Chang, L. Zhao, and G. Wang, "Optimal Coverage Scheme Based on Genetic Algorithm n Wireless Sensor Networks." *Control Decision* 1(22), pp. 1289–1292, 2007.

[21] G. J. Min, and H. Y. Hee, "A Target Coverage Scheduling Scheme Based on Genetic Algorithms n Directional Sensor Networks." *Sensors*, Sensor 11, pp. 1888–1906, 2011.

[22] Jafar, M, Moghaddam, H, Kalam, A, Nowdeh, SA, Ahmadi,A, Babanezhad, M and S. Saha. "Optimal Sizing and Energy Management of Stand-Alone Hybrid Photovoltaic/Wind System Based on Hydrogen Storage Considering LOEE and LOLE Reliability Indices Using Flower Pollination Algorithm." *Renewable Energy*, vol. 135, pp. 1412–1434, 2019.

[23] A. Draw. "On the Performances of the Flower Pollination Algorithm – Qualitative and Quantitative Analyses." *Applied Soft Computing*, vol. 34, pp. 349–371, 2015.

[24] L. A. Moncayo–Martínez and E. Mastrocinque. "A Multi-Objective Intelligent Water Drop Algorithm to Minimize Cost of Goods Sold

and Time to Market in Logistics Networks." *Expert Systems with Applications*, vol. 64, pp. 455–466, 2016.

[25] F. Xi, Y. Pang, G. Liu, S. Wang, W. Li, C. Zhang,and Z. L. Wang. "Self-Powered Intelligent Buoy System by Water Wave Energy for Sustainable and Autonomous Wireless Sensing and Data Transmission." *Nano Energy*, vol. 61, pp. 1–9, 2019.

[26] A. Siddiqua, M. A. Shah, H. A. Khattak, I. U. Din,and M. Guizani. "iCAFE: Intelligent Congestion Avoidance and Fast Emergency Services." vol. 90, pp. 366–375, 2019.

[27] P. R. Srivastava. "A Cooperative Approach to Optimize the Printed Circuit Boards Drill Routing Process Using Intelligent Water Drops." *Computers & Electrical Engineering*, vol. 43, pp. 270–277, 2015.

[28] A. Khosravi, J. J. G. Pabon, R. N. N. Koury, and L. Machado. "Using Machine Learning Algorithms to Predict the Pressure Drop During Evaporation of R407C." *Applied Thermal Engineering*, vol. 133, pp. 361–370, 2018.

[29] L. Chen, X. Tian, Y. Li, C. Yang, L. Lu, Z. Zhou, and Y. Nie. "An AIE Dye Based Smartphone and LDA Integrated Portable, Intelligent and Rapid Detection System as Trace Water Indicator and Cyanide Detector." *Dyes and Pigments*, vol. 166, pp. 1–7, 2019.

[30] M. Adhikar and T. Amgoth. "An Intelligent Water Drops-Based Workflow Scheduling for IaaS Cloud." *Applied Soft Computing*, vol. 77, pp. 547–566, 2019.

[31] S. Elsherbiny, E. Eldaydamony, M. Alrahmawy, and A. E. Reyad. "An Extended Intelligent Water Drops Algorithm for Workflow Scheduling in Cloud Computing Environment." *Egyptian Informatics Journal*, vol. 19, no. 1, pp. 33–55, 2018.

[32] Z. Jiao, K. Ma, Y. Rong, P. Wang, H. Zhang, and S. Wang. "Path Planning Method Using Adaptive Polymorphic Ant Colony Algorithm for Smart Wheelchairs." *Journal of Computational Science*, vol. 25, pp. 50–57, 2018.

[33] M. Qiuet, M , Z. Ming , Z, J. Li, J, J. Liu , J, G. Quan ,G and Y. Y. Zhu. "Informer Homed Routing Fault Tolerance Mechanism for Wireless Sensor Networks." *Journal of Systems Architecture*, vol. 59, no. 4–5, pp. 260–270, 2013.

[34] A. A. H. Hassan, W. M. Shah, A. H. H. Habeb, M. F. I Othman, and M. N. Al-Mhiqani. "An Improved Energy-Efficient Clustering Protocol to Prolong the Lifetime of the WSN-Based IoT." *IEEE Access*, 8, pp. 200500–200517, 2020.

[35] M. N. Khan, H. U. Rahman, M. A. Almaiah, M. Z. Khan, A. Khan, M. Raza, M. Al-Zahrani, O. Almomani, and R. Khan. “Improving Energy Efficiency with Content-Based Adaptive and Dynamic Scheduling in Wireless Sensor Networks.” *IEEE Access*, 8, pp. 176495–176520, 2020.

[36] C. Ting and C. Liao. “ A Memetic Algorithm for Extending Wireless Sensor Network Lifetime.” *Informational Science* vol. 180, no. 24, pp. 4818–4833, 2010.

[37] J. Zhang, F. Ren, S. Gao, H. Yang, and C. Lin. “Dynamic Routing for Data Integrity and Delay Differentiated Services in Wireless Sensor Networks.” *IEEE Transactions on Mobile Computing*, vol. 14, no. 2, pp. 328–343, 2015.

[38] S. Begum, N. Tara, and S. Sultana. “Energy-Efficient Target Coverage in Wireless Sensor Networks Based on Modified Ant Colony Algorithm.” *International Journal of Ad hoc, Sensor & Ubiquitous Computing (IJASUC)* vol. 1, no. 4, pp. 29–36, 2010.

[39] G. Dong, X. Fu, H. Li, and P. Xie. “Cooperative Ant Colony Genetic Algorithm Based on Spark.” *Computers and Electrical Engineering*, vol. 60, pp. 66–75, 2017.

[40] Y. Jian, and Y. Li. “Research on Intelligent Cognitive Function Enhancement of Intelligent Robot Based on Ant Colony Algorithm” *Cognitive Systems Research*, vol. 56, pp. 203–212, 2019.

[41] Z. H. Jia , ZH, Y. Wang, Y, C. Wu, C ,Y. Yang , Y, X. Zhang, X and H. Chen, H 2019, “Multi-objective energy-aware batch scheduling using ant colony optimization algorithm.” *Computers and Industrial Engineering*, vol. 131, pp. 41–56, 2019.

[42] H. Shah-Hosseini. “Intelligent Water Drops Algorithm: A New Optimization Method for Solving the Multiple Knapsack Problem.” *International Journal of Intelligent Computing and Cybernetics.*

[43] Wu, X & Zeng, F 2018, ‘Design of electro-hydraulic servo loading controlling system based on fuzzy intelligent water drop fusion algorithm’, Computers & Electrical Engineering, vol. 71, pp. 485–491.

[44] P. Datta and B. Sharma. A Survey on IoT Architectures, Protocols, Security and Smart City Based Applications. In 2017 8th *International Conference on Computing, Communication and Networking Technologies (ICCCNT)* (pp. 1–5). IEEE, 2017, July.

# 10

# Soft Open Points in Active Distribution Systems

**Md Abu Saaklayen[1], Xiaodong Liang[1], Sherif O. Faried[1], and Massimo Mitolo[2]**

[1]The University of Saskatchewan, Saskatoon, SK S7N 5A9, Canada
[2]Irvine Valley College, Irvine, CA 92618, USA
E-mail: vpu975@mail.usask.ca; xil659@mail.usask.ca; sherif.faried@usask.ca; mmitolo@ivc.edu

## Abstract

In this chapter, the Soft Open Point (SOP), an emerging power electronics device in distribution networks, is introduced. SOPs can be installed to replace the normally open points (NOPs) in distribution networks to achieve the flexible connection between feeders and the upgrade from radial to closed-loop configuration. Under normal operation, SOPs can provide active power flow control, reactive power compensation, and voltage regulation; under fault conditions, SOPs offer post fault supply restoration. In this chapter, the basic concept of the SOP and its benefits are explained; a comparison between SOPs and other power electronic devices used in distribution networks is provided; the modeling method of SOPs, along with their placement and sizing in active distribution networks are reviewed.

**Keywords:** Active distribution systems, Normally Open Points, voltage regulation, Soft Open Points, service restoration.

## 10.1 Introduction

Environmental concerns, technological innovation, and government new policies have led to increasing integration of renewable energy distributed generation (DG) [1]. Besides the undeniable benefits, DGs, along with energy storage, electric vehicles, and demand-side resources, have made the operation of distribution networks more complex and challenging. Traditional distribution networks are usually operated in an open loop, which makes it difficult to accommodate volatile outputs of DGs [2, 3]. A distribution network may occasionally become overloaded due to different types of loads and intermittent renewable energy sources.

Currently, voltage issues are mitigated by using conventional reactive power compensation devices, including on-load tap changer (OLTC), voltage regulators (VR), capacitor banks, and direct scheduling of dispatchable distributed generators. However, due to the slow response and the discrete voltage regulation offered by conventional compensation devices, it is rather difficult to regulate voltages precisely when the power generation of DGs and the load fluctuate in active distribution networks (ADNs) [4]. In addition, due to the intermittent power generation output of renewable energy-based DGs, the control of dispersed DGs can be difficult for distribution system operators. The regulating capability of DGs cannot solve the centralized optimization problem. In this situation, the SOP, an emerging power electronics device, can be used as an alternative to traditional measures to solve overloading, voltage violations, and supply restoration problems in a distribution system with high DG penetration.

## 10.2 Basic Concept of SOP

The SOP can be installed to replace the NOPs in distribution networks to achieve a flexible connection between feeders [5]. Upgrading a distribution network from its original radial structure to a normally closed-loop configuration has attracted research interest. A closed-loop configuration is advantageous over a radial one because the load can be balanced between feeders, resulting in a better voltage profile and the improved reliability of the system [6]. However, the loop configuration increases the risk of wide-area failures because any single network fault can be quickly propagated over a broad region. Thus, more complicated and expensive protection schemes are required for such interconnected network configurations [7, 8]. The SOP is proposed to combine the benefits of both radial and loop/mesh operated networks while avoiding the drawbacks of both [8].

The basic configuration of SOPs consists of two back-to-back voltage source converters (VSC) (Figure 10.1). Two VSCs are located between the feeder endpoints and are connected through a common dc link. The main

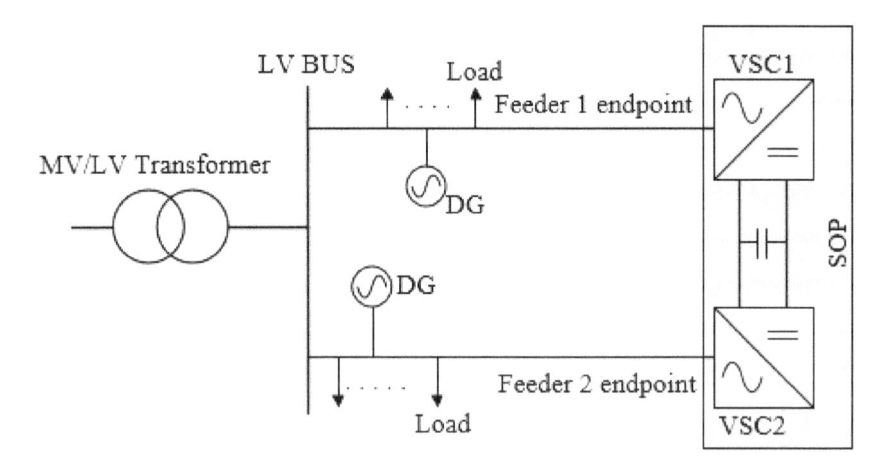


**Figure 10.1** Basic configuration of distribution networks with a SOP.

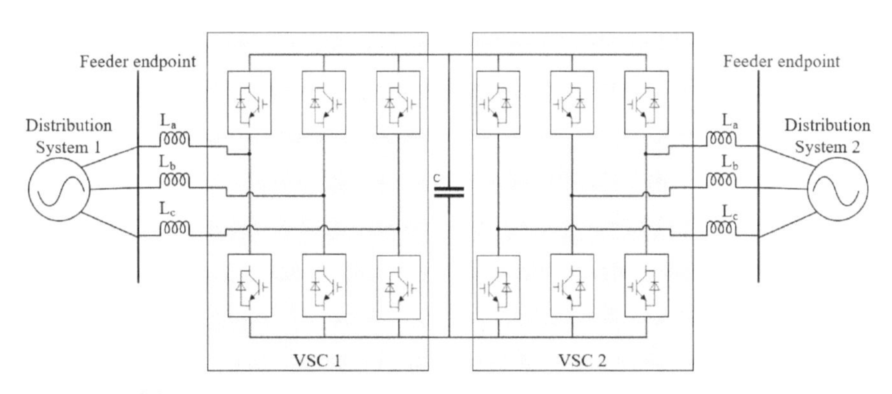


**Figure 10.2** Main circuit topology of the back-to-back VSC-based SOP [12, 13].

circuit topology of the back-to-back VSCs is shown in Figure 10.2, which consists of a dc capacitor to provide an energy buffer and reduce the dc side voltage ripple, and of two two-level three-phase insulated gate bipolar transistor (IGBT)-based VSCs to generate voltage waveforms using pulse-width-modulation (PWM). Each VSC terminal is connected to a series filter; the inductances of the filter are indicated with $L_a$, $L_b$, and $L_c$ for phases A, B, and C, respectively. The inductances provide the high-frequency harmonic attenuation, limit the rising rate of short circuit currents, and facilitate the power flow control. The line resistance and reactance between the filter and the feeder endpoint are ignored due to the short distance [9].

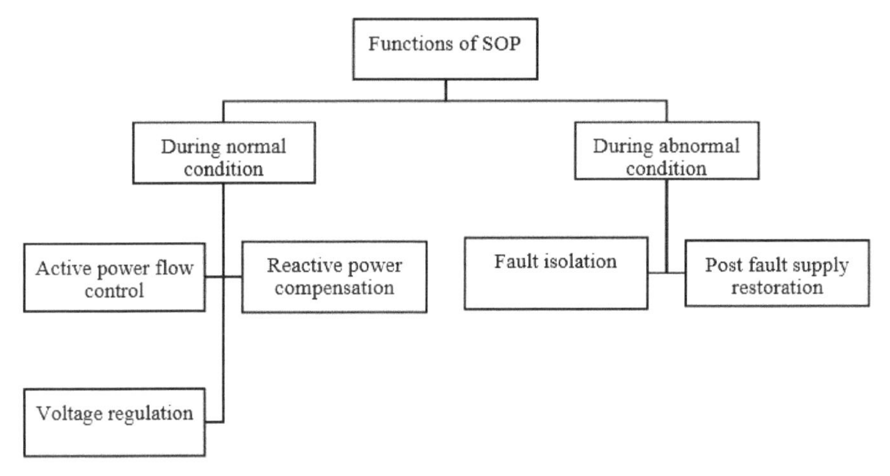


**Figure 10.3** Basic functions of SOPs.

Under normal operating conditions, SOPs can offer active power flow control, reactive power compensation, and voltage regulation; under abnormal operating conditions, SOPs can also enable post-fault supply restoration.

The safety hazards caused by the frequent shifting of conventional tie switches can be eliminated through SOPs. SOPs can significantly improve flexibility, expand the controllability, enhance the operation control speed, and increase DG hosting capacity in active distribution networks [10, 11]; the voltage profile and the economy and reliability of the power supply can also be significantly improved [11]. Therefore, the deployment of SOPs can improve the system operation, and facilitate a large penetration of renewable energy sources in distribution networks. The basic functions of SOPs are shown in Figure 10.3.

### 10.2.1 Benefits of SOPs

SOPs, also known as "loop power flow controllers", "series controllers" and "DC links", are multifunctional power electronic devices (Figure 10.4). Compared to mechanical switches, the most important advantages of SOPs are given below:

1. Continuous and dynamic power flow control between connected feeders.
2. SOPs can be used for dynamic voltage support because the injection and absorption of reactive power can be independently controlled at both VSC terminals.

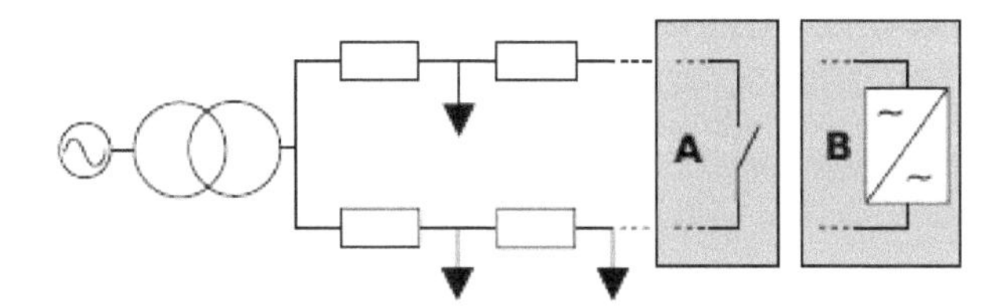


**Figure 10.4** Single line diagram of a simple distribution system, where option A represents a NOP connection and option B represents a SOP connection [8].

3. SOPs enhance the power quality of distribution networks by mitigating voltage imbalance, sags, flickers, and low order harmonics [14].
4. SOPs can be used to connect any feeders irrespective of angular or rated voltage differences between them. SOPs can connect the feeders that are supplied by different substations, which is not possible with mechanical switches.
5. SOPs can disconnect the fault-side feeder within a few milliseconds, and therefore limit the fault current propagation [9, 15, 16].
6. SOPs have the capability of instantaneous fault-current control, and thus, can limit the increase in fault magnitude. Incorporation of SOPs does not require modifications of the existing network's protection assumptions, or upgrades of protection devices, which is generally the case for a closed-loop operation [17].
7. The integration of SOPs reduces the energy loss and increases the hosting capacity of the distribution network [18, 19].

A comparison between SOPs and NOPs is shown in Table 10.1.

**Table 10.1** Comparison of emulation and nomograms

| | NOPs (Tie switch) | SOPs |
|---|---|---|
| Type | Mechanical structure | Fully controlled power electronic device |
| Number of terminals | 2 | 2 and more |
| Number of status | 2 | Multiple status |
| Operation mode | Open-loop | Closed-loop |
| Control capability | Control on/off of feeders | Control active power and reactive power of feeders |
| Fault handling | Fault isolation | Low voltage ride through, loads transfer, uninterrupted power supply. |

## 10.3 Comparison of SOPs with other Power Electronic Devices in Distribution Systems

When the power flow is primarily unidirectional, on-load tap changers enable the automatic voltage control in medium voltage distribution networks. However, DGs in distribution networks require a bi-directional power flow, which can be realized through the implementation of power electronics devices. Large-scale DGs are mostly connected at the medium-voltage level because the cost of the related equipment is lower than that at the high-voltage level. The introduction of DGs at the medium-voltage level produces a significant effect on the voltage profile.

Power electronics technology has been mostly employed in transmission networks in the form of flexible alternating current transmission system (FACTS) devices. Power electronics devices in distribution networks did not attract significant interest until the increased deployment of DGs. In addition to SOPs, commonly used power electronics devices in distribution networks include distribution static synchronous compensators (D-STATCOM), distribution unified power flow controllers (D-UPFC), and smart transformers (ST). The typical connection diagram of these devices is shown in Figure 10.5.

The D-STATCOM is used at the distribution voltage level for active and reactive power compensation, and it is typically connected in parallel with the load. It has advantages over conventional capacitors because it prevents resonances and performs a continuous compensation at a near-unity power factor. Compared to other devices, the cost of implementing a D-STATCOM in active distribution networks is low, but its incapability of feeder load balancing and post fault supply restoration has limited its application [21].

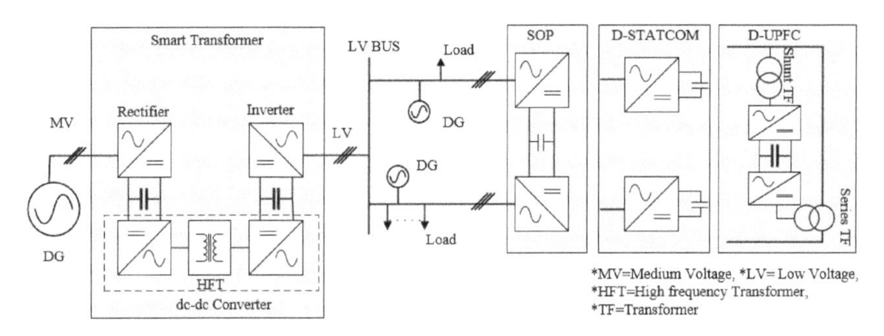


**Figure 10.5** Power electronic devices in distribution networks for control of voltage and power flow [12, 16, 21].

**Table 10.2** Power electronics devices for network compensation in distribution systems [21], [16].

| Functions | D-STATCOM | D-UPFC | Smart Transformer | SOP |
|---|---|---|---|---|
| Feeder connection | None | Direct (synchronous) | Direct (synchronous) | DC-Link (asynchronous) |
| Real power exchange | No | Yes | Yes | Yes |
| Reactive power support | Yes | Yes | Yes | Yes |
| Load balancing | No | Yes | Yes | Yes |
| Fault isolation capability | No | No | No | Yes |
| Post fault restoration | No | Yes | Yes | Yes |
| Connected number of feeders | One | One or more | One or more | Two or more |
| No of converters | One | Two | One to three | Two or more |
| Bidirectional power flow | No | No | Yes | Yes |

The D-UPFC is a third-generation power electronics device and is a combination of STATCOM and static synchronous series compensator (SSSC). The D-UPFC implements the power flow regulation in distribution systems by controlling the line's active and reactive power under steady-state conditions. The D-UPFC operates well under balanced conditions but will not work if disturbances or faults occur on the source site. The UPFC concept was proposed in 1995 by L. Gyugyi with Westinghouse [22].

The smart transformer is a power electronics-based transformer, also known as solid state transformer (SST), whose concept was first introduced in 1968 by McMurray [23]. With the reactive power injection, the smart transformer can support the voltage profile in the medium voltage grid under critical conditions and solve the grid congestion by acting on a single-phase basis. As an energy router, smart transformers can manage both energy flow and information flow between local grids and data centre's [24].

Table 10.2 summarizes the functions of the D-STATCOM, D-UPFC, smart transformer, and SOP. In literature, the back-to-back voltage source converter (B-t-B-VSC), unified power flow controller (UPFC), and SSSC are also classified as different types of SOP [13].

As shown in Table 10.1, SOPs can perform most of the functions and can replace other power electronic devices to minimize the cost. In addition, the SOP has fault isolation capability, making it a perfect device for service restoration.

## 10.4 Principle and Modeling of SOPs in Active Distribution Networks

The SOP has its own control system to achieve flexible control of active and reactive power transmitted through its terminal. A single line diagram of a medium voltage distribution network with an SOP connected at the remote ends of two feeders is shown in Figure 10.6. The two voltage source converters, VSC1 and VSC2, are connected via a common DC bus. The active and reactive power provided by VSC1 and VSC2 is P1, Q1, and P2, Q2, respectively. The power delivered by an SOP can be controlled in the four quadrants of the power chart, as shown in Figure 10.7. The x- and y-axis in Figure 10.7 are active power (P) and reactive power (Q), respectively; the positive values indicate that the voltage source converter delivers power, whereas the negative values indicate that the voltage source converter absorbs power. The circles represent the capacity (i.e., the maximum apparent powers S1 and S2) for the two voltage source converters. Each voltage source converter can operate in any region of the four quadrants within its capacity limit [25].

The controllable variables for SOPs are three-phase active and reactive power outputs of the back-to-back converters. Control modes of converters of SOPs are determined according to the operation state of a distribution system. The overall control modes of SOPs under various operating conditions are shown in Table 10.3. Under normal conditions, the PQ-VdcQ control is usually selected as the control mode of SOPs, where one converter controls the flow of active power and reactive power of the connected feeder,

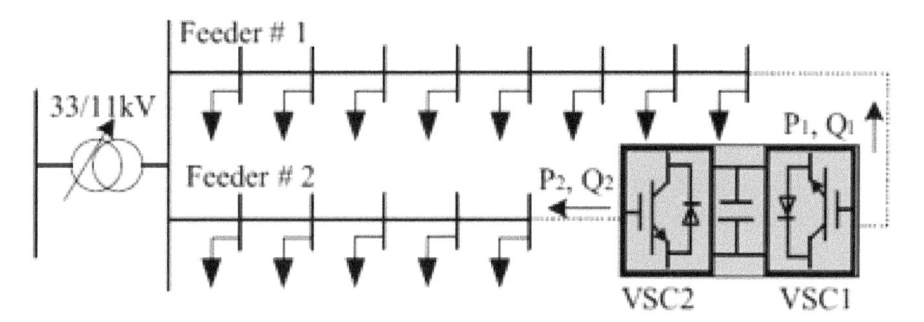


**Figure 10.6** A distribution network with a SOP at the remote ends of two feeders [25].

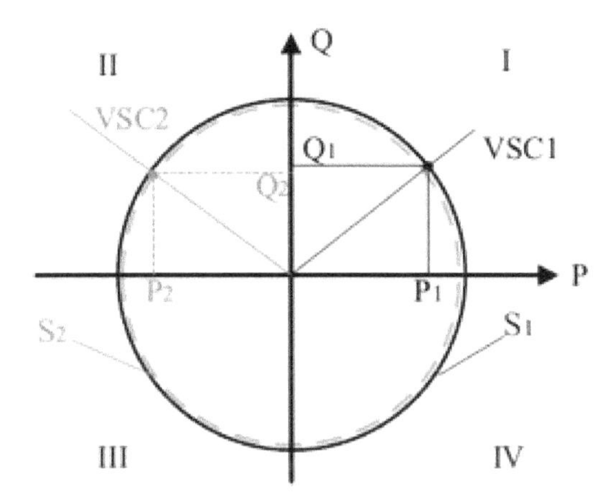


**Figure 10.7** An example of a SOP's operating point: active and reactive power provided by a SOP with two voltage source converters with the same rating [25].

**Table 10.3** Control modes of the back-to-back voltage source converter–based SOP [20, 26].

| Control Mode | VSC1 | VSC2 | Applicable Scenarios |
|---|---|---|---|
| 1 | PQ/PV$_{ac}$ | V$_{dc}$Q | Normal operation |
| 2 | PQ/PV$_{ac}$ | V$_{dc}$V$_{ac}$ | Normal operation |
| 3 | V$_{ac}\theta$ | V$_{dc}$Q/ V$_{dc}$V$_{ac}$ | Faults on VSC1 side |
| 4 | V$_{dc}$Q/ V$_{dc}$V$_{ac}$ | V$_{ac}\theta$ | Faults on VSC2 side |

and the other converter is responsible for the control of DC voltage and reactive power output. Under abnormal conditions, the SOP is operated in the restoration control mode for the electrical supply restoration. For a fault that occurred in the feeder connected to VSC1, the Vac$\theta$-VdcQ control mode is chosen for the SOP to restore power, and the voltage source converter on the faulty side (VSC1) is switched to Vac$\theta$ control after the fault isolation. VSC1, therefore, provides voltage and frequency support to perform the load restoration [20, 26].

When there is a fault in the connected feeders, the SOP quickly deactivates itself or enters a low-voltage ride-through (LVRT) process. The faulty part of the distribution system is then identified and isolated with the aid of breakers/remote control switches, through a pilot protection scheme. After the isolation of the faulty section, the LVRT process ends. To restore the power in the outage area, the control mode of the SOP is then switched to the restoration control mode [20, 27]. The flow chart of the supply restoration strategy using SOPs is shown in Figure 10.8.

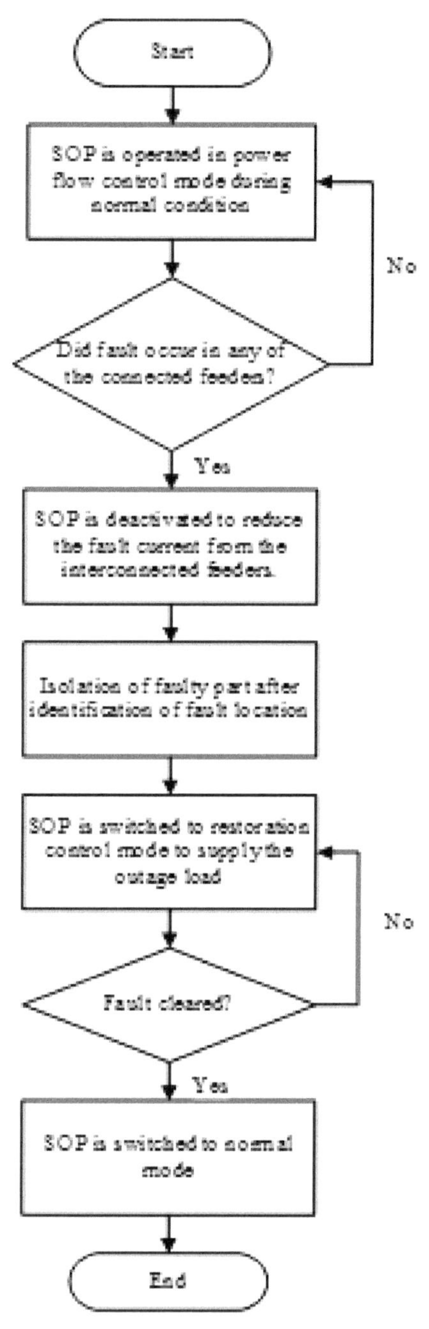


**Figure 10.8** Flow chart for the supply restoration strategy using SOPs [27].

### 10.4.1 Mathematical Modeling of SOPs in Active Distribution Networks

The three-phase power of SOPs can be controlled independently, providing that the three-phase SOP consists of three single-phase SOP modules. The DC isolation enables the converters of the SOP to independently control reactive power outputs within their capacity constraints. The active power is not an independent variable because the sum of the active power values should be equal to zero. Referring to Figure 10.1, the mathematical model of SOPs to solve the optimization problem can be formulated as follows [25, 27, 28]:

***SOP active power constraint:***

$$P_1^{SOP} + P_2^{SOP} + P_1^{SOP-loss} + P_2^{SOP-loss} = 0 \tag{10.1}$$

***SOP internal power loss equations:***

$$P_1^{SOP-loss} = A_L^{SOP} \sqrt{(P_1^{SOP})^2 + (Q_1^{SOP})^2} \tag{10.2}$$

$$P_2^{SOP-loss} = A_L^{SOP} \sqrt{(P_2^{SOP})^2 + (Q_2^{SOP})^2} \tag{10.3}$$

***SOP capacity constraint:***

$$\sqrt{(P_1^{SOP})^2 + (Q_1^{SOP})^2} \tag{10.4}$$

$$\sqrt{(P_2^{SOP})^2 + (Q_2^{SOP})^2} \tag{10.5}$$

***SOP reactive power constraint:***

$$Q_{1,\min}^{SOP} \leq Q_1^{SOP} \leq Q_{1,\max}^{SOP} \tag{10.6}$$

Here,

| | | |
|---|---|---|
| $P_1^{SOP}/Q_1^{SOP}$ | : | Active/reactive power injection by SOP through VSC1 at feeder 1 |
| $P_2^{SOP}/Q_2^{SOP}$ | : | Active/reactive power injection by SOP through VSC2 at feeder 2 |
| $P_1^{SOP-loss}/P_2^{SOP-loss}$ | : | Active power losses of SOP connected at feeders 1 and 2 |

| | | |
|---|---|---|
| $S_1^{SOP}/S_2^{SOP}$ | : | Capacity limits of SOP connected at feeders 1 and 2 |
| $Q_{1,\min}^{SOP}/Q_{1,\max}^{SOP}$ | : | Upper/lower limit of reactive power provided by SOP connected at feeder 1 |
| $Q_{2,\min}^{SOP}/Q_{2,\max}^{SOP}$ | : | Upper/lower limit of reactive power provided by SOP connected at feeder 2 |
| $A_L^{SOP}$ | : | Loss coefficient of SOP |

## 10.5 Classification of Existing SOP Configurations

SOPs can be primarily divided into two categories: AC-SOP and DC-SOP. Figures 10.9 and 10.10 show the classification and the connection diagrams of existing SOP configurations [16]; 2T means two-terminal, NMG means networked microgrid, and ES means energy storage.

### 10.5.1 Two-Terminal Soft Open Points

The two-terminal SOP is the basic configuration, which contains two back-to-back converters. Its application can be classified into two types: connecting two feeders and connecting two NMGs.

An SOP is normally installed between two radial medium- or low-voltage distribution feeders integrated with large-scale DGs. To achieve the maximum load restoration during an outage, it is better to place the SOP at the end of the feeders in a radial type of distribution network.

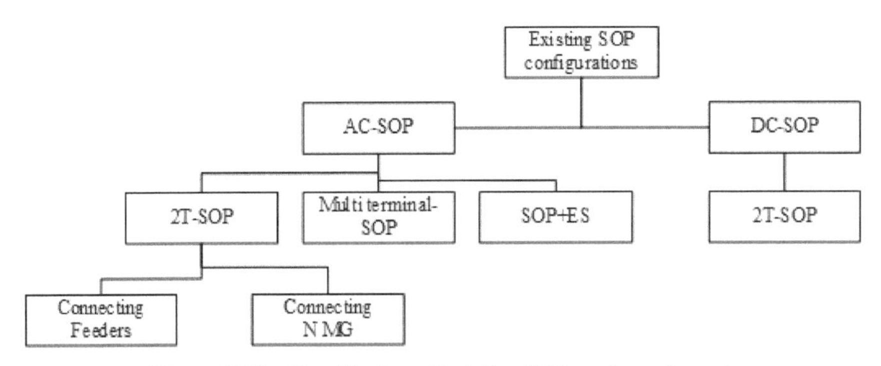


**Figure 10.9** Classification of existing SOP configurations.

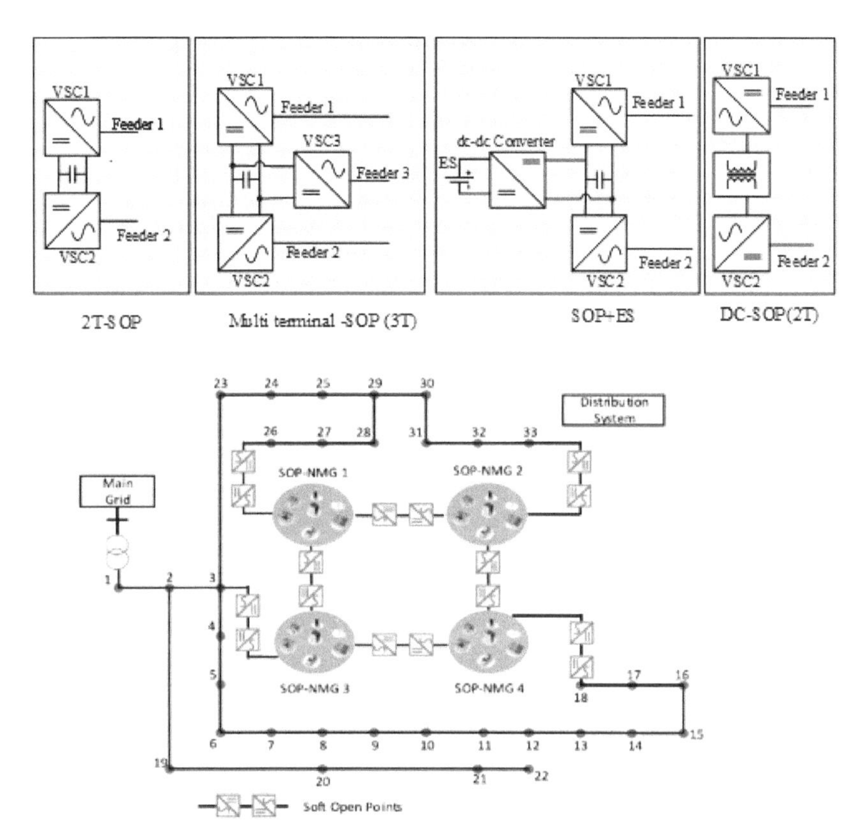


**Figure 10.10** Connection diagram of existing SOP configurations [16, 20, 29].

SOPs can also be used to connect microgrids (MGs), so that during load restoration, SOPs can continuously control and regulate the power flow and the voltage in a more flexible manner than traditional switches. The concept of SOP-based NMGs is proposed in [29], where NMGs refer to MG clusters. Each SOP is fully controlled by power electronics devices and can be used to continuously adjust the exchanged power flow among MGs and regulate the voltage profile. An example is given in Figure 10.10, where each SOP interconnects two MGs or connects one MG to a distribution network.

## 10.5.2 Multi-Terminal Soft Open Points

The purpose of multi-terminal SOPs is to connect more than two feeders in distribution networks to balance the power flow by regulating active and reactive power. Multiterminal SOPs reduce power losses and increase the

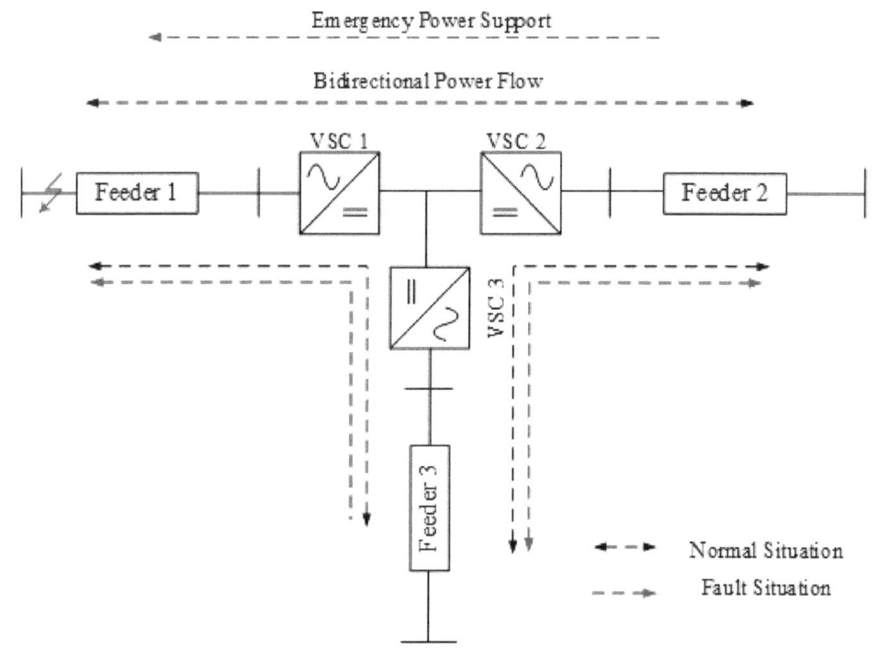


**Figure 10.11** Power flow of a three-terminal SOP under various operating conditions [20].

renewable energy hosting capacity of a distribution network, as well as restore more power during outages compared to two-terminal SOPs [18, 20].

In Figure 10.11, the power flow of a three-terminal SOP under normal conditions (black dotted line) and fault conditions (red dotted lines) is illustrated. Due to the intermittent nature of DGs and loads, the power flow among feeders is bi-directional. When a fault occurs, the power flow becomes unidirectional from the normal feeder to the faulty feeder through the SOP. The other two feeders still maintain a bi-directional power flow, depending on the power generation and the load demand.

### 10.5.3 Soft Open Points with Energy Storage

SOPs can be integrated with energy storage to improve the economy of distribution systems and increase the maximum hosting capacity of DGs [30]. Energy storage is needed for distribution networks due to the fluctuating nature of the connected renewable energy-based DGs and loads [31].

The battery storage can be connected to an SOP through a dc-dc converter, and the energy is absorbed or released through this converter. The

**Table 10.4** Contol modes of DC-SOPs [32]

| Control Mode | VSC1 | VSC2 | Applicable Scenarios |
|---|---|---|---|
| 1 | $P_{DC}$ | $V_{ac}\theta$ | Normal operation |
| 2 | $V_{ac}\theta$ | $P_{DC}$ | Normal operation |
| 3 | $V_{dc}$ | $V_{ac}\theta$ | Fault on VSC1 side |
| 4 | $V_{ac}\theta$ | $V_{dc}$ | Fault on VSC2 side |

configuration of SOPs with energy storage can place the energy storage in a single location rather than disperse the storage in the distribution network; the system efficiency can be, therefore, improved. To solve the optimization problem for such system configurations, operating constraints on the energy storage system should be considered, together with the SOP's constraints.

### 10.5.4 DC Soft Open Points

The DC-SOP is based on a bi-directional DC-DC converter designed with a high-frequency transformer and two voltage source converters. A DC-SOP connects two DC feeders, and similar to the AC-SOP, it is placed at the end of DC feeders. The DC-SOP offers the electrical isolation between DC feeders and performs as a DC circuit breaker to isolate faults occurring along the medium-voltage DC power grid. The DC-SOP can control active power flow through it, and provide voltage regulation to improve the voltage profile of the distribution system [32]. The control modes of a DC-SOP under normal and fault conditions are summarized in Table 10.4.

## 10.6 Planning for Sizing and Placement of SOPs in Distribution Networks

The goal of optimal planning is to minimize annual costs (i.e., operation, maintenance, electricity purchase) and power losses. The placement of SOPs and other power electronics devices in active distribution networks mainly depends on:

1. Geographical distance, weakly connected areas, and the node voltage control ability.
2. Branch currents and the maximum amount of load that can be supplied by the SOP after the fault isolation.

The location and capacity of SOPs affect power transfer between feeders and the capacity of DGs. Therefore, the installation of DGs and SOPs in a distribution network should be planned together.

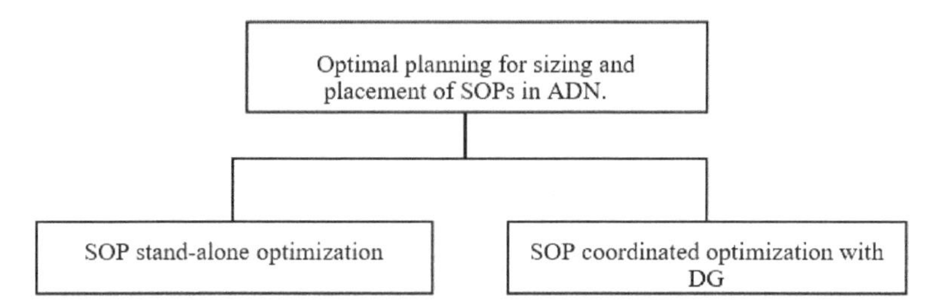


**Figure 10.12** Planning of SOP placement in active distribution networks.

Figure 10.12 summarizes the optimal placement planning strategy of SOPs, which can be mainly classified into two categories: 1) optimal sizing and sitting of SOPs in an existing distribution network (i.e., SOP stand-alone optimization); 2) coordinated sizing and sitting of SOPs along with other active elements (i.e., DGs, capacitors, energy storage such as battery banks) in a distribution network (i.e., SOP coordinated optimization).

The capacities and optimal installation sites of SOPs are proposed in [11] based on DGs characteristics and the changes of the network topology. A large-scale mixed integer non-linear programming (MINLP) model is used to obtain the optimal solution, and a mixed integer second order cone programming (MISOCP) model is adopted to solve the problem.

As a case study, this method has been applied to an 11 kV distribution system of the Taiwan Power Company (Figure 10.14 and Table 10.5). After introducing SOPs, the cost of annual energy losses of the system was reduced by 23.6% [11].

In [10], a novel multi-objective planning method is used to place SOPs in distribution networks to minimize both annual total costs and annual power losses. The SOP integrated with an energy storage system is proposed in [31] for the economic planning of distribution networks, considering the time sequence characteristics of the load and the fluctuation of DGs.

To mitigate the voltage violation risk due to forecasting errors of renewable power generation, a two-stage robust optimization model is formulated for the SOP allocation to minimize the total cost of SOPs investment and the network operation [33]. The widely used K-means clustering method is incorporated to obtain several representative scenarios, considering output variations of renewable power generation throughout the whole year. The power flow betweenness index (PFBI) and the voltage deviation index (VDI) are calculated for all possible locations to find the optimum place for the SOP. The PFBI is defined as the contribution of a line to power transfer between

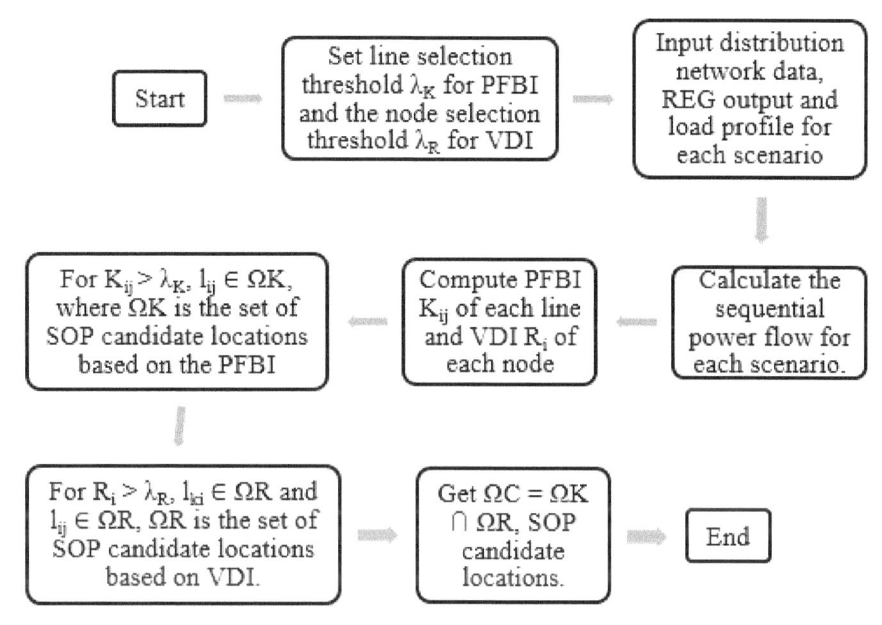


**Figure 10.13** Process flow diagram to select candidate locations of SOPs in a distribution system based on PFBI and VDI [33].

power sources and loads; VDI is defined as the deviation of the voltage of a node with respect to its rated voltage. The SOP should be related to a node with a greater VDI, which indicates more chances to have voltage violations at that node.

The process flow diagram to select the candidate locations of SOPs is shown in Figure 10.13, where $K_{ij}$ is the average PFBI of line *ij*, $R_i$ is the average VDI of node *i*, $l_{ki}$, and $l_{ij}$ are the branches with node *i* as the starting node, and the ending node, respectively. The detailed formula to calculate PFBI and VDI can be found in [33].

In the above models, the demand-response criterion is not included. In [34], by considering the characteristics of DGs and the demand response, the planning model of SOPs in active distribution networks is established through the combination of planning and operation. Based on a comparative

**Table 10.5** The location and capacity of SOPs [11].

| Locations | 5–55 | 7–60 | 12–72 | 20–83 | 40–42 |
|---|---|---|---|---|---|
| Capacity of SOPs (KVA) | 600 | 600 | 600 | 600 | 200 |

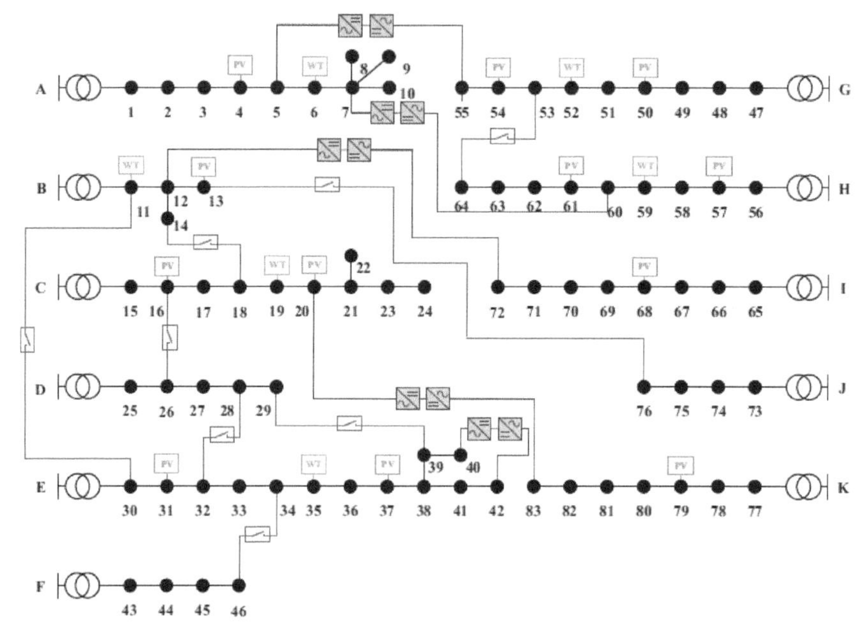


**Figure 10.14** Locations of SOPs in Taiwan power company distribution system [11].

analysis, it is found that this planning model can maximize economic benefits and achieve higher feasibility.

### 10.6.1 SOP Coordinated Optimization

Coordinated planning of SOPs and DGs can improve the performance of a distribution system. In the literature, the coordinated allocation of SOPs is investigated under both balanced and unbalanced conditions. The review in both cases follows.

#### 10.6.1.1 SOP Coordinated Optimization in Balanced Distribution Networks

A coordinated optimal placement method of DGs, capacitor banks, and SOPs in distribution networks is recommended in [35]. To solve the optimization problem, a bi-level programming model is used, where the upper level minimizes annual costs (i.e., operation and maintenance, and electricity purchase) and the lower level minimizes power losses. In [36], to reduce power losses, improve voltage profile, and manage congestion, a fast covariance matrix adaptation evolution strategy with an increasing population size (IPOP-CMA-ES) algorithm is proposed to obtain the optimal allocation of DGs and SOPs.

In [37], a new two-stage coordinated optimization framework is developed for the planning of the battery storage integrated with the SOP by considering demand response and conservation voltage reduction (CVR) schemes. The optimization framework is shown in Figure 10.15. The objective of the proposed methodology is to minimize the total investment and operating costs of battery storage and SOPs. It is found that the implementation of the demand response and the conservation voltage reduction schemes can simultaneously improve the economic benefits, reliability, flexibility, and environmental performance of a distribution network.

#### 10.6.1.2 SOP Coordinated Optimization in Unbalanced Distribution Networks

Asymmetric parameters of lines, unbalanced loads, and DGs can cause voltage imbalance in distribution networks [38]. Unbalanced voltages can hinder the normal operation of power equipment and increase power losses [39]. The

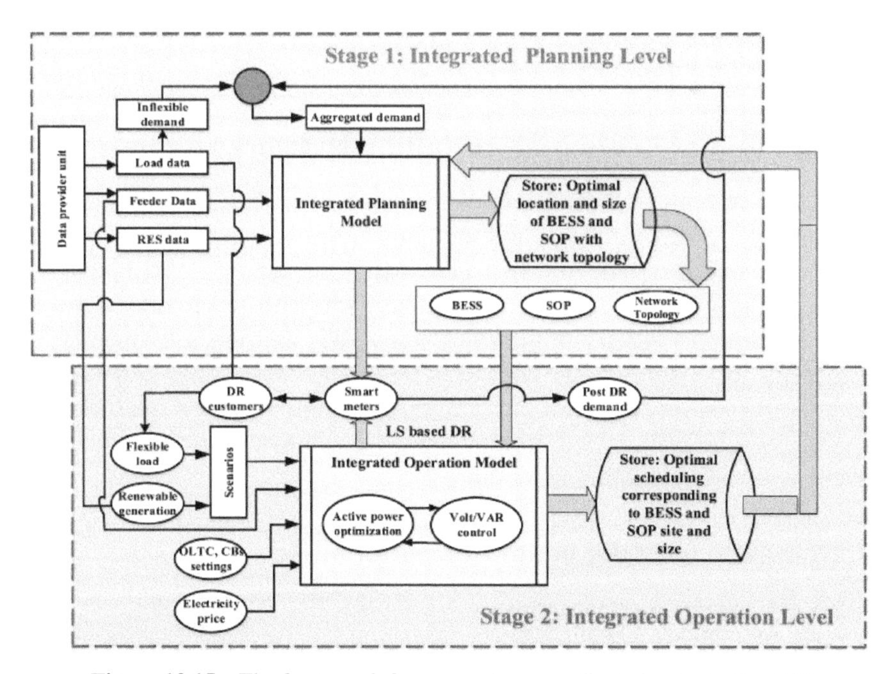


Figure 10.15 The framework for a two-stage coordinated optimization [37].

active and reactive power at each phase of the DG converter and SOPs can be controlled independently [40], which can improve the load unbalances in a distribution network.

The above mentioned research on planning problems of DGs and SOPs is primarily focused on symmetrical distribution networks, where the power and line parameters of each phase are assumed identical [35, 37]. Considering the curtailment of active and reactive power support of DGs, the sizing and placement of inverter-based DG units are analysed in [41], but this optimization model only considers investment and operation costs, and disregards the effect of power losses and unbalance. Also, in [42], the optimal sizing and placement of DG units are studied without considering the influence of system unbalance.

To address this research gap, a bi-level optimization for DGs and SOPs planning, which incorporates power control in an unbalanced distribution network, is proposed in [43].

The upper–level model is for optimal sizing and placement of DG and SOP units to minimize the total cost of the unbalanced distribution network, whereas the lower-level model considers the capability of DG and SOP units in terms of the power loss reduction and voltage unbalance improvement. A MISOCP model is used for the optimal planning of DG units and SOPs by incorporating active management of the converter.

## 10.7 Operation of SOPs in Distribution Networks

The operation control of traditional distribution networks is commonly obtained with the control of DGs and auxiliary devices, including on-load tap changers, capacitors, and tie switches [44]. This type of control has limitations and insufficient precision; in fact, it is difficult to provide high-precision real-time operation optimization when DGs and loads frequently fluctuate. In this case, through the control of their power electronics devices, SOPs offer fast, flexible, and accurate power flow control and optimization [8, 45, 46]. In this section, we will review SOP's operations under both normal and abnormal operating conditions.

### 10.7.1 Operation of SOPs under Normal Conditions

Under normal operating conditions, SOPs ensure the load balance in a network by monitoring feeders in real-time and accurately controlling their active power transmission. SOPs can also provide reactive power compensation to feeders according to the demand to reduce voltage fluctuations and improve the DG hosting capacity of distribution networks [18]. Under normal operating conditions, the SOP operates in the power flow control mode.

#### 10.7.1.1 Control Block Diagram for Power Flow Control Mode Operation of SOPs

The objective of the power flow control mode in SOPs is to deliver the required P and Q and regulate the voltage within the acceptable range and balance the load between the connected feeders. In literature, for the control of three-phase voltage source converters, the traditional vector control method in the synchronous direct-quadrature (d-q) frame is adopted. Reference [47] proposes a dual closed-loop current-controlled strategy to operate SOPs for the feeder's power flow control under normal conditions. In [9, 15, 48], the two voltage source converters of the SOP operate in the current source mode. Such a current-control strategy has advantages, including: 1) it provides a decoupled control of active and reactive power components, and 2) it inherently limits the current of voltage source converters during a network fault.

The overall control structure of the SOP consists of the following three main components: 1) outer power control loop; 2) inner current control loop; and 3) phase-locked loop (PLL) [9].

**Outer Power Control Loop**

In the outer power control loop (Figure 10.16), one converter operates with the P–Q control scheme. The errors of active power and reactive power are transformed into the reference d-q current components, $i^*{}_d$ and $i_q$, through proportional-integral (PI) controllers (the superscript asterisk denotes the

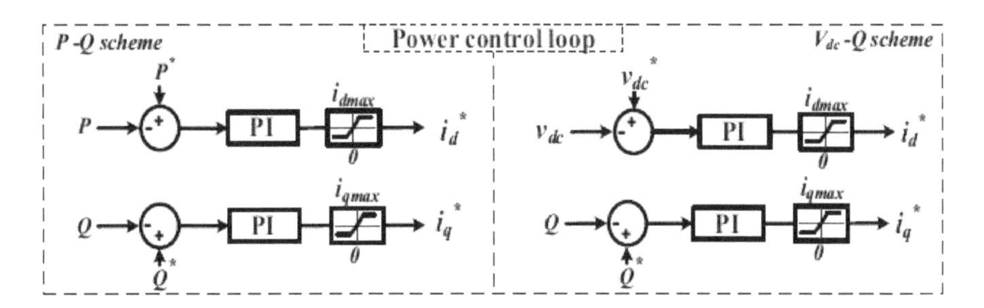


**Figure 10.16** Block diagram of the power control loop of the SOP in the power flow control mode[9].

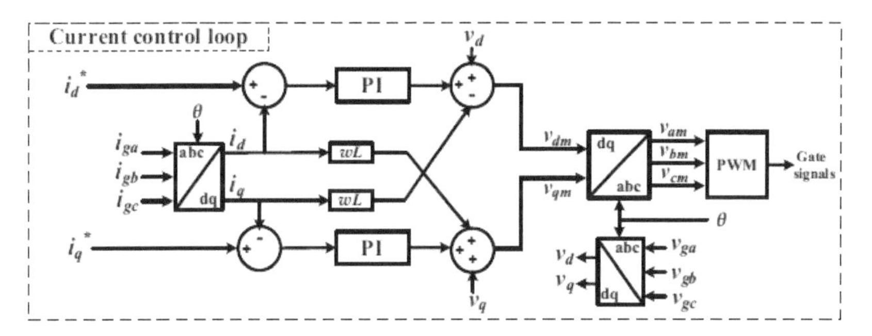


**Figure 10.17** Current control loop block diagram of the SOP in the power flow control mode [9].

reference values). The other converter operates in the $V_{dc}$-Q control scheme by maintaining a constant dc side voltage for a stable and balanced active power through the dc link. Dynamic limiters for $i^*_d$ and $i^*_q$ are inserted to enable overcurrent limitation during faults and disturbances.

**Inner Current Control Loop**

The inner control loop is shown in Figure 10.17, where the reference converter's d-q voltage, $V_{dm}$, and $V_{qm}$, are determined through PI controllers by considering d-q current errors. The cross-coupling inductance L is the inductance between the converter terminal and the feeder endpoint (i.e., the filter inductance). The L remains constant when the power flow control mode is used for the operation of SOPs. To achieve a good dynamic response, the voltage feed-forward and current feedback compensations are used [49]. Gate signals for IGBTs are obtained through the PWM, once $V_{dm}$ and $V_{qm}$ are transformed into the converter terminal voltage by means of the Park's transformation [50].

**Phase-Locked Loop (PLL)**

To synchronize the output voltage of the voltage source converter with the voltage of ac networks, the PLL plays a vital role. The block diagram of the PLL control is shown in Figure 10.18.

A PLL control topology based on the p-q theory is used in [51]. By using the sum of the products of the feedback signals, $f_\alpha$ and $f_\beta$, and input $\alpha$-$\beta$ voltages transformed through Clark's transformation [50], the variation of the angular frequency $\Delta\omega$ is calculated as follows:

$$\Delta\omega = V_\alpha f_\alpha + V_\beta f_\beta \tag{10.7}$$

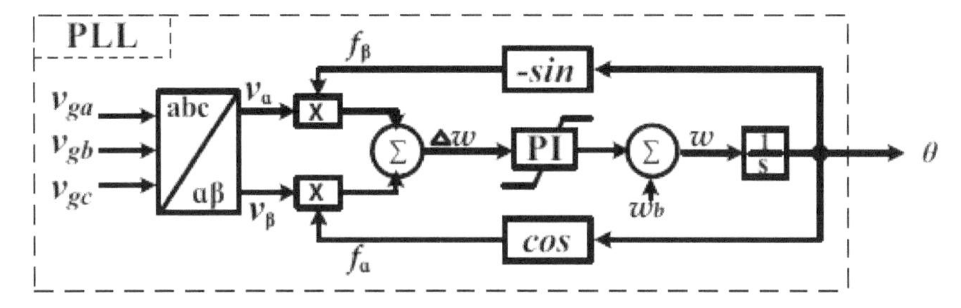


**Figure 10.18** PLL block diagram of the SOP in the power flow control mode [9].

**Table 10.6** Optimization objectives of SOPs under normal conditions

| Optimization objectives | Optimal objective functions | Ref. |
|---|---|---|
| Voltage profile improvement, feeder load balancing, and energy loss minimization | Optimal dispatch of the SOP's active and reactive power | [25] |
| | Minimization of the feeder's load unbalance condition and the total power loss | [52] |
| Voltage profile improvement | Minimization of voltage deviations | [53] |
| Voltage profile improvement and feeder load balancing | Minimizing the loss of active power and maintaining the feeder load balancing | [8] |
| Mitigating unbalance conditions within three phases | Minimization of the total power loss, and voltage and current unbalance | [28] |

The PLL output angle $\theta$, with a frequency $f = 2\pi\omega$, is then obtained using a PI controller, a feedback compensation of the base angular frequency $\omega_b$ and an integrator.

SOPs can significantly facilitate the economic operation of distribution networks by optimizing active power flow and providing reactive power support. A summary of optimization objectives of SOPs under normal conditions is presented in Table 10.6.

### 10.7.2 Operation of SOP during Abnormal Condition (Supply Restoration)

Network reconfiguration, or island division, is generally executed for the supply restoration of traditional distribution networks. After the detection and isolation of faults, the topology structure of distribution networks is reconfigured to restore power to the outage area by changing the switch status. The installation of SOPs in distribution networks not only can perform the supply restoration but also delivers reactive power to reduce losses and avoid frequent switching actions. During supply restoration, it is essential to coordinate properly between switching operations of control modes at the faulty terminal of SOPs and the distribution automation system.

#### 10.7.2.1 Control Block Diagram for Supply Restoration Mode of SOPs

Under fault conditions, the LVRT capability of SOPs ensures the uninterrupted power supply to important loads; the converter of SOPs on the faulty side acts in the voltage source mode to supply the required voltage, along with the stable frequency, whereas the converter connected with unfaulty feeders is in the normal current source mode of operation. After the fault clearance, the converter goes back to the normal current source mode. Reference [9] investigates the supply restoration mode under balanced and unbalanced fault conditions, the transition from power flow control mode to supply restoration mode, and from supply restoration mode to power flow control mode. The control block diagrams for supply restoration and transition mode are shown in Figures 10.19 and 10.20, respectively.

#### 10.7.2.2 Supply Restoration Approaches Using SOPs

The SOP-based supply restoration method in distribution systems is investigated in [54], where the tie switch is replaced by a two-terminal SOP. Once a

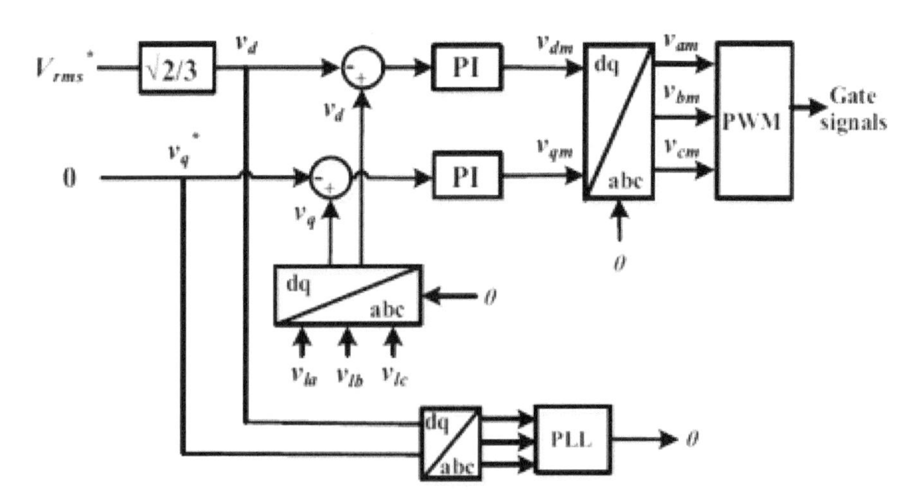


**Figure 10.19** The control block diagram of the interface voltage source converter for the supply restoration control mode [9].

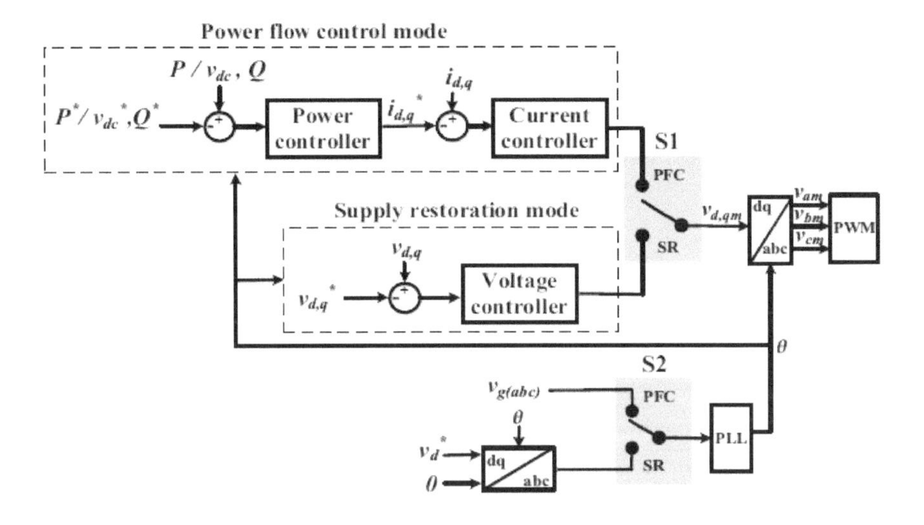


**Figure 10.20** The control mode transition system [9].

fault occurs, the active power of the nonfault terminal of the SOP is optimized to restore as many curtailed loads as possible. The performance of supply restoration employing the joint scheme of the network reconfiguration and the SOPs is the most effective combination. However, in [54], the model treats the maximum restored loads as a single objective, and the positive effect of SOPs on the system's operational state is not taken into account.

The influence of different load shedding schemes on supply restoration is investigated in [55], which also incorporates SOPs. A joint optimization model to minimize load shedding, operation times of the tie switch, and network losses is established in [56], which optimizes the switch status and active power transmitted by SOPs. A supply restoration model through the coordination of multiple SOPs is proposed in [27], and it is verified that the coordination of multiple SOPs can expand the range of supply restoration and enhance the self-healing capability of distribution networks. A restoration model for distribution networks by including routing repair crew, mobile battery vehicles, and SOP-connected-NMGs is proposed in [29]. Results showed that SOP-connected-NMGs can restore more load than traditional switch-based NMGs and improve the system resiliency.

Most studies in supply restoration of distribution networks using SOPs focus on the coordination of SOPs, tie switches, and the development of control strategies of SOPs. However, the impact of uncertainties from DGs and loads is not considered. An SOP-based islanding partition model of distribution networks is established in [57], considering the time-series characteristics of DGs, energy storage systems, and loads. The proposed model coordinates control strategies among DGs, energy storage systems, and SOPs in the island formation, and takes the failure duration into account to restore more loads. The uncertainties of the output of non dispatchable DGs are considered to ensure the robustness and feasibility of the proposed method.

A novel bi-level service restoration model using three-terminal SOP is proposed in [20] for optimal service restoration. In the proposed model, the objective of the upper level is to minimize the load loss risk, while the lower level handles uncertainties of load and DGs by adopting the interval robust optimization method.

A summary of objectives for service restoration using SOPs in distribution networks is shown in Table 10.7.

**Table 10.7** The summary of objectives for service restoration using SOPs in distribution networks

| Objectives | Schemes | Uncertainties/unbalanced effect consideration | Ref. |
|---|---|---|---|
| Minimizing load shedding | Coordination of SOPs and tie switches | Uncertainties of DGs and loads in [20] | [20], [54], [55] |
| | Coordination of routing repair crew (RRCs), mobile battery carried vehicles (MBCVs), and networked microgrids (NMGs) formed by SOPs. | - | [29] |
| | Coordinated operation of DGs, energy storage and SOPs | Time-series characteristics of DGs, energy storage and load | [57] |
| Minimizing load shedding, operation time/number of tie switch and network losses | Coordination of SOPs and tie switches | – | [56] |
| Minimizing load shedding and network losses | Coordination of multiple SOPs and tie switches | – | [27] |
| Minimizing load shedding and voltage unbalance | Coordination of multi-terminal SOPs and tie switches | Uncertainties of DGs and loads in an unbalanced distribution network | [58] |

## 10.8 Conclusion and Future Research Direction

An extensive literature review of SOP is presented in this chapter. SOPs are used to minimize network losses, regulate the voltage profile to maintain a good power quality for customers, reduce the outage time through fast response, and increase the flexibility of distribution networks. Most published papers investigated the economic planning and operation of SOPs, but SOPs' application with microgrids in distribution systems is barely explored. In addition, the control and communication architecture of SOPs in distribution systems are not thoroughly studied. The future research directions on SOPs in distribution networks should include:

1. The planning and operation of SOPs in microgrid formation are hardly investigated in the literature. The application of SOPs in microgrid formation and supply restoration is a promising future research direction, specifically, by interconnecting community microgrids or networked-microgrids through SOPs to improve system resiliency and reduce costs.
2. Most of the previous research has been conducted under balanced operating conditions. The performance of SOPs should also be investigated under unbalanced operating conditions.
3. The detailed control and communication architecture of SOPs with response time should be further investigated.
4. Theeconomic and financial benefits of SOPs with respect to other power electronics devices should be evaluated.
5. The capability of SOPs has been investigated in very few real large-scale distribution systems [11, 59]. SOPs need to be verified in more real systems to assess their true capability.

## References

[1] V. B. Pamshetti, S. Singh, A. K. Thakur, and S. P. Singh. "Multi-Stage Coordination Volt/VAR Control with CVR in Active Distribution Network in presence of Inverter-Based DG Units and Soft Open Points." IEEE Transactions on Industrial Applications, vol. 57, no. 3, pp. 2035–2047, May–June 2021.

[2] E. Du et al. "Managing Wind Power Uncertainty through Strategic Reserve Purchasing." IEEE Transactions Power System, vol. 32, no. 4, pp. 2547–2559, Jul. 2017.

[3] S. Liu, P. X. Liu, X. Wang, Z. Wang, and W. Meng. "Effects of correlated photovoltaic power and load uncertainties on grid-connected microgrid

day-ahead scheduling." IET General Transmission Distribution, vol. 11, no. 14, pp. 3620–3627, Sep. 2017.

[4] M. B. Liu, C. A. Canizares, and W. Huang. "Reactive Power and Voltage Control in Distribution Systems with Limited Switching Operations." IEEE Transactions Power System, vol. 24, no. 2, pp. 889–899, May 2009.

[5] P. Li, H. Ji, J. Zhao, G. Song, F. Ding, and J. Wu. "Coordinated Control Method of Voltage and Reactive Power for Active Distribution Networks Based on Soft Open Point." IEEE Transactions on Sustainable Energy, vol. 8, no. 4, pp. 1430–1442, Oct. 2017.

[6] T.-H. Chen, W.-T. Huang, J.-C. Gu, G.-C. Pu, Y.-F. Hsu, and T.-Y. Guo. "Feasibility Study of Upgrading Primary Feeders from Radial and Open-Loop to Normally Closed-Loop Arrangement." IEEE Transactions. Power System, vol. 19, no. 3, pp. 1308–1316, Aug. 2004.

[7] N. Okada, H. Kobayashi, K. Takigawa, M. Ichikawa, and K. Kurokawa. "Loop Power Flow Control and Voltage Characteristics of Distribution System for Distributed Generation Including PV System." in Proceedings of 3rd World Conference onPhotovoltaic Energy Conversion, 2003, May 2003, vol. 3, pp. 2284–2287 Vol.3.

[8] J. M. Bloemink and T. C. Green. "Increasing Distributed Generation Penetration Using Soft Normally-Open Points." in IEEE PES General Meeting, Minneapolis, MN, Jul. 2010, pp. 1–8.

[9] W. Cao, J. Wu, N. Jenkins, C. Wang, and T. Green. "Operating Principle of Soft Open Points for Electrical Distribution Network Operation." Applied Energy, vol. 164, pp. 245–257, 2016.

[10] S. Zhang, L. Zhang, K. Li, H. Zhang, J. Lyu, and H. Cheng. "Multi-objective Planning of Soft Open Point in Active Distribution Network Based on Bi-level Programming," in 2019 IEEE Innovative Smart Grid Technologies - Asia (ISGT Asia), Chengdu, China, May 2019, pp. 3251–3255.

[11] C. Wang, G. Song, P. Li, H. Ji, J. Zhao, and J. Wu. "Optimal Siting and Sizing of Soft Open Points in Active Electrical Distribution Networks." Applied Energy, vol. 189, pp. 301–309, Mar. 2017.

[12] S. Sun, W. Cong, Y. Sheng, M. Chen and Z. Wei. "Distributed Power Service Restoration Method of Distribution Network with Soft Open Point." 2020 IEEE/IAS Industrial and Commercial Power System Asia (I&CPS Asia), 2020, pp. 1616–1621.

[13] Y. Chen, J. Sun, X. Zha, Y. Yang, and F. Xu. "A Novel Node Flexibility Evaluation Method of Active Distribution Network for SNOP

Integration." IEEE Journal of Emerging and Selected Topics in Circuits and Systems, vol. 11, no. 1, pp. 188–198, Mar. 2021.

[14] J. M. Bloemink and T. C. Green. "Increasing Photovoltaic Penetration with Local Energy Storage and Soft Normally-Open Points," in 2011 IEEE Power and Energy Society General Meeting, San Diego, CA, Jul. 2011, pp. 1–8.

[15] H. Hafezi and H. Laaksonen. "Autonomous Soft Open Point Control for Active Distribution Network Voltage Level Management." in 2019 IEEE Milan PowerTech, Milan, Italy, Jun. 2019, pp. 1–6.

[16] K. S. Fuad, H. Hafezi, K. Kauhaniemi, and H. Laaksonen. "Soft Open Point in Distribution Networks." IEEE Access, vol. 8, pp. 210550–210565, 2020.

[17] E. Romero-Ramos, A. Gómez-Expósito, A. Marano-Marcolini, J. M. Maza-Ortega, and J. L. Martinez-Ramos. "Assessing the loadability of active distribution networks in the presence of DC controllable links." IET General Transmission Distribution, vol. 5, no. 11, p. 1105–1113, 2011.

[18] M. A. Abdelrahman, C. Long, J. Wu, and N. Jenkins. "Optimal Operation of Multi-Terminal Soft Open Point to Increase Hosting Capacity of Distributed Generation in Medium Voltage Networks." in 2018 53rd International Universities Power Engineering Conference (UPEC), Glasgow, Sep. 2018, pp. 1–6.

[19] L. J. Thomas, A. Burchill, D. J. Rogers, M. Guest, and N. Jenkins. "Assessing Distribution Network Hosting Capacity with the Addition of Soft Open Points." in 5th IET International Conference on Renewable Power Generation (RPG) 2016, London, UK, 2016, p. 32 (6.)

[20] W. Liu et al. "A Bi-level Interval Robust Optimization Model for Service Restoration in Flexible Distribution Networks." IEEE Transactions Power Systems, vol. 36, no. 3, pp. 1843–1855, May 2021.

[21] J. M. Bloemink and T. C. Green. "Benefits of Distribution-Level Power Electronics for Supporting Distributed Generation Growth." IEEE Transactions Power Delivery, vol. 28, no. 2, pp. 911–919, Apr. 2013.

[22] W. Chen, X. Lou, X. Ding, and C. Guo. "Unified data-driven Stochastic and Robust Service Restoration Method Using Non-Parametric Estimation in Distribution Networks with Soft Open Points." IET General Transmission Distribution, vol. 14, no. 17, pp. 3433–3443, Sep. 2020.

[23] W. McMurray. "Power Converter Circuits Having A High Frequency Link." June 23,1970. Accessed: May 21, 2021. [Online].

[24] M. Liserre, G. Buticchi, M. Andresen, G. De Carne, L. F. Costa, and Z. X. Zou. "The Smart Transformer: Impact on the Electric Grid and Technology Challenges." IEEE Industrial Electrons Magazine, vol. 10, no. 2, pp. 46–58, Jun. 2016.

[25] C. Long, J. Wu, L. Thomas, and N. Jenkins. "Optimal Operation of Soft Open Points in Medium Voltage Electrical Distribution Networks with Distributed Generation." Applied Energy, vol. 184, pp. 427–437, 2016.

[26] P. Li, G. Song, H. Ji, J. Zhao, C. Wang, and J. Wu. "A Supply Restoration Method of Distribution System Based on Soft Open Point." in 2016 IEEE Innovative Smart Grid Technologies - Asia (ISGT-Asia), Melbourne, Australia, Nov. 2016, pp. 535–539.

[27] P. Li, J. Ji, H. Ji, G. Song, C. Wang, and J. Wu. "Self-Healing Oriented Supply Restoration Method Based on the Coordination of Multiple SOPs in Active Distribution Networks." Energy, vol. 195, p. 116968, Mar. 2020.

[28] P. Li et al. "Optimal Operation of Soft Open Points in Active Distribution Networks Under Three-Phase Unbalanced Conditions." IEEE Transactions Smart Grid, vol. 10, no. 1, pp. 380–391, Jan. 2019.

[29] T. Ding, Z. Wang, W. Jia, B. Chen, C. Chen, and M. Shahidehpour. "Multiperiod Distribution System Restoration with Routing Repair Crews, Mobile Electric Vehicles, and Soft-Open-Point Networked Microgrids." IEEE Transactions Smart Grid, vol. 11, no. 6, pp. 4795–4808, Nov. 2020.

[30] H. Ji, C. Wang, P. Li, J. Zhao, G. Song, J. Wu. "Quantified Flexibility Evaluation of Soft Open Points to Improve DG Penetration in and., Applied Energy, vol. 218, pp. 338–348, 2018.

[31] Z. Ding and C. Wang. "Planning of Soft Open Point Integrated with ESS in Active Distribution Network." 2021 3rd Asia Energy and Electrical Engineering Symposium (AEEES), 2021, pp.427–432.

[32] Z. Zhu, D. Liu, Q. Liao, F. Tang, J. Zhang, and H. Jiang. "Optimal Power Scheduling for Medium Voltage ACDC Hybrid Distribution Network." Sustainability, vol.10, no. 2, p.318, Jan. 2018.

[33] P. Cong, Z. Hu, W. Tang, C. Lou, and L. Zhang. "Optimal Allocation of Soft Open Points in Active Distribution Network with High Penetration of Renewable Energy Generations." IET General Transmission Distribution, vol. 14, no. 26, pp. 6732–6740, Dec. 2020.

[34] J. Lu, H. Yang, Y. Wei, and J. Huang. "Planning of Soft Open Point Considering Demand Response." in 2019 IEEE Sustainable Power and Energy Conference (iSPEC), Beijing, China, Nov. 2019, pp. 246–251.

[35] Z. Lu, S. Chen, C. Ying, H. Shaowei, and T. Wei. "Coordinated Optimal Allocation of DGs, Capacitor Banks and SOPs in ADN considering Dispatching results." 9th International Conference on Applied Energy, ICAE2017, pp. 21–24 August 2017, Cardiff, UK.
[36] M. Yin and K. Li. "Optimal Allocation of Distributed Generations with SOP in Distribution Systems." in 2020 IEEE Power Energy Society General Meeting (PESGM), Aug. 2020, pp. 1–5.
[37] V. B. Pamshetti and S. P. Singh. "Coordinated Allocation of BESS and SOP in High PV Penetrated Distribution Network Incorporating DR and CVR Schemes." IEEE System Jornals, pp. 1–11, 2020.
[38] E. Prieto-Araujo, A. Junyent-Ferre, G. Clariana-Colet, and O. Gomis-Bellmunt. "Control of Modular Multilevel Converters under Singular Unbalanced Voltage Conditions with Equal Positive and Negative Sequence Components." IEEE Transactions Power System, vol. 32, no. 3, pp. 2131–2141, May 2017.
[39] K. Ma, R. Li, and F. Li. "Quantification of Additional Asset Reinforcement Cost from 3-Phase Imbalance." IEEE Transactions Power System, vol. 31, no. 4, pp. 2885–2891, Jul. 2016.
[40] S. M. Fazeli, H. W. Ping, N. B. A. Rahim, and B. T. Ooi. "Individual-Phase Control of 3-Phase 4-Wire Voltage–Source Converter." IET Power Electron., vol. 7, no. 9, pp. 2354–2364, 2014.
[41] M. Bazrafshan, N. Gatsis, and E. Dall'Anese. "Placement and Sizing of Inverter-Based Renewable Systems in Multi-Phase Distribution Networks." IEEE Transactions Power System, vol. 34, no. 2, pp. 918–930, Mar. 2019.
[42] M. M. Othman, W. El-Khattam, Y.G. Hegazy, and A.Y. Abdelaziz. "Optimal Placement and Sizing of Distributed Generators in Unbalanced Distribution System Using Supervised Big Bang Crunch Method." IEEE Transactions Power System,, vol. 30, no. 2, pp. 911–919, March 2015.
[43] J. Wang and Q. Wang. "Coordinated Planning of Converter-Based DG Units and Soft Open Points Incorporating Active Management in Unbalanced Distribution Networks." IEEE Transactions Sustainable Energy, vol. 11, no. 3, pp. 2015–2027, July 2020.
[44] A. Gabash and P. Li. "Active-Reactive Optimal Power Flow in Distribution Networks with Embedded Generation and Battery Storage." IEEE Transactions Power System, vol. 27, no. 4, pp. 2026–2035, Nov. 2012.

[45] W. Cao, J. Wu, and N. Jenkins. "Feeder Load Balancing in MV Distribution Networks Using Soft Normally-Open Points." IEEE PES Innovative Smart Grid Technologies, Europe, 2014, pp. 1–6.
[46] W. Cao, J. Wu, N. Jenkins, C. Wang, and T. Green. "Benefits Analysis of Soft Open Points for Electrical Distribution Network Operation." Applied Energy, Volume 165, 2016, pp. 36–47.
[47] N. R. Chaudhuri, R. Majumder, B. Chaudhuri, and J. Pan. "Stability Analysis of VSC MTDC Grids Connected to Multimachine AC Systems." IEEE Transactions Power Delivery, vol. 26, no. 4, pp. 2774–2784, Oct. 2011.
[48] Q. Qi, J. Wu, and C. Long. "Multi-Objective Operation Optimization of an Electrical Distribution Network with Soft Open Point." Applied Energy, Volume 208, 2017, pp. 734–744.
[49] J. M. E. Huerta, J. Castello-Moreno, J. R. Fischer, and R. Garcia-Gil. "A Synchronous Reference Frame Robust Predictive Current Control for Three-Phase Grid-Connected Inverters." IEEE Transactions Industrial Electrons, vol. 57, no. 3, pp. 954–962, Mar. 2010.
[50] M. A. Salam and S. Sethulakshmi. "Control for Grid Connected and Intentional Islanded Operation of Distributed Generation." in 2017 Innovations in Power and Advanced Computing Technologies (i-PACT), Vellore, Apr. 2017, pp. 1–6.
[51] L. G. B. Rolim, D. R. D. Costa, and M. Aredes. "Analysis and Software Implementation of a Robust Synchronizing PLL Circuit Based on the pq Theory." IEEE Transactions Industrial Electrons, vol. 53, no. 6, pp. 1919–1926, Dec. 2006.
[52] H. Ji. "An Enhanced SOCP-Based Method for Feeder Load Balancing Using the Multi-Terminal Soft Open Point in Active Distribution Networks." Applied Energy, vol. 208, pp. 986–995, 2017.
[53] Y. Zheng, Y. Song, and D. J. Hill. "A General Coordinated Voltage Regulation Method in Distribution Networks with Soft Open Points." International Journal of Electrical Power & Energy Systems., vol. 116, p. 105571, Mar. 2020.
[54] P. Li, G. Song, H. Ji, J. Zhao, C. Wang, and J. Wu. "A Supply Restoration Method of Distribution System Based on Soft Open Point." in 2016 IEEE Innovative Smart Grid Technologies - Asia (ISGT-Asia), Melbourne, Australia, Nov. 2016, pp. 535–539.
[55] Z. Xie, B. Han, G. Li, C. Xu, and C. Jiang. "Load Shedding Method of Power Restoration of Distribution Network Based on SOP." in 2019 IEEE Innovative Smart Grid Technologies - Asia (ISGT Asia), Chengdu, China, May 2019, pp. 3997–4001.

[56] Z. Li, Z. Tang, W. Chao, H. Zou, X. Wu, and C. Lin. "Multi-objective Supply Restoration in Active Distribution Networks with Soft Open Points." in 2018 2nd IEEE Conference on Energy Internet and Energy System Integration (EI2), Beijing, Oct. 2018, pp. 1–5.

[57] H. Ji, C. Wang, P. Li, G. Song, and J. Wu. "SOP-Based Islanding Partition Method of Active Distribution Networks Considering the Characteristics of DG, Energy Storage System and Load," Energy, vol. 155, pp. 312–325, Jul. 2018.

[58] J. Wang, N. Zhou, and Q. Wang. "Data-Driven Stochastic Service Restoration in Unbalanced Active Distribution Networks with Multi-Terminal Soft Open Points." International Journal of Electrical Power & Energy Systems, vol. 121, pp.106069 (1–15), Oct. 2020.

[59] Q. Hou, J. Zheng, and N. Dai. "Application of Soft Open Point for Flexible Interconnection of Urban Distribution Network." in 2019 IEEE PES Asia-Pacific Power and Energy Engineering Conference (APPEEC), Macao, Macao, Dec. 2019, pp. 1–5.

# 11

# Future Advances in Wind Energy Engineering

**Biswajit Mohapatra**

Einstein Academy of Technology and Management. Odisha

## Abstract

Wind energy is a renewable and non-conventional energy source, which is the result of uneven heating up of the earth surface by the sun. It is a clean, inexhaustible, indigenous energy resource. In recent years, wind energy can become one of the most important and promising sources of renewable energy provided with additional transmission capacity and better means of maintaining system reliability. Some of the recent advances in the field of wind energy engineering have resulted in better reliability, controllability, low cost, and less maintenance. Some areas in which advances have taken place are a development of variable speed wind turbines, airborne wind energy, offshore floating concepts, smart rotors, wind-induced energy harvesting devices, blade tip-mounted rotors, unconventional power transmission systems, multi-rotor turbines, alternative support structures, and modular high voltage direct current generators, innovative blade manufacturing techniques, diffuser-augmented turbines, and small turbine technologies.

**Key Words:** Airborne wind energy, variable speed wind turbines, offshore floating concepts, blade tip-mounted rotors, diffuser-augmented turbines, multi-rotor turbines.

## 11.1 Introduction

Wind power can be counted as one of the cleanest forms of power available on earth to date. By developing appropriate technologies to exact wind power from the wind we can decrease our dependency on power generated from non renewable sources of energy. Some of the major advantages of wind power are Sustainable energy source, reduction of pollution which is linked with health issues. Due to developments in the field of wind energy generation technologies peoples feel that it will reduce carbon emission to the atmosphere but it comes with many challenges also. The important challenge is the cost-effectiveness in comparison with other renewable sources, especially from the solar energy sector. There are several emerging technologies that are arising that could fulfill the future requirement of power. Within the approaching topics we going to visit and discuss all those technologies thoroughly and their implications also.

### 11.1.1 Airborne Wind Energy

From the start of the 2000s, industries are making research on offshore installations. These locations are far enough from the coast where the wind recourses are heavily available with stronger wind speeds which are very regular in nature, which allows invariable use and appropriate plan for the production of maximum power after conversion. For achieving this condition the scientist has discovered a new way of extracting power from wind energy that is Airborne Wind Energy (AWE). AWE adopts the principle of extracting power from wind at increased altitudes. AWE is an umbrella name given to the idea to convert wind energy into electricity with the help of an autonomous kite or unmanned aircraft, linked to the underside by one or more tethers [3]. The essential principle was introduced by the seminal work of Loyd [1] during which he analyzed the utmost energy which is prepared to be theoretically extracted with AWESs supported tethered wings. AWE systems can be classified into different types depending on their implementation which is shown in Figure 11.1.

#### 11.1.1.1 Ground-Gen airborne wind energy systems

In Ground-Generator Airborne Wind Energy Systems (GGAWES) production of electricity takes place by utilizing the aerodynamic forces which is been experienced by the aircraft and this energy is transmitted through a rope to the underside. As previously anticipated, GG-AWESs are often

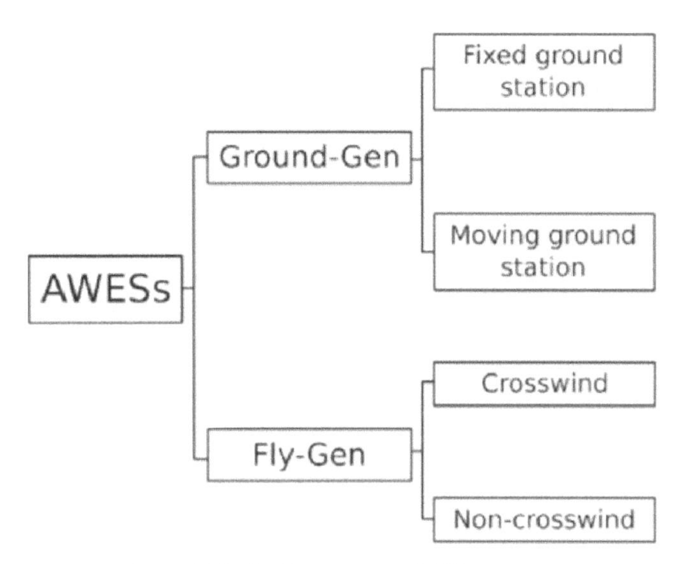


**Figure 11.1** Different types of AWE systems.

distinguished in devices with fixed or moving-ground-station. Fixed-ground-station GG-AWES (or Pumping Kite Generators) technology is the most promising sector in which almost all academic and private companies are showing their interest. Extraction of energy from the wind takes place in two phases, one is the generation phase during which energy is produced and another one is the recovery phase during which a small amount of energy is consumed. (Figure 11.2).

The principle of operation of this mechanism says that the ropes, which feel the traction forces, are wound on a winch that is connected to motor-generators axles. In the generation model, the aircraft is propelled in the forward direction which in turn produces a lift force and consequently, traction (unwinding) force is exerted on the ropes that tend to rotate the electrical generators. For a maximum generation, the aircraft should have a crosswind flight (Figure 11.2a) with a jig–jack path which looks like numerical eight. As compared to a non-crosswind flight (with the aircraft remaining in a static position in the sky), strong apparent wind was induced on the aircraft that increases the pulling force on the rope. In the recovery phase (Figure 11.2b) motors rewind the ropes bringing the aircraft back to its original position from the underside, which possesses a positive balance. The gross energy produced in the generation phase should be larger than the energy spent

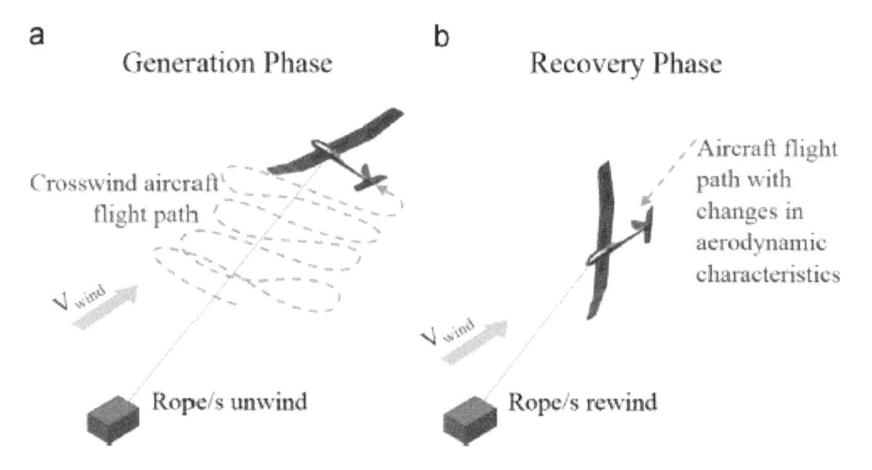


**Figure 11.2** Extraction of energy from wind.

within the recovery phase. Moving-ground-station GG–AWES are generally a more complex system that aims at providing always a positive power flow which makes it possible to simplify their connection to the grid.

Basically, there are two types of moving-ground-station GGAWES:

1. **Vertical axis generators** are the generators that are fixed at the ground stations on the periphery of the rotor of a large electric generator with a vertical axis. In this case, the aircraft forces make the ground stations rotate together with the rotor, which in turn transmits torque to the generator.
2. **Rail generators** (closed loop rail or open loop rail) where ground stations are integrated on rail vehicles and electric energy is generated from vehicle motion. In these systems, energy generation looks like a reverse operation of an electric train.

#### 11.1.1.2 Fly-Gen airborne wind energy systems

In Fly-Gen AWESs, power is generated on the aircraft when it is in the air only and this power is transmitted to the ground through special cables which are connected to the aircraft. Conversion of energy in the case of FG-AWESs is done by specially designed turbines that are connected to the aircraft. Only crosswind generation are analyzed because it has been observed that it can influence one or two orders of magnitude on top of noncrosswind generation [1]. AWESs concept which exploits crosswind power has therefore a strong competitive advantage over non crosswind concepts in terms of obtainable power and, the economics of the complete system.

## 11.2 Offshore Floating Wind Concepts

Some innovations are been made on the floating wind concept and the most important out of them is the floating network establishment. The floating structures are not fixed on the base of the seafloor but they are supported by semi-submersible, tension leg or spar platforms, and these platforms are kept stable by different mooring and anchoring systems. The most important innovation which has taken place in the field of floating wind concept is the creation of the floating web network to support the platforms. Different types of floating wind structures has been evolved from the concepts of fixed structure technology which is shown in Figure 11.3, for this reasons we have to optimize the potential of the floating structures. The optimization is only possible if we integrate the design of the platform with the turbine design. As we go in the more detailed design of the turbines used in floating platforms which will include downwind rotors, high tip speed ratio operation and possibly two bladed rotors would have an impressive impact on the value of the platform.

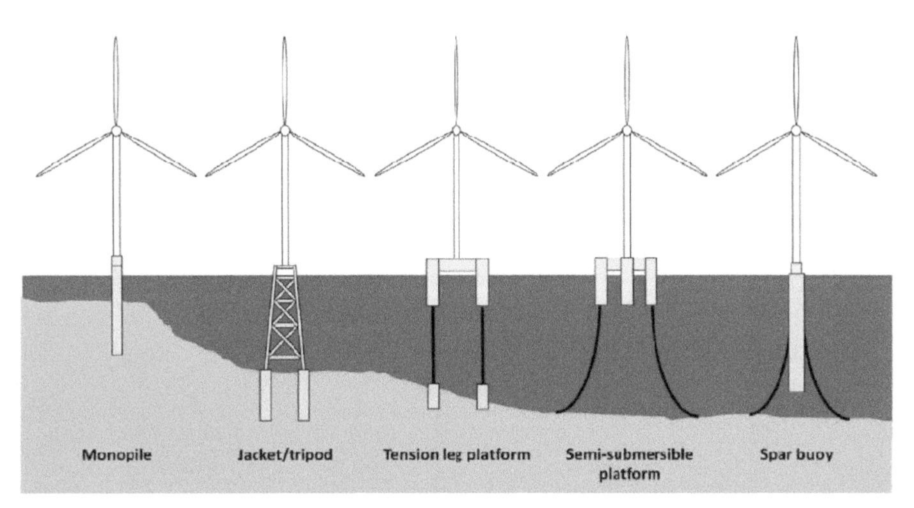


**Figure 11.3** Different types of floating wind structures.

We are lacking behind in finding some ideas which will be cost-effective for floating type structure as compared to fixed structure because of the techniques which we considered such as: catenary moored semi-submersible platforms [2], the strain leg platform seems to be smaller and lighter structures, but we require a well designed structure which can increase the level of stress on the tendon and anchor system(see e.g. Ref. [3] for design considerations); and also the spar-buoy design [4], which is more suited to deeper waters ($>$ 80 m) [5].

In trouble conditions such as dept of (60–300 m) with higher wind speeds which requires higher cabling and mooring costs. But the floating design has an advantage over the bottom mounted design due to less sensitivity with the extent of water depth. Floating turbine design resembles less transportation, installation, and assembly cost because these structures are assembled at the ground and directly transferred offshore. Floating structures should be designed in such a manner that they can survive in extreme conditions also; if not then it will have an impact on the capital and operational cost.

From the practical point of view, we can think about placing VAWT on the floating platform but the Technical readiness level (TRL) is very low. By using large turbines in the case of fixed offshore wind farms results in lowing the cost so floating offshore projects can also have low cost by the use of large turbines. Theoretically, turbines may be scaled up to a minimum of 20 MW, though future scaling-up difficulties may restrict development to around 10 MW.

### 11.2.1 Floating hybrid energy platforms

It is becoming difficult to produce maximum energy from offshore sites both in terms of keeping the cost minimum and more eco-friendly. These two factors come into the picture because the industry is going to explore the potential of wind in the Atlantic region within the next decade. For this reason, the researchers are looking towards hybrid platforms from which maximum energy can be extracted at a low cost.

The best example of a hybrid system is the Blackbird system, which consists of a storage base anchored Uniaxial Hybrid vertical axis wind turbine (VAWT) in association with a wave energy converter (WEC) which is placed on a submerged tension leg buoy platform. The unit exhibits a perfect combination of VAWT, WEC, water level generator which is floating in a single line attached to each other through cables and fixed to the seafloor through a drilled anchor system.

In the above discussion, we have discussed the wind-wave system similarly wind-wind system also exists which uses a multi converter which can be placed on the same platform as that of other systems, one such exam of the wind-wind system is SCDnezzy [6]. Multi turbine platforms can be placed in the floating water but still, debates are going on to place an airborne systems in floating platforms. If some technical solutions could be found out in which the AWE can work with the conventional wind turbines on the same platform then it would become very beneficial in terms of reduction in the cost of energy.

The benefits of going for a hybrid platform lie with the fact that different forms of energy are produced on the same platform, and the overall production of energy is also increased by sharing the common infrastructure. Some analysis says that the sites which are been chosen for wave energy production has also great potential for wind energy. Another benefit of using hybrid platforms is that the output energy variation is very less due to a combination of wave and wind energy.

Hybrid platforms are stable but the floating platforms only having a wind turbine are very unstable but the employment of the STC (Spar-Torus-Combination) helps to stabilize the platform, which increases the performance. Many projects which are in the pre commercial stage are placed in water which has a high cost of return as compared to conventional offshore systems. The offshore system will automatically lean fox maximizing the cost of energy as time passes.

Floating hybrid wind turbines face some challenges as compared to the single floating technique, as the system is a little bit complex in nature. One in every of the challenges is the shortage of information regarding hybrid dynamics and interaction with the floater. For these devices to figure there must be a transparent synergy between technologies. If we want to increase the potential of hybrid devices then we have to keep the values as like that of single floating devices, i.e. of the order of 10 MW.

## 11.3 Smart Rotors Technology

If we increase the size of the blade then it is necessary to take into consideration some blade/rotor concepts which are been utilized to adjust the blades or rotors according to the variable flow of wind, adjustment of turbulence spot on the blade, shearing stress on the blade, etc. In some blades where the length is greater than 70 m there, we cannot identify the optimal operation point because there is a variable inflow situation though out the length of the blade. If optimal blade setting was achieved, this can reduce the load on the wind farm, helps in getting constant power. Those devices which can help in achieving these goals are been categorized under smart rotor technology [7].

### 11.3.1 Passive and active control systems

Passive systems are those systems that are not controlled by operators or automatic systems. The load is been segregated within the blade span by bend twist coupling (BTC), or has been allocated at specific regions of the blade. According to these specifications, carbon fiber composites material and 3-D printing technology are essential for the manufacturing these types of blades which will be for the long run.

One of the passive systems is adding low drag vortex generators to the blades of the wind turbines which reduce the blade erosion by decreasing the drag force by preventing the flow separation of the wind. The inclusion of passive devices takes place from the starting of the design process and by the use of modernized multi disciplinary techniques and optimization methods it can also be incorporated into the existing blades.

Another type of active control mechanism used is Circulation control (CC). In this mechanism, some slots are made on the surface of the blade from which compressed air is produced. This compressed are is being utilized by the blade to dynamically control the performance of the blade in terms of controlling the lift and drag force which ultimately results in control of output power.

### 11.3.2 Degree of development, challenges and potential of smart rotors

The scope of smart rotor technology is to increase the capacity of large wind turbines (>20 MW). Increasing the size of the turbine seems to be a good choice for reducing the levelized cost of energy because the investment in this seems to be a little part of the per-unit capital investment for the installed capacity of the plant. By integrating the passive load alleviation technology with active control and rotor as the sensor will reduce the weight of the system. Some of the advantages of applying this technique are that it reduces the energy cost, increases the length of the blade which results in an increase of swept area which increases the capacity of the turbine. Both the active and passive control can be combined and can be applied to VAWTs and HAWTs. There's a need for the measurement of wind flow to the active drive system so that the smart rotor technology can be implemented effectively [8]. The measurement of wind flow can be achieved by using nacelle-mounted cup anemometers and wind vanes or spinner-mounted sonic anemometers [9]. By using the CFD code we can design the 3D model of the rotor where the approaching wind effect can be demonstrated. The data which are been collected from this model can be utilized as an input to the smart rotor technology [10].

In the case of the CC system, we require a highly efficient pumping system and variable speed fans which are the most important challenge for this system. Fan technology is a well developed technology but it is been designed to work with constant loads only. To develop an economical design with less mass and elliptical aerofoils, with accurate placement of the slots for air delivery is the challenge faced by the manufacturers. By the use of carbon in the blades with the elliptical structure, the issue of flexibility and less mass can be overcome. However, the cost of the rotor system which has to be incorporated in the CC system is unknown but a rough calculation of energy cost says that benefit will be there if we consider that the rotor cost jumps up to four times. An approximate calculation says that the pumping power has a share of less than 10% of the overall power generated by the system. If the size of the rotor is increased without changing the load demand then the system can produce more power as compared to the standard rotor. Advanced technologies are been developed but more work has to be done in the field of structural design. CC system technology is heavily patented, whether or not many patents are generic. Since research during this field is especially performed by the industry it's difficult to properly estimate its TRL (2–3 within the property right, maybe higher). The IPC (Individual pitch control) and BTC techniques are tested under a windy condition which should give a TRL of around 4. The TRL of BTC depends upon the way in which it has been designed but in some cases it goes up to 5–7 because there are some specific techniques for manufacturing which can be used but the constraints are the manufacturing should be in large scale. Exciting manufacturing processes don't look good till now; several components need to be made by hand due to inadequate industrial manufacturing technology. If an answer for the challenges is found, the TRL evolution can follow a median to possibly mean if there's a disclosure of a number of the steer required. Most of the technologies which are been discussed will take a decade for enhancing the capability but out of them, BTC can reach the market within 5 years. Till now the turbines which are been produced have a capability to produce 12 MW so we can expect an upliftment of about 10 to 50 MW of power. These technologies have to compete with each other to get more funding which will ultimately going to increase the TRL of these technologies. Some of the smart rotor technologies are been developed by industries, but many more concepts are at the primary level of research. Now it is difficult to say that which technology is going to run for the long term.

## 11.4 Wind Turbine with TIP Rotors

This conceptual technology consists of wind turbines where the standard torque transmission by gearbox and generator is substituted by a fast-rotating rotor/generator mounted on the tip region of every blade (see Figure 11.4).

While conventional turbines extract power at a free wind speed of around 10 m/s by conversion of torque, the tip-rotor converts power at around 70 m/s. The concept will be designed for both two- or three-bladed turbines [11]. The efficiency of the tip rotors to convert power may well be near 100% at low tip speed ratios, because the usual Betz limitation of 59% doesn't apply for the moving tip rotor [12]. This can be because it's the thrust of

**Figure 11.4** Fast rotating rotor.

the tip rotors which is providing the reaction torque to the first rotor causing them to extract power as they move. It can therefore be shown that within the idealized Betz model case [13], the ratio of the ability extracted by a tip rotor to it extracted by the first rotor is $C_p/ C_t$ = (1-a), where $C_p$ is the power coefficient of the tip rotor, $C_t$ is its thrust coefficient and a is that the axial induction factor from idealized actuator disc theory. If the tip rotor were itself optimized, then a = 1/3. This is able to imply the general efficiency of the tip rotor to be the optimal efficiency of the first rotor multiplied by (1-1/3), i.e. (16/27) × (1-1/3) = 0.395. If the overall efficiency were determined by each rotor operating at the Betz limit, i.e. (16/27) (16/27) =0.351, it may be seen that this configuration has the potential to slightly exceed this value. If we calculate the efficiency in this manner then the speed of the rotor should be decreased and the size of the tip rotor would be increased which will not satisfy the purpose for which these types of systems are designed. So in a practical point f view, we have to sacrifice efficiency so that the weight would be less and speed would be more [14]. Siemens has demonstrated that the use of direct drive generators for e-propellers for aircraft can reach a performance of 5kW/kg [15]. This demonstration has opened a path for thinking about a system of 10 MW with two blades with two generators placed at the tip of the blades, the weight of the blade is comparatively high than the generators. We require a drive train for a 10MW wind turbine whose weight is very large, by using this technology this weight issue is been resolved which helps in decreasing the capital cost for the plant. Another factor that adds to the cost reduction and weight reduction is that this mechanism doesn't use a shaft and gearbox.

Two main challenges associated with high-speed rotor is noise and erosion of blade due to more friction with the air. The aerodynamics and aeroelastic of this technology are more complicated as compared to other turbines, so more research has been done on the material used for manufacturing the blade and the shape of the blade. The tip rotor concept can be applicable for large turbines because they face less challenge related to centrifugal for on the rotor. These concepts can also be applicable in offshore farms because there the noise problem and visibility issues can be ignored.

The development in this technology is very less so no prototypes are been made or tested practically to date. The TRL of such an idea is thus 1–2 with possible scalability of the identical order of magnitude as for a standard 10 MW offshore turbine. As the TRL for this technology is very slow for which investors are hesitating to invest in this technology. The technology which is been discussed above is the beginning level of reach that to be an academic point of view. A detailed research is required so that public funds can be utilized in this project to face the challenges faced by wind generators.

## 11.5 Multi Rotor Wind Turbine

For increasing the efficiency and decrease the load on the turbine we can replace the single rotor setup with multiple rotor setup MRS (Figure 11.5). This innovative technique can scale up the production up to 20 MW with the same rotors and can reduce the overall cost of the plant. In this system, there are provisions for individual rotor control which allows the system to readily respond to uneven windy conditions which helps in not affecting the efficiency of the system. By the use of MRS technology, the problem of structural and materialistic requirements is been solved which is required for increasing the capacity of the plant. The size of the rotor is optimized by considering the operational and maintenance perspective not the aerodynamics.

**Figure 11.5** Multiple rotor setup (MRS).

One of the most important advantages of this technology is standardization [16]. We can produce small rotors in an industrial manner which can reduce the cost but the process which is been adopted for the manufacturing of the large turbines needs customization. In addition to that the failure of one rotor does not after the production of energy from the whole system. One commercial project named EU INNWIND, which is an MRS of 20MW capacity with 45 rotors each having a rating of 440 KW is been compared with a single turbine 20 MW project [15, 16] and it was found that the MRS project produces 8% higher than single rotor plant. Design details are required for optimizing the cost required for operation and maintenance in a reliable manner. Reliability studies have shown that, while there is often an oversized maintenance advantage in having much smaller components [17, 18], avoidance of the utilization of jack-up vessels is critical to possess effective design to avoid minor failures. Modeling an MRS is a challenging job and needs more development. We have to consider the aerodynamic behavior of the MRS which seems to need more research [19]. The designs which are been made in the multi rotor system for turbulent wind conditions and the aeroelastic behavior are at their initial steps. The active control and structural vibration alleviation techniques in the design increase the challenge for the system. As the no. of components is increasing in the designed system we have to think about the reliability and stability of the system.

Some works are going on to develop three or seven rotor systems for small-scale applications [20]. In this field, the latest project is been developed by European company Vestas with four rotors and a capacity of 900 KW, with a TRL of 5–6. Overall we can say that this technology can reach up to a capacity of 20 MW with an average TRL of 2–4. More work is needed in the field of modeling and system design so that it can serve in the long term. Since this technology is been demonstrated commercially, we can expect an industrial investment in further research in this technology so that it can be brought up to the maturity level, also some public funding can be utilized to solve some of the fundamental challenges related to design and control.

## 11.6 Diffuser Augmented Wind Turbines

In the below section we are going to discuss a turbine whose name is been coined due to its structure which looks like a diffuser funnel that collects the wind and concentrates them to one point for which it is been called Diffuser Augmented Wind Turbines (DAWTs), also called wind lens or shrouded wind turbines. The diffuser is modified by adding a broad ring or brim around the exit point and an inlet shroud at the doorway, thus creating a "lens effect". This design increases diffuser performance and it has been demonstrated as producing increased power compared to straightforward turbines, for a given turbine diameter and wind speed [21]. So we can reduce the LOCE as compared to an un-shrouded machine with the condition that the cost of the diffuser should be lesser as compared to the manufacturing cost of the large rotor used to produce the same amount of energy. This remains a serious challenge. Some other benefits of using this technology are the reduction of noise and also the decrease in visibility of it is being embedded with a building structure. The development of DAWTs is at its semi commercial level in Japan which can produce only a few kilowatts of power. The latest development which occurred in the field of DAWTs is at Kyushu University from single DAWTs to multiple rotors shown in Figure 11.6, with a maximum tested the power of 100 kW [22]. Multi-rotor DAWTs show good performance as compared to MRS [22]. One of the challenges faced by the diffuser augmented system is the structural loading and also the question

**Figure 11.6** Multi-rotor DAWT.

of whether the system has the potential to compensate larger power gains with lightweight rotor whether single or multiple with a nacelle system. This technology is costlier than non ducted technology but it is beneficial for small markets. The TRL of this technology is around 5–6 and can be scaled up to 1 MW. These technologies are applicable for small-scale applications and we don't see any large investment coming into these technologies. We can expect public funding for these technologies which will be used in urban applications.

## 11.7 Other Small Wind Turbine Technologies

The following non mainstream concepts are briefly reviewed, namely, wind turbines supported maglev, innovative vertical axis, and a cross axis design:

1. Railroading (also MagLev) wind turbines use full-permanent magnets to try and do to eliminate friction through levitation of the blades [23] (see Figure 11.7(a)). This technology is a form of advanced (TRL = 8) for little power applications (∼ kW). One of the most important challenges for this technology is it is not economically viable and also it is not suitable for an increase in capacity [24].
2. VAWTs, as per the name it has a vertical shaft around which the rotor rotates. Some of the latest applications which is been evolved from an offshore applications are VAWTs which is shown in Figure 11.7(b) suffers from low rotational speed which causes high torque, increase in mass, and cost of the system. One technique used to avoid low rotational speed is a secondary rotor which may use the tip speed of the VAWT instead of addressing very low shaft speeds and really high torque.

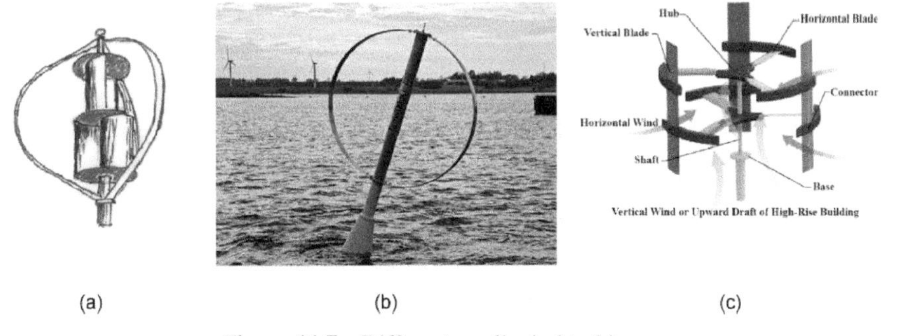


**Figure 11.7** Different small wind turbines.

3. The Cross Axis turbine (CAWT) has the capability to extract energy from wind flowing in both horizontal as well as vertical directions as shown in Figure 11.7(c). The CAWT isn't an appropriate concept but if some extraordinary designs are been made for this concept then it could be very useful to be used in urban areas.

Almost all these technologies are for low power applications or for low wind areas such as urban areas which can be integrated into the structure of buildings. The cost of this technology can be reduced if efficient designs are developed, manufacturing cost reduces, material cost decreases, and also less installation cost. Some legal challenges are also present in implementing these technologies.

## 11.8 Wind Induced Energy Harvesting from Aeroelastic Phenomena

Another way of producing energy from wind is from vibration which is been produced by a flow of air inside the mechanical systems which are designed specially to experience a large amount of oscillations that can be used to extract energy. These systems should be attached with appropriate energy conversion devices to produce energy. These technologies can't be used for large-scale generation but can be used for some applications where the instruments require their own power supply such as wireless sensors or structural health monitoring systems. For small-scale applications, these technologies are very cost-effective.

Among fluid-structure-interaction phenomena [25], those considered suitable for energy harvesting applications include: 1. dynamic instability of classical flutter [26], 2. interference between vortex-induced vibrations (VIV) [27, 28], and, 3. dynamic instability or galloping.

The devices which are been designed as per a flutter contain a flat plate of specific length and are suspended elastically to oscillate when the wind comes in contact with that plate. The flat plate is allowed to oscillate in two motions which is mechanically called heaving and pitching motion. Extraction of energy takes place in heaving motion due to less damping effect. A linear generator or solenoids are connected to produce energy (Figure 11.8). Figure 11.9 a show the VIV and galloping-based device which is having a circular hollow body of specific length connected to a spring and translational system. Due to ocean current vibration of the cylinder taking place and the translational system convert it to rotational motion which is connected to a generator to

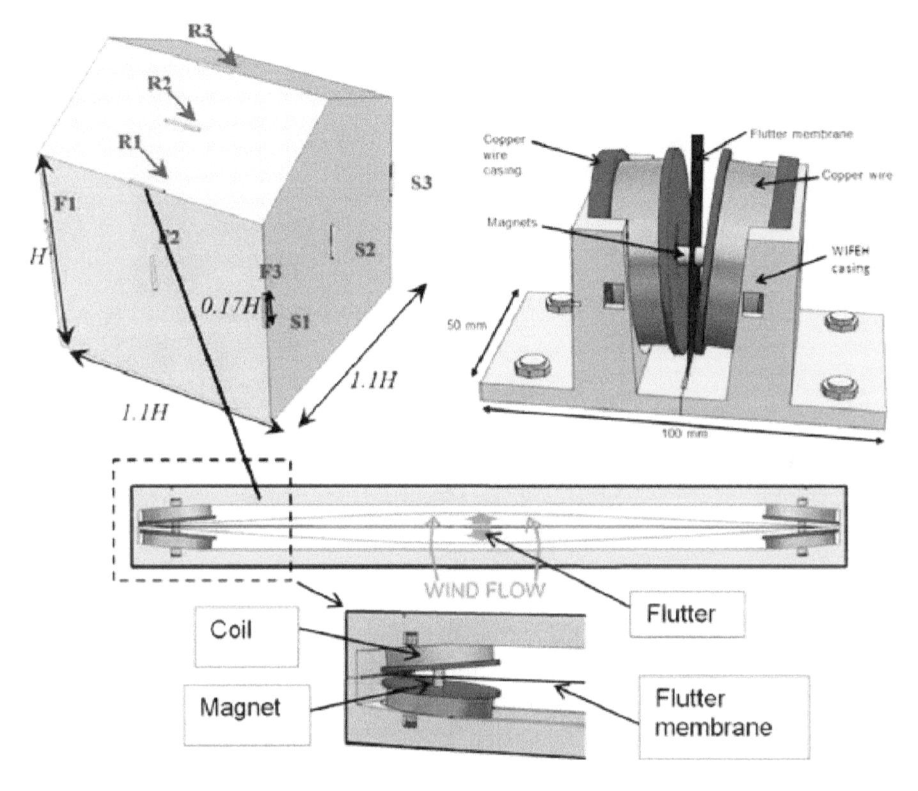


**Figure 11.8** Wind induced energy harvesting aeroelastic phenomena.

produce energy, the same concept can be utilized in the wind in which the vortex created by the flow of wind produces the vibration of the cylinder. For perfect execution of the above technologies for energy harvestmen following points should be taken into consideration:

1. the speed of the wind at which device starts working(cut-in velocity), should be as low as possible;
2. the amplitude of vibration, should be as high as possible;
3. the loss level, which decides the amount of energy passed to the next conversion device;

We have to use electromagnetic transducers for high vibrations at low frequencies otherwise we can use piezoelectric transducers for higher frequencies. Research shows that if we take flapping foils with full passive motion has an efficiency of about 30 to 35% [29], which is very high as

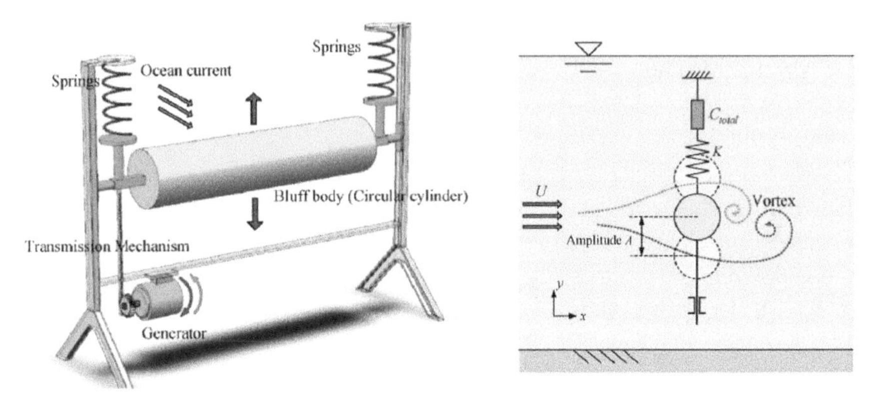


**Figure 11.9** VIV and galloping device.

compared to present used technology having efficiency below 5% with low TRL and the power obtained per unit length is only 2 W/m and the foil has a wide of 10 cm with a flow of wind at 10 m/s. There are some conceptual projects.

One such project is the piezo-tree, the structure resembles the shape of a tree and it contains piezoelectric leaves which vibrate or floats with the wind. This project faces the challenge of low output power and collection of power from each leaf, so reach should be carried out on this project to improve the aerodynamic property of the structure, reduction in the mass of the project so that the motion of the leave can be increased.

Considering the real time application perspective both the devices which contain a single oscillatory system or array of them can be adopted, but a little bit of investigation is required in case of multiple systems installation from the point of view of interference. VIV excitation system is the only system which has a narrow operational range and is not suitable for variable flow of wind. For this reason, Vortex Corporation is developing a bladeless nominal power device called Vortex Tacoma [30]. Those devices which are been developed by the use of VIV and galloping technology show that they can have a wide operational range. Some of the recent studies show that the combination of both technologies can be utilized for harvesting power from low energy systems. The best example of this type of project is "PiezoTsensor" which is been funded by EU Space Agency [31]. Due to the small opening in these technologies whatever R&D taking place in this sector is limited to academic level only and not at the commercial level. The main challenge in this field is to channel more and more funding into these small openings which can be developed into a promising sector for energy conversion.

## References

[1] U. Ahrens, M. Diehl, R. Schmehl. "Airborne Wind Energy Berlin." *Heidelberg: Springer Berlin Heidelberg*; 2013. https://doi.org/10.1007/978-3-642-39965-7.

[2] H. K. van, E. Dietrich, J. Smeltink, K. Berentsen, S. M. V. der, R. Haffner, et al. "Study on Challenges in the Commercialization of Airborne Wind Energy Systems Draft Final Report." *Brussels: European Commission*; 2018. https://doi.org/ 10.2777/87591.

[3] Y. Liu, S. Li, Q. Yi, and D. Chen. "Developments in Semi-Submersible Floating Foundations Supporting Wind Turbines: A Comprehensive Review." *Renewable Sustainable Energy Reviews*, 2016;60: pp. 433–449. https://doi.org/10.1016/j.rser.2016.01.109

[4] E. E. Bachynski, and T. Moan. "Design Considerations for Tension Leg Platform Wind Turbines." *Mar Struct* 2012;29: pp. 89–114. https://doi.org/10.1016/j.marstruc.2012.09. 001.

[5] Z. Cheng, K. Wang, Z. Gao, T. Moan, and C. Z. Cheng. "A Comparative Study on Dynamic Responses of Spar-Type Floating Horizontal and Vertical Axis." *Wind Turbines* 2017;20: pp. 305–323. https://doi.org/10.1002/we.2007.

[6] J. Azcona, D. Palacio, X. Munduate, L. González, T. A. Nygaard. "Impact of Mooring Lines Dynamics on the Fatigue and Ultimate Loads of Three Offshore Floating Wind Turbines Computed with IEC 61400-3 Guideline." *Wind Energy* 2017;20: pp. 797–813. https://doi.org/10.1002/we.2064.

[7] Z. Gao, T. Moan, L. Wan, and C. Michailides. "Comparative Numerical and Experimental Study of Two Combined Wind and Wave Energy Concepts." *Journal of Ocean Engineering and Science* 2016;1: pp. 36–51. https://doi.org/10.1016/J.JOES.2015.12.006.

[8] T. K. Barlas, G. J. V. D. Veen, and V. K. Gam. "Model Predictive Control for Wind Turbines with Distributed Active Flaps: Incorporating Inflow Signals and Actuator Constraints." *Wind Energy* 2012;15: pp. 757–771. https://doi.org/10.1002/we.503.

[9] A. Cooperman, and M. Martinez. "Load Monitoring for Active Control of Wind Turbines." *Renewable Sustainable Energy Reviews* 2015;41: pp. 189–201. https://doi.org/10.1016/j.rser. 2014.08.029.

[10] M. Zhang, B. Tan, and J. Xu. "Smart Fatigue Load Control on the Large-Scale Wind Turbine Blades Using Different Sensing Signals."

*Renewable Energy* 2016;87: pp. 111–9. https://doi.org/10.1016/J.RENENE.2015.10.011.

[11] A. G. Siemens. "Press Release: Siemens Develops World-Record Electric Motor for Aircraft" 2015: 3www.siemens.com/press/en/pressrelease/press=/en/pressrelease/2015/corporate/pr2015030156coen.htm, [accessed on 14 November 2018].

[12] P. Jamieson. "Innovation in Wind Turbine Design." *second ed.* Wiley; 2018.https:// doi.org/10.1002/9781119975441.

[13] P. Jamieson, M. Branney, K. Hart, P. K. Chaviaropoulos, G. Sieros, and S. Voutsinas, et al. "Innovative Turbine Concepts - Deliverable 1.33 of the EU INNWIND Project". 2015.

[14] P. J. Tavner, J. Xiang, and F. Spinato. "Reliability Analysis for Wind Turbines." *Wind Energy* 2007;10: pp. 1–18. https://doi.org/10.1002/we.204 ER.

[15] P. J. Tavner, F. Spinato, G. V. Bussel, and E. Koutoulakos. "The Reliability of Different Wind Turbine Concepts, with Relevance to Offshore Applications." *Proceur Wind Energy Conferance Brussels*, 31st march – 3rd april 2008, 2008.

[16] P. Chasapogiannis, J. M. Prospathopoulos, S. G. Voutsinas, and T. K. Chaviaropoulos. "Analysis of the Aerodynamic Performance of the Multi-Rotor Concept." *Journal of Physics Conference Series* 2014;524:012084. https://doi.org/10.1088/1742-6596/524/1/012084.

[17] U. Göltenbott. "Aerodynamics of Multi-Rotor Wind Turbine Systems Using Diffuseraugmentation PhD Thesis Kyushu University; 2017." https://doi.org/10.15017/ 1807035.

[18] Y. Ohya, J. Miyazaki, U. Göltenbott, and K. Watanabe. "Power Augmentation of Shrouded Wind Turbines in a Multirotor System." *Journal Energy Resource Technology* 2017;139: p. 051202. https://doi.org/10.1115/1.4035754.

[19] U. Göltenbott, Y. Ohya, S. Yoshida, and P. Jamieson. "Aerodynamic Interaction of Diffuser Augmented Wind Turbines in Multi-Rotor Systems." *Renewable Energy* 2017;112: pp. 25–34. https://doi.org/10.1016/j.renene.2017.05.014.

[20] S. V. Kozlov, E. A. Sirotkin, and E. V. Solomin. "Wind Turbine Rotor Magnetic Levitation." 2016 *2nd* International *Conference on Industrial Engineering, Applications and Manufacturing* ICIEAM 2016 - Proceedings 2016: pp. 5–8. https://doi.org/10.1109/ICIEAM.2016.7911477.

[21] H. Elahi, M. Eugeni, P. Gaudenzi, H. Elahi, M. Eugeni, and P. Gaudenzi. "A Review on Mechanisms Forpiezoelectric-Based Energy Harvesters." *Energies* 2018;11: p. 1850. https://doi.org/10.3390/en11071850.

[22] A. I. Aquino, J. K. Calautit, and B. R. Hughes. "Evaluation of the Integration of the Windinduced Flutter Energy Harvester (WIFEH) into the Built Environment: Experimental and Numerical Analysis." *Applied Energy* 2017;207: pp. 61–77. https://doi.org/10.1016/j. apenergy.2017.06.041.

[23] B. Zhang, B. Song, Z. Mao, W. Tian, and B. Li. "Numerical Investigation on VIV Energy Harvesting of Bluff Bodies with Different Cross Sections in Tandem Arrangement." *Energy* 2017;133: pp. 723–36. https://doi.org/10.1016/j.energy.2017.05.051.

[24] L. Pigolotti, C. Mannini, and G. Bartoli. "Experimental Study on the Flutter-Induced Motion of Two-Degree-of-Freedom Plates." *Journal of Fluids Structures* 2017;75: pp. 77–98. https://doi.org/10.1016/j.jfluidstructs.2017.07.014.

[25] L. Caracoglia. "Modeling the Coupled Electro-Mechanical Response of a Torsionalflutter-Based Wind Harvester with a Focus on Energy Efficiency Examination." *Journal of Wind Engineering and Industrial Aerodynamics* 2018;174: pp. 437–50. https://doi.org/10.1016/j.jweia.2017.10.017.

[26] W. T. Chong, W. K. Muzammil, K. H. Wong, C. T. Wang, M. Gwani, and Y. J. Chu, et al. "Cross Axis Wind Turbine: Pushing the Limit of Wind Turbine Technology with Complementary Design." *Applied Energy* 2017;207: pp. 78–95. https://doi.org/10.1016/j.apenergy.2017.06.099.

[27] U. S. Paulsen. "Deliverable D5.1 - Sizing of a Spar-Type Floating Support Structure." 2013.

[28] L. Caracoglia. "Modeling the Coupled Electro-Mechanical Response of a Torsionalflutter-Based Wind Harvester with a Focus on Energy Efficiency Examination." *Journal of Wind Engineering and Industrial Aerodynamics,* 2018;174: pp. 437–450. https://doi.org/10.1016/j.jweia.2017.10.017.

[29] P. Jamieson, and M. Branney. "Multi-Rotors; A Solution to 20 MW and Beyond? Energy S. Watson, et al." *Renewable and Sustainable Energy Reviews* 113 (2019) 109270 19 Procedia 2012;24: pp. 52–59. https://doi.org/10.1016/J.EGYPRO.2012.06.086

[30] Bladeless Vortex. 14 may story of a tech startup – vortex bladeless aero-generator accessed https://vortexbladeless.com/story-vortex-bladeless-tech-startup/; 2018 [accessed on 29 October 2018].

[31] PiezoTsensor 2016 accessed www.piezotsensor.eu[accessed on 29 October 2018].

# Index

**A**
Active distribution systems 157, 173, 177, 253
Airborne wind energy 287, 288, 290

**B**
Blade tip-mounted rotors 287
Blockchain 135, 145, 146, 149

**C**
Coverage Area 238, 248
Cyber-attack 13, 121, 136, 142, 151, 152

**D**
Daily uncertainties 85, 110
DC Microgrid 5, 135, 136, 139, 140
DC Power Flow Controllers 21, 31, 72, 77
DC/DC Converters 21, 31, 58, 59, 60, 62
Dc-link Capacitor 185, 188, 199, 202
Differential Protection Scheme 135, 136, 147
Diffuser-augmented turbines 287

**H**
HVDC Transmission Systems 21, 22, 23, 24, 38, 62

**I**
IGBT 34, 44, 188, 194, 196, 255

**L**
Lifespan of WSN 238, 244

**M**
Microgrid formation 157
Microgrid topology 157, 162
Multi objective indices 84, 85, 91
Multi rotor turbines 287

**N**
New optimization algorithms 85
Normally Open Points 253

**O**
Objectives 85, 126, 157, 159, 162, 176, 279
Offshore floating concepts 287
Optimal integration 84, 98, 105, 110
Optimal planning 157, 162, 267, 272
Optimization 84, 157, 171, 242

**P**
Photovoltaic distributed generator 84

Power Converters 5, 9, 48, 70, 77, 188, 207
Power Electronics 1, 9, 21, 38, 253, 259, 272
PV Inverter 185, 187, 189, 196

**R**
Relay 136
Reliability 4, 42, 149, 159, 174, 185
Renewable distributed generation 84, 114, 183

**S**
Sensor 16, 17, 237, 240, 241, 242, 243, 278
Service restoration 260, 278, 279
Smart distribution grids 83, 111
Smart grid 1, 3, 5, 8, 13, 118, 119
Soft Open Points 253, 264, 265, 267
Switch placement 157, 172, 174, 176, 182
System modelling 157

**T**
Thermal Stress 194, 198

**V**
Variable speed wind turbines 287
Voltage regulation 13, 23, 58, 253, 256

**W**
Wind turbine distributed generator 84
WSN 240, 243, 244, 245, 246

# About the Editors

**Kolla Bhanu Prakash** is working as Professor and Research Group Head for A.I & Data Science Research group at KLEF. He published 80 research papers in International, National journals and Conferences. He published Scopus and SCI publications of 80 with H-Index of 15 and total citations are 450. He is reviewer for several international journals including IEEE, Elsevier, Springer, Wiley and Taylor and Francis and served as reviewer, keynote speaker and TPC member for several International conferences. His research interests include deep learning, data science and quantum computing. He received Best Researcher Award. He authored 12 books with International reputed publishers like Elsevier, Springer, River, CRC, Degryuter and Wiley. He also published 7 patents and 1 Copyright. He is guiding 8 Ph.D scholars. Dr.K.Bhanu Prakash is IEEE Senior Member, Fellow-ISRD, Treasurer - ACM Amaravathi Chapter, India, LMISTE, MIAENG, SMIRED. He is Series Editor – "Next Generation Computing & Communication Engineering" – Wiley publishers. He is series editor - "Industry 5.0: Artificial Intelligence, Cyber-Physical Systems, Mechatronics and Smart Grids" – CRC publishers.

**Dedicated to** My Parents, Sri. Kolla Narayana Rao & Smt. Kolla Uma Maheswari
And my Wife, Mrs. M. V. Prasanna Lakshmi

**Sanjeevikumar Padmanaban** (Member'12–Senior Member'15, IEEE) received the bachelor's degree in electrical engineering from the University of Madras, Chennai, India, in 2002, the master's degree (Hons.) in electrical engineering from Pondicherry University, Puducherry, India, in 2006, and the PhD degree in electrical engineering from the University of Bologna, Bologna, Italy, in 2012. He was an Associate Professor with VIT University from 2012 to 2013. In 2013, he joined the National Institute of Technology, India, as a Faculty Member. In 2014, he was invited as a Visiting Researcher at the Department of Electrical Engineering, Qatar University, Doha, Qatar, funded by the Qatar National Research Foundation (Government of Qatar). He continued his research activities with the Dublin Institute of Technology, Dublin, Ireland, in 2014. Further, he served an Associate Professor with the Department of Electrical and Electronics Engineering, University of Johannesburg, Johannesburg, South Africa, from 2016 to 2018. Since 2018, he has been a Faculty Member with the Department of Energy Technology, Aalborg University, Esbjerg, Denmark. He has authored over 300 scientific papers.

S. Padmanaban was the recipient of the Best Paper cum Most Excellence Research Paper Award from IET-SEISCON'13, IET-CEAT'16, IEEE-EECSI'19, IEEE-CENCON'19 and five best paper awards from ETAEERE'16 sponsored Lecture Notes in Electrical Engineering, Springer book. He is a Fellow of the Institution of Engineers, India, the Institution of Electronics and Telecommunication Engineers, India, and the Institution of Engineering and Technology, U.K. He is an Editor/Associate Editor/Editorial Board for refereed journals, in particular the IEEE SYSTEMS JOURNAL, IEEE Transaction on Industry Applications, IEEE ACCESS, *IET Power Electronics*, *IET Electronics Letters,* and Wiley-*International Transactions on Electrical Energy Systems,* Subject Editorial Board Member—*Energy Sources—Energies Journal*, MDPI, and the Subject Editor for the *IET Renewable Power Generation*, *IET Generation, Transmission and Distribution*, and *FACTS* journal (Canada)

**Dedicated to** my wife, son, and daughter.

**Mitolo** has been awarded the Knighthood in the *Order of Merit of the Italian Republic* for merit acquired by the nation in recognition of his scientific work.

Sir Massimo received the Ph.D. in Electrical Engineering from the University of Napoli "Federico II", Italy. He is a *Fellow* of IEEE *"for contributions to the electrical safety of low- voltage systems"*, and a *Fellow* of the Institution of Engineering and Technology (IET) of London (UK). He is a registered Professional Engineer in the state of California and in Italy.

Sir Massimo is currently a Full Professor of Electrical Engineering at Irvine Valley College, Irvine, CA, USA, and a Senior Consultant in the matter of failure analysis and electrical safety with Engineering Systems Inc., ESi. Professor Mitolo's research and industrial experience are in analysis and grounding of power systems, and electrical safety engineering.

Dr. Mitolo has authored over 170 journal papers and the books "*Electrical Safety of Low-Voltage Systems*"(McGraw-Hill),"*Laboratory Manual for Introduction to Electronics: A Basic Approach*" (Pearson),"*Analysis of Grounding and Bonding Systems*" (CRC Press), "*Electrical Safety Engineering of Renewable Energy Systems*" (John Wiley & Sons).

Sir Massimo is currently the deputy *Editor-in-Chief* of the *IEEE Transactions on Industry Applications*. He is active within the Industrial and Commercial Power Systems Department of the IEEE Industry Applications Society (IAS) in numerous committees and working groups.

The recipient of the IEEE Region 6 *2015 Outstanding Engineer Award*, Dr. Mitolo has earned nine *Best Paper Awards*, numerous recognitions, among which are the IEEE *Ralph H. Lee I&CPS Department Prize Award*, the IEEE I *&CPS 2015 Department Achievement Award*, and the *James E. Ballinger Engineer of the Year 2013 Award* from the Orange County Engineering Council.

**Dedication** To my Mother, Fernanda Alagia Mitolo, unfailing source of love and support.